# WALTER ROTHSCHILD

*The Man, the Museum and the Menagerie*

Miriam Rothschild

Published by the Natural History Museum, London

*To the memory of*
*Alix de Rothschild*
*for whom this book was written*

Hardback edition published by Balaban Publishers, 1983
First paperback edition published by the Natural History Museum, London, 2008

ISBN 978 0 565 09228 3

Printed by Anthony Rowe Limited

Note
Please note that this text was written in 1983, when the Natural History Museum was know as the British Museum Natural History, and the Natural History Museum Tring was know as Tring Museum. Please also note that some of the appendices have been omitted in this edition and that it includes only a selection of the images appearing in the original edition.

# Contents

| | | |
|---|---|---|
| Author's Preface | | v |
| Acknowledgements | | viii |
| Lionel Walter Rothschild—principal events | | xi |

CHAPTERS 1—35

| | | |
|---|---|---|
| 1 | Walter Rothschild | 1 |
| 2 | Tring is Fairyland | 3 |
| 3 | Emma, Walter's Mother | 9 |
| 4 | Emma—The iron apronstrings | 19 |
| 5 | Walter's Father—The *Eminence Blanche* | 25 |
| 6 | Walter's Father—Nobody wanted to touch Lloyd George | 41 |
| 7 | Because I am so happy, Mama | 52 |
| 8 | I have nearly driven Althaus out of his senses | 57 |
| 9 | Who is the pink and gold boy in the corner? | 70 |
| 10 | I am not a Lieutenant, I am a Captain | 80 |
| 11 | Please ask Walter | 86 |
| 12 | Rozsika, Walter's sister-in-law | 94 |
| 13 | My Museum | 100 |
| 14 | The rump of our only old male is rather dicky ... | 110 |
| 15 | The fellow is always right | 120 |
| 16 | Walter's Curators—Ernst Hartert and the birds | 128 |
| 17 | Walter's Curators—Karl Jordan and the butterflies | 143 |
| 18 | Walter's Collectors | 154 |
| 19 | Walter's Collectors—A.F.R. Wollaston and N.C. Rothschild | 170 |
| 20 | Walter's Collectors—William Doherty | 177 |
| 21 | Walter's Collectors—The King of Bulgaria is coming on Friday ... | 181 |
| 22 | Hurrah! We are off ... | 185 |
| 23 | The Giants | 197 |
| 24 | The Rothschilds and animals | 206 |
| 25 | The Great Row | 215 |

26 Catherine wheels 224
27 If His Majesty's Government will send me a message ... 233
28 Dear Lord Rothschild 249
29 What has become of two ostriches ... 270
30 A figure in the background 283
31 You will be painted blue and yellow and exposed in the High Street 287
32 The Primrose Way—The truth about the Rothschilds 295
33 The birds cross the Atlantic 302
34 Home Farm 305
35 Candle ends 313

BIBLIOGRAPHICAL NOTES 319

BIBLIOGRAPHY
Books 350
Newspapers, Journals, Catalogues and Reports 355

APPENDIX 1
Synopsis of species and subspecies of plants and animals named in
honour of Walter Rothschild 361

FAMILY TREES 362

INDEX 366

# Author's Preface

When a biography of the second Lord Rothschild was first mooted someone remarked — not unkindly — that it was impossible to write a readable book about a secretive man whose life had consisted for fifty odd years of a 14-hour day of virtually incomprehensible work — technical descriptions of the tail end of insects — punctuated at intervals by brief, flaming rows.

It is true that Walter's pathological secrecy, the difficulty of normal communication imposed by his speech problems — and crippling lack of voice-control — that he never escaped from his mother's sheltering roof and solicitous care, and above all the fact that the great bulk of his private papers were destroyed or lost, make a conventional biography well nigh impossible. Studies of his *entourage* and his family — especially his parents, Natty and Emma — become imperative, since otherwise every approach is blocked by a huge silent question mark and we are confronted with a series of single episodes — strung lightly together on a network of several million words about birds' and butterflies' wings.

Nonetheless, the temptation remains to treat Walter as an isolated figure — like the Laocoon or Beowulf — since he was cut off from the rest of humanity by his eccentric and confusing blend of creativity and childishness, the massive contradictions in his nature, and his total lack of self-confidence coupled with a peculiar brand of enthusiastic megalomania. Furthermore his museum, his vast collections and the impressive combined output of himself and his curators, could provide ample material for several volumes. In his own case the loss of all his personal papers was perhaps less frustrating than the destruction of his father's correspondence, although it also involved the loss of the Tring Museum diaries and ledgers (see Candle Ends, p. 313). We still possess the letters recording the day to day business of his curators (lodged in the library of the British Museum of Natural History) which — although they tell us little if anything about Walter himself — give one a fair idea of the scope of his zoological activities. They would, in fact, take about two years to re-read, working a normal

8-hour day! One of his collectors, Albert Meek, wrote him some 500 letters describing his butterfly hunts in the Solomon Islands.

Despite the erection of his ivory tower, Walter was, at one time or another, engaged in several different unconnected fields of activity. Those who are concerned with his Edwardian and Jewish background, and with his family and his role in the evolution of the Balfour Declaration, may well find the chapters describing the evolution of giant tortoises or expeditions he sent to the Galapagos and Solomon Islands rather puzzling. Similarly naturalists who are intrigued by the black beards of mountain gorillas or the beaks of Darwin's finches, rather than the birth of the State of Israel, and to whom the names of Weizmann and Sokolow and Samuel mean nothing at all, may well marvel at the double life this strange character seems to have pursued. Colonel Richard Meinertzhagen, who declared in his diary that his friend Walter Rothschild had no use for women because they bored him, would also have been surprised at his own lack of insight.

To a great extent this book is based on personal recollections, since I shared for eight months of every year (from 1908 to 1935) my grandmother's hospitable roof at Tring, together with my mother, sisters and brother and my uncle, Walter Rothschild. I also worked sporadically for about forty years in his museum, both before and after his death.

This circumstance has imposed restrictions upon me as a biographer, for there are certain confidences I cannot disregard. Thus I feel compelled to respect Walter's desire to keep the identity of the blackmailing couple secret — and I must reflect in silence upon the fact that retribution on an unimaginable scale descended on the surviving member of this unsavoury pair.

The scientific sections of this book are founded on the forty-one volumes of *Novitates Zoologicae* (the Tring Museum periodical) (see also Walter's complete bibliography in Vol. 41, 1938) the Tring Museum correspondence and papers (mostly unclassified) and that portion of the Lane Library relating directly or indirectly to the Tring Museum and Walter Rothschild. Of this not inconsiderable pile of papers Walter's own letters to his curators (about 300 survived) have been used most extensively. Similarly the 320 letters from Walter to Albert Gunther in the British Museum (Natural History) are quoted liberally.

The chapters concerned with the Balfour Declaration owe much to Leonard Stein's brilliant book *The Balfour Declaration* (1961), *The Papers and Letters of Chaim Weizmann* (Vols VII-X) (1975-77), Isaiah Friedman's *The Question of Palestine 1914-1918* (1937) and Herbert Samuel's *Memoirs* (1945).

The chapters relating to Walter Rothschild's family are based on letters and papers in the Rothschild Archives, the Lane Library and the Rothschild Family Tree compiled by the third Lord Rothschild (privately printed, London 1973, 1981).

The bibliography contains references to published and unpublished sources of information (printed books, printed sources other than books, and manuscript sources). In addition bibliographical notes (to which the numbers in the text refer) are supplied for each chapter.

First names have been used freely for members of the family, since the number of Rothschilds is confusing in this context. "Rothschild" without a first name has been reserved for Walter only. In addition to the Family Tree there are brief biographical notes in the index concerning Walter Rothschild's relatives.

I am especially indebted to Professor Richard Davis for guiding me through the jungle of the Rothschild Archives and for his patient help and advice with my text. Maldwyn Rowlands, Chief Librarian at the British Museum of Natural History, was infinitely generous both with time and knowledge, and Phyllis Thomas, Walter's librarian, responded nobly and patiently over the years to relentless and persistent questioning.

Miriam Rothschild

# Acknowledgements

There is, inevitably, a lack of space in which to acknowledge all the help I have received in compiling this motley volume. Sadly, Alix de Rothschild, who encouraged me to begin it, insisted that I finish it, and discussed it chapter by chapter during her long and harrowing illness, died before it went to press. But at least she knew that I had chosen the ending she preferred.

I would like to single out several individuals to whom I am particularly indebted. First and foremost the two professional graphologists, one in the UK, Mrs. Joan Cambridge, and the other (who wishes to remain anonymous) in Switzerland, who analyzed the handwriting of Walter and all the members of his immediate family; Miss Edith Evans, Miss Carmen Wheatley and Mr. Nicholas Wollaston who examined various documents on my behalf, abstracted letters and manuscripts at Libraries in London, Cambridge, Oxford and Edinburgh; Mrs. Rozsika Parker, the Hon. Emma Rothschild, Mrs. Anne Scott and Colonel Robert Traub, who read the entire manuscript and suggested various alterations and corrections, all of which I gratefully acknowledge. In addition there are specialists in various fields in which Walter Rothschild was permanently or transiently involved, who have kindly read and criticized the relevant chapters. For this invaluable and infinitely good-natured and kindly assistance I have to thank Sir Isaiah Berlin, Sir Cyril Clarke, Professor Richard Davis, Professor Vince Dethier, the Dowager Lady Egremont, Professor Gottfried Fraenkel, the late Sir John Foster, Professor Roy Foster, Dr. Rudolph Freudenberg, Miss Ada Jordan, Sir Anthony Lambert, Mr. Noel Machin, Professor Albert Neuberger, Lord Rothschild, Dr. David Snow, Professor Richard Southwood, Mr. Ken Slavin, Dr. David Stoddart, Professor Mayir Vereté and Professor John Vincent.

Letters from the Royal Archives are reproduced by gracious permission of Her Majesty the Queen. I must also acknowledge the receipt of much factual information, and permission to consult private correspondence and various files, quote from hitherto unpublished letters and diaries, and reproduce illustrations and letters, given by Sir Geoffrey Agnew, Mrs. Catherine Althaus, the American Museum of Natural History (Dr. Thomas Nicholson and Dr. Lester Short), the Anglo Jewish Archives (Mr. Joseph Monk), the Earl of

Balfour, the Bibliothèque Nationale (Mlle Annie Angremy and Monsieur Georges Le Rider), the Board of Deputies of British Jews, the British Museum, the British Museum (Natural History) (Dr. R.H. Hedley, Dr. I. Galbraith, Miss Pamela Gilbert, Dr. Derek Goodwin, Miss Dorothy Norman, Dr. Gordon Sheals, Mr. Frans Smit, Mr. Michael Walters), Colonel William Behrens, Mrs. Georgina Blakiston, Miss Maria Brassey, Mr. Michael Bucks, Professor Arthur Cain, the Central Zionist Archives (Dr. Michael Heymann), Professor A. Chapeau, the Dowager Marchioness of Cholmondley, Lady Churston, the Earl of Crawford, Mme Marianne de Glasyer, Lord Egremont, Viscount Esher, the late Monsieur Bertrand Gille, Mr. Bob Grace, the Duc de Gramont, Mrs. Lizzie Ann Hastings, Dr. T.E.B. Howarth, *the Jewish Chronicle*, the Earl Lloyd George of Dwyfed (and the Beaverbrook Foundation and National Library of Wales for permission to quote from the David Lloyd George letters), Mr. C. Marshall (Legal and General Assurance Society Ltd). Professor Ernst Mayr, the National Army Museum (Mr. Aubrey Bowden, Mr. P. Buckley), Lord Newton, Mr. Leonard Rance, Dr. R. Robson, Dr. Ruth Rogers (Kress Library of Business and Economics, Harvard University), the Earl of Rosebery, Baron Edmond de Rothschild, Mr. Edmund de Rothschild, Baroness Elie de Rothschild, Mrs. James de Rothschild, Lord & Lady Rothschild, the Rothschild Archives (Mr. Gershom Knight, Mrs. Yvonne Moss), Mr. Cosmo Russell, the Marquis of Salisbury, the Earl of Scarborough, Dr. Theresa Searight, the Scottish Record Office (Dr. James Galbraith, Dr. Frances Shaw), Mr. H. Silver (Sun Alliance Assurance Co.), Miss Mary Smith, Miss Phyllis Thomas, *The Times* Newspaper, the Weizmann Archives (Dr. Nehama Chalom, Dr. Thelma Jaffe).

I am greatly indebted to the various photographers who have spent endless time and trouble on both the colour and black and white plates, but in particular I would like to thank Mr. John Haywood and Miss Jill Morley Smith for the preparation of the majority of the black and white photographs and in many cases for improving considerably upon the faded originals, and also for the Colour Plates IVa and XI. I am also greatly indebted to Monsieur Jean Besancenot for Colour Plate V, Mr. Peter Green who supervised the photography of the specimens in the British Museum (Natural History) both at Tring and in London (Colour Plates VI-X), Mr. Eric Hosking for Colour Plate XII, Mr. Edward Leigh (for Colour Plates I, II & IVc), and Mr. Tom Scott (for Colour Plate III).

I profited vastly from informative and stimulating discussions with Mme. Diane Benvenuti, Sir Isaiah Berlin, Professor Richard Davis, Professor Roy Foster, Sir Terence Morrison-Scott, Mr. Thomas Pakenham, the late Baroness Hilda de Rothschild, Professor Mayir Vereté and Professor John Vincent.

I must thank Miss José Beer for endless patience and forebearance while typing and retyping the manuscript, for the care she bestowed on the bibliographical notes and the bibliography. I was enormously impressed by the skill with which the Curwen Press printed the colour and black and white plates and would like to acknowledge their help and advice. Lastly I am infinitely grateful to Dr. Miriam Balaban without whose enthusiasm, breathtaking drive and sheer professional skill this book would most certainly never have been published.

# Lionel Walter Rothschild, 2nd Baron Rothschild of Tring 1868-1937

| | |
|---|---|
| 1868 | Born 8th February |
| 1872-73 | Family moved to Tring Park |
| 1878 | Began collecting for his "Museum", aged 7 |
| 1878-86 | Educated at home with a tutor. Converted shed into his first museum. |
| 1886-87 | Autumn 1886 to July 1887: Bonn University |
| 1887-89 | Magdalene College, Cambridge<br>Sent Palmer to the South Seas to collect birds |
| 1889 | Museum built<br>Entered N.M. Rothschild & Sons, New Court, the family banking firm<br>2nd Lieutenant in Bucks Yeomanry (Royal Bucks Hussars) |
| 1890 | Elected to represent Manchester on Board of Deputies of British Jews |
| 1892 | Museum opened to public<br>E. Hartert appointed curator |
| 1893 | 1st volume of *Avifauna of Laysan* published<br>K. Jordan appointed curator of invertebrates |
| 1894 | *Novitates Zoologicae* (the Museum publication) issued (continued as Editor until his death) |
| 1894-1925 | Papers on Giant Land Tortoises |
| 1895 | Revision of the Papilios |
| 1897 | Sent expedition to the Galapagos Islands. (This was one of the more important of the large collecting expeditions he sponsored) |
| 1898 | Hon. PhD of Giessen University |
| 1898-1913 | Monograph of *Charaxes* |
| 1899 | Review of the Ornithology of the Galapagos Islands<br>Appointed Trustee of the British Museum (1899-1937) |
| 1899-1910 | MP for Aylesbury (as a Liberal and Liberal Unionist) |

| | |
|---|---|
| 1900 | Monograph on the genus *Casuarius* |
| 1903 | Publication of Revision of the Lepidopterous family Sphingidae |
| 1906 | Revision of the American Papilios |
| 1907 | Monograph on Extinct Birds |
| 1907-08 | Library and Lepidoptera Halls added to Tring Museum |
| 1908 | Resigned from N.M. Rothschild & Sons |
| 1911 | Elected Fellow of the Royal Society |
| 1915 | Succeeded to the title on the death of his father |
| 1915-17 | Balfour Declaration negotiations |
| 1917 | 31st October: Balfour Declaration approved by Cabinet<br>1st November: Rothschild sends congratulations to Weizmann and Sokolow re Declaration<br>2nd November: Balfour Declaration in the form of a letter dated 2nd November 1917, received by him at Tring |
| 1930 | Hartert retired owing to ill health |
| 1931 | President Zoological Section (D) of the British Association |
| 1932 | Sale of collection of Birds to the Natural History Museum of New York |
| 1935 | Death of his mother<br>Fracture of femur in walking accident<br>Moved to Home Farm from Tring Park |
| 1936 | Gift of the Tring Museum collections and library to the Trustees of the British Museum |
| 1937 | Died 27th August |

## PRINCIPAL COLLECTIONS 1893-1937

### General collection for public display

2000 complete mounted mammals (including a Quagga, 13 Gorillas, 228 Marsupials, 24 Echidnas, etc.) plus 200 heads and 300 pairs of antlers

2400 mounted birds (including 62 Birds of Paradise, 520 Hummingbirds, 62 Cassowaries, etc.)

680 reptiles (including 144 giant tortoises)

914 fishes

A representative collection of Invertebrates

## Students' Department

1400 mammal skins and skulls
300,000 bird skins (all but 4470 sold to New York Natural History Museum)
200,000 birds' eggs
300 dried reptiles
2¼ million set Lepidoptera (thousands of types)
300,000 beetles (sold or given away except for types and paratypes of two-thirds of the known species of Anthribidae)
Library: approximately 30,000 books (including Audubon's *Birds of America*, Merian's *Insects of Surinam*, Moses Harris's paintings and drawings for the Aurelians, etc., etc.)

### Some Tring Museum contributions based on a study of the collections

| | |
|---|---|
| 1894-1939 | *Novitates Zoologicae* (Tring Museum Periodical: 40 vols, approx. 23,000 pages, 600 plates) |
| 1895 | Trinomial nomenclature adopted, linked to a new concept of subspeciation based on geographical representation |
| 1896 | Biological species clearly described and defined |
| | First clear statement that speciation is the joint product of mutations and isolation |
| 1896-1905 | Geographical isolation described: a prerequisite for the building up of an intrinsic isolating mechanism for speciation |
| 1903 | Monograph: *Revision of the Lepidopterous Family Sphingidae* |
| 1903-22 | Monograph: *Die Vögel der Paläarktischen Fauna* |
| 1911 | Genetic theory of mimetic polymorphism presented |
| 1916 | Sympatric species defined |
| 1931 | Account of work on geographical variation and speciation in "The pioneer work of the Systematist." |

# CHAPTER 1

# Walter Rothschild

IT IS NOT EASY to be born. The average man is squeezed out into the world with blood to lubricate his passage and wild shrieks of anguish to speed him on his way. If, as an additional complication, he arrives on the scene with a theoretical silver spoon in his mouth, the life awaiting him beyond the draw-sheet is fraught with extra hazards. Walter Rothschild, however, was born in a serene silence. It was before the days of anaesthesia and relaxing pre-natal exercises, but his mother was a woman of deep sustaining faith and a will of iron. At the end of 48 hours of labour there was nothing for her son to hear beyond the muffled clatter of hoofs and the rumble of wheels passing over the straw spread in the street below. While on the one hand the presence of the much envied silver spoon imposes certain grave restraints on the child of good fortune, it ultimately affords great possibilities, confusing alternatives and an agreeable vista of ease and indolence amid delicious surroundings. Walter was confronted with none of these problems: from the early age of seven he knew exactly what he was going to do. He said so and he did it.

There is no record of where or why or how the first meeting with Alfred Minall took place. Minall worked as a joiner for a firm which helped in the rebuilding of Walter's home, Tring Park. He was also, apparently, a highly skilled taxidermist. During the conventional afternoon outing, flanked on one side by a German governess and on the other by a nursemaid pushing a pram — in case he got tired — Walter seems to have spent a never-to-be-forgotten enthralling hour watching Minall skin a mouse. Probably on this occasion he saw for the first time a small collection of stuffed animals in Minall's cottage.

Nursery tea was attended by both his parents and during a lull in the conversation he suddenly stood up and made, for a seven-year-old, a long and crystal clear pronouncement: "Mama, Papa. I am going to make a museum and Mr. Minall is going to help me look after it."

Throughout manhood Walter suffered from a crippling impediment in his speech which rendered normal conversation virtually impossible, and there is no reason to suppose that during childhood he was free of this particular handicap. But on this occasion he neither hesitated nor groped for words.

His audience was understandably impressed.

Fifty years later he had amassed the greatest collection of animals ever assembled by one man, ranging from starfish to gorillas. It included 2¼ million set butterflies and moths, 300,000 bird skins, 144 giant tortoises, 200,000 birds' eggs and 30,000 relevant scientific books. He, and the two collaborators he had selected as assistants, described between them 5000 new species and published over 1200 books and papers based on the collections. A century later over 100,000 people visit his museum every year. In 1880, when Walter was twelve years old, Minall duly became his first curator.

# CHAPTER 2

# Tring Is Fairyland

THE "ODD MAN" at Tring Park, dressed in a white boiler suit for the purpose, prepared twenty-four large scuttles overnight to assist the housemaids in laying the fires the following morning. These were placed in strategic positions in little serving rooms around the house so that no maid need carry a heavy scuttle — with a shovel attached — more than a few yards. There were five housemaids and they laid the first fire at 5 am and the last, Lady Rothschild's* bedroom fire, at 7.30. That one, and only that one, was lit by the head housemaid herself. The bottom of each scuttle was filled with small lumps of coal; there followed two little logs and two bundles of short split sticks, cut to precisely the same length, then a bundle of thin brown twigs of the type used in the construction of besom brooms, then a couple of handfuls of curly, fresh wood shavings, two neatly folded sheets of newspaper, a box of matches, two or three home-made spills and a single black velvet, fingerless glove. On the top of the scuttle was a removable tray which contained an empty compartment for yesterday's ashes and clinkers, a brush, a pot of grate blacking, a pot of brass polish and a leather, and, folded neatly across the top, a coarse linen checked cloth, eventually spread by the maid on the hearth, upon which she knelt like a prayer rug. The head housemaid was a stout lady, tightly corseted, and she found kneeling a trial .... These scuttles aroused in the grandchildren an unassuaged, dangerous longing to light fires — so enticing did they look — and were in some strange way the very essence of Tring Park. You knew as you looked at the scuttles that a damp log was unimaginable. The fires lit effortlessly — it was never necessary to hold the paper across the grate to get the flames to draw. The ritual was, like everything else at Tring, a little severe and impersonal, but the essence of quality, of excellence, always interesting.

Emma fell in love with her corner of the Chiltern Hills at first sight. It was a wonderful love affair which lasted, serene and untroubled, until she died at the age of 91 in her bedroom overlooking the beechwoods along the crest of the ridge. She gave her heart to Tring Park, like a queen to her country, and in

* Emma, Walter's mother.

3

return it brought her great happiness. If only one could be sure that, during her long, orderly, narrowed and melancholy widowhood — for she never recovered her *joie de vivre* after Natty's* death — she looked back over the sixty years of her reign with a sense of great achievement, untroubled by the fear of change and disintegration threatening Tring Park. But no-one knew how she envisaged the future. She never complained or explained. Possibly she was content to leave it to her daughter-in-law Rozsika. In one single characteristic Emma resembled her elder son, for although outwardly at ease, accomplished, dignified, with impeccable manners which came from the heart not the head, and all the social graces, yet inwardly she was reserved and shy — even diffident — and, like Walter, found communication extremely difficult, and revelations which contained even the hint of an emotional undertone, well nigh impossible.

Tring Park was a Wren house, and it stood in a seventeenth century deer Park miraculously untouched by time,[1] at the foot of the Chiltern Hills. The fish ponds had at some period been drained, but otherwise all remained as it was when Charles II — so we are told — made love to Nell Gwynne at Elinors the smaller house on the estate. The Black Boy had perfect taste and flair, for no more beautiful spot could be found anywhere in England in 1873 than the ridge of hanging beech woods dropping down into rolling parkland, with the Vale of Aylesbury spread out in a peaceful emerald patch-work, meeting the misty grey skyline beyond the tiny wedding-cake-like turrets of Mentmore in the middle distance.

The Rothschild connection began in the eighteen-thirties when Nathan Mayer Rothschild,** the founder of the English branch of the family business, and then 56 years old, rented Tring Park as a summer residence from its owner, Mr. Kay.[2] But a quarrel ensued when a dog belonging to one of his grooms pulled down a deer in the Park, and the owner threatened not to renew the lease. Walter's younger brother Charles, in his scrap album of humorous mementoes, preserved a card which was addressed, in 1834, to N.M. Rothschild, Esq., reading as follows: "A haunch of venison from Tring Park. With Mr. Kay's compliments." Was it sent before or after the Great Squabble? We do not know.

Uncle Anthony de Rothschild's daughter, Constance, and Emma were close friends, and the former was pressing Emma, who had married in 1867, to find a home near hers in the vicinity of Aston Clinton. Quite apart from her sincere affection for her cousin, a trusted married friend in the neighbourhood would

---

* Nathan Mayer, 1st Lord Rothschild, Walter's father, died in 1915.
** Walter's great grandfather.

provide Constance with a suitable chaperone and a little more freedom, which at that moment was sadly lacking. Natty and Emma were very much in love and longing to find a place of their own, especially near Aylesbury, which he had represented in Parliament[3] since the age of 25. Furthermore Lionel Rothschild and his wife — with whom in the early days the young couple shared not only a London residence but Bentley Priory — were greatly loved and respected, but remorselessly demanding and exacting parents. Suddenly, in 1872, Tring Park came onto the market. Walter's father — we know this from a letter written by Emma to Constance, and recopied in her leather bound locked notebook — would not have dared broach the subject himself and certainly had not the means to buy Tring.[4] No doubt Connie urged her father Sir Anthony to make the suggestion. In 1842 Mayer Amschel (Nathan Mayer's fourth son) had acquired a property at Mentmore.[5] Anthony had bought Aston Clinton[6] estate about the same time, and Lionel himself had acquired Halton and various farms around Aylesbury, Hulcott and Bierton. Later Baron Ferdinand, who decided to settle in England in 1859, built Waddesdon, seven miles from Aylesbury. Therefore the purchase of Tring fitted in well with the family's interest, both political and rural, in the district. In May 1872 Lionel duly bought Tring Park for £230,000 — a large sum of money for a country house in those days. It is doubtful if the future owner played a major role in the considerable alterations carried out at this stage, although it seems odd if Lionel did not seek his son's views, as even while at Cambridge Natty was consulted over a wide variety of issues, particularly the farm and garden, and often tendered terse advice,[7] unasked for. Certainly Emma — aged 29 and both competent and energetic — was not invited to consider the decoration or furnishing — not even for Baby Evelina's nursery. She was merely shown patterns of the chintz once it had been selected. It appears that Tring Park was lent to the young couple, complete in every detail, rather like the loan of a canteen of silver. From a letter written to her mother-in-law it would appear that Emma's first glimpse of Tring occurred the day she moved in! It is hard to believe that she and Natty had not contrived to ride over while staying at Aston Clinton, for they were both keen riders — a mere three miles distant — for at least one surreptitious visit. This is her first impression of Tring:[8]

*Tring Park [No date]*

*My dear Aunt,*
*[for Charlotte was Emma's aunt before she became her mother-in-law]*

*I write in a great hurry to give these lines to Sales. (Natty has not yet arrived.) I am so bewildered by all I see that I cannot find words to express to you how much I admire this beautiful house and its perfect arrangements. Everything is charming and combines elegance with comfort. I am writing in my bedroom which is so magnificent I can hardly believe myself to be its occupant. I have often longed to possess the carpet which is described in the Arabian Nights and to be transported to fairyland, but I have obtained that and without the carpet, for Tring is indeed quite fairyland and I cannot thank you enough for all the trouble and care you have bestowed on this house and for you and Uncle Lionel and your kindness in allowing us to live here. The dining room is a perfect success and the pictures look as if they had always been placed there. The blue room is too lovely and I could not imagine the beauty when you showed me the paper and the chintz at M-----! It would take too long to enumerate all I admire but I see a predominance of my favourite Chinese and Indian ornaments — one prettier than the other. Thousand thousand thanks! I hope my inadequately expressed words will not lessen the assurance of my deeply felt gratitude. The childrens' rooms are too beautiful and Baby [Evelina] looks at the pictures with delight. Walter was not sick and seems none the worse for his journey. "Walter thanks you very much for furnishing all so nicely." This is his own message....*

*Your loving*
*EMMA*

Although Emma, who was born and educated in Frankfurt, was quinquelingual and spoke English, French and German perfectly, not only without a trace of accent, but with every idiom in its right context, in the early days of her marriage her letters in English were a trifle uneasy — but even allowing for the need to extol her mother-in-law and appear overwhelmed by her generosity, there is a truthful ring about her enthusiasm.

For obvious reasons she dwells in this letter upon the house itself, but what Emma really loved was the park with the fallow deer under the trees — for Mr. Kay's deer were still there, as indeed they had been continuously since Charles Gore's[9] day — and the mauve and white violets in the short grass, the bluebell wood and the rooks, which nested in trees touching the roof and cawed of unnumbered springs as nowhere else. Natty was also a great lover of nature: a sentiment which he and his wife shared and which they passed on to both their sons. Later she wrote:[10]

*Here we have bright sun and a pleasant breeze .... We are out nearly all day and the children enjoy themselves exceedingly. Tring Park looks beautiful; the grass and the trees are luxuriantly green and the air is perfumed with the smell of hay and the lime trees. The hay has not suffered much and they are speedily getting in the last crop in the Park .... Excepting a walk to the P.O. I've not left the Park precincts which are more entertaining than anything beyond.*

Emma was never an outdoor gardener and there were merely a few token flowerbeds and stone urns filled with cascading fuchsias and blue lobelias in front of the French windows. A cunningly concealed haw-haw divided the house and lawn from the rolling chalk grassland. She loved the illusion of living *in* the Park at the foot of the Chiltern Hills. It was the most beautiful place in the world.

Natty, it must be acknowledged, made a fearful mess of the subsequent alterations to Tring Park. In enlarging it and improving its comfort he entirely spoilt the outside of the house and produced a solid block of red and white masonry with a slate grey roof, vaguely reminiscent of a large hospital. Some unkind people likened it to a well windowed Victorian prison, but Loulou (Lewis) Harcourt, a friend of the family[11] thought it very pretty and exceedingly comfortable. It certainly resembled no other country house in England. Underneath its curious façade the Wren structure is far from lost; it could easily have been rescued. A dozen or more paper knives were cut from the rejected timber and adorned with a silver inlay engraved "From oak beams, Old Tring Park."

The interior decoration and furnishing at Tring was a weird and wonderful mixture of the very beautiful and immensely ugly. Lord Newton, in his diary,[12] described it laconically as "not attractive". But it was in the grand manner; only the arrangement was sometimes horrific. There were so many fine things in the house, from the *Morning Walk*\*, and the parcel gilt Orpheus Cup[13] to the Spanish leather wallpaper and the delicately gold-cased Mezuzah hanging on Lady Rothschild's brass bedstead; furthermore the rooms were so lofty and airy and well proportioned that with a little imagination and a scraping away of dark, shiny, brown Victorian paint all the incongruity could have been eliminated. Except perhaps for the *décor* of the smoking room .... This was added at a much later date by Natty himself. At intervals throughout his life Natty was subject to extraordinary lapses of artistic taste, due to the deeply sentimental streak in his nature, which occasionally overwhelmed him, swamping his clear vision and normal sense of values. He bought *Squire Hallet and his wife*, he bought *Mrs. Lloyd,*[14] and shortly before he died he acquired the original water colours of Blake's *Songs of Innocence*\*\* (who was not at that time

---

\* Undoubtedly the most beautiful picture ever painted by Thomas Gainsborough, bought from Agnew (who amused Emma by calling her *Vanda coerulea* "Horchids") in 1884 for £10,000 ( a complicated deal, so this is an approximate figure). Natty was very disappointed because she could not accept Gainsborough's *Two shepherd boys with dogs fighting* — Emma could not bear dog fights. Natty rightly felt it was a wonderful picture, and offered it to her as a birthday present.

\*\* There was a hint that it was his intention to give the *Songs of Innocence* to Lady Gosford (in 1915) as a present. Natty had a long and close friendship with this delightful woman and several years before this we find him writing to Haldane for a copy of his translation of Schopenhauer as "Lady Gosford is going to Switzerland and I believe she has been unable to obtain a copy".[15]

a fashionable artist), yet he also fell in love with the amazing group of pseudo-Greek marble figures — kitsch, and very bad kitsch too — which he installed over the fireplace in the Smoking Room at Tring. Round the walls were matching, bas-relief white marble plaques, of nude sexless figures (short hair was the only feature clearly distinguishing men from women) each one set in a pink marble surround. He had seen the originals somewhere in Italy, and had them copied at immense cost, and sent to England. He installed his own superb Louis XV desk near a window and surrounded it with leather sofas and armchairs, Limoges plates and palm trees in Chinese cache-pots. The family used it as a sitting room. In his old age Walter was still occupying one of these immense leather arm chairs, with a soda water siphon on a table beside him and a pile of Wild West magazines on the floor — the Art Nouveau pseudo-Victory of Samothrace still balanced gracefully on one toe on the prow of her ship, eternally holding her bronze olive branch aloft. With six plate glass windows, pink and white marble walls, and rich, highly polished chestnut reddish-brown wooden doors, with elaborate brass fittings, it must have been unequalled anywhere in the world as an anti-cosy sitting room.[16] Richard Meinertzhagen — a friend of Walter's, one of the best ornithologists of the day and an ardent Zionist — wrote in his diary:[17] "Whenever I go to Tring I feel I must dress in my best clothes. The house is luxuriously furnished, and contains so many works of art and lovely things that I feel I must dress for them. I always enjoy going there. Lady Rothschild is a wonderful old lady with many reminiscences as clear as crystal." But this belonged to the future .... On the day on which the family moved into Tring Park and the enthusiastic letter was dispatched to his mother, Natty was in his mid thirties, and the happiest and most creative years of his life lay immediately ahead. His son was only five years old. Emma already knew with profound conviction that she and Walter would never want to live anywhere else but in Fairyland, and they never did.

*Dear mama*

# CHAPTER 3

# Emma, Walter's Mother

THE ROTHSCHILDS are a truly incestuous family. It was always assumed by the world at large that marriages were arranged between cousins in order to keep the family fortune intact, readily available and secret, and there is little doubt that this was an obvious and practical advantage of such unions. (Among themselves it was said — not without a grain of truth — that the wives' marriage settlements consisted of worthless South American railway shares that had proved a nuisance to their men folk.) But usually the explanation was quite otherwise — the cousins fell in love with one another. Walter's irreverent nieces[1] remarked that it was a wise man who knew which Rothschild was his own father....

Emma and Natty's romance followed a slightly different course from usual, for Emma (whose father was head of the Frankfurt branch of the Rothschild family and had seven daughters but no male heir) had already decided to accept the Baron James's offer of marriage — although the thought of living in Paris did not greatly appeal to her — when she and Natty discovered they had fallen madly in love. This was the real thing! There was only one possible solution as far as Emma was concerned, and she never experienced a moment's doubt. It was not without reason that Lord Rosebery — who was then Prime Minister and married to her cousin Hannah — exclaimed: "Emma, you have the courage of a lion!"[2] This courage sprang from a confident belief that she knew what was right and what was wrong, and Emma all her life uncompromisingly and forcefully followed these deep convictions. In this instance duty and desire were for once indissolubly linked. Eventually the shocked and sorrowful Baron James had to be content with the next best — her younger sister Thérèse — while Natty and Emma lived happily ever after. Their engagement, which was very brief, caused momentary consternation and required most skilful handling in order to prevent family discord. It was either tactfully ignored by everyone in current correspondence, or — what is more likely — the letters referring to it were immediately destroyed. Emma was doubtful about her reception at Gunnersbury (her future father-in-law's house) once Natty had announced their decision, and was immensely relieved and sincerely grateful when Baroness Lionel — her aunt Charlotte — received her and her mother with open arms. On her return to Frankfurt on September 21st 1866 she wrote her aunt and uncle a six page letter.[3] Had one not known of the underlying drama, this letter would not have attracted any attention, and no doubt for this reason it was not destroyed; a sigh of relief escapes from between the lines:

> *Our passage was excellent, but though we did not suffer from the usual discomfort we were very sick at heart. I can assure you Mama and I talked unceasingly of all the Dear Ones we had just left and I cannot tell you how sad we feel to have parted from you. Pray let me thank you for all your kindness and love — my heart was too full when at Gunnersbury to permit me to express all the deep thankfulness I felt for your affectionate welcome. I only hope to show you my gratitude one day by true filial devotion, and trust I may not prove quite unworthy of the numerous proofs of tenderness you have shown me. My words very inadequately express all I feel but I claim your kind indulgence and hope you will see in these lines the tender and affectionate heart which dictates them.*

Natty, under his shy, gruff exterior and disconcertingly abrupt and forthright manner, was an exceedingly affectionate and sentimental man. In his wallet he always kept one of Emma's childhood letters,[4] written to his mother, and a love poem she wrote for him. His most treasured possession was a small gold heart — her first present — which he carried on his watch chain. He never tired of explaining to his granddaughter that the tiny little, almost invisible stones — worn and flattened by forty years of persistent knocking against gold links — spelled DEAREST by virtue of the first letter of their names: Diamond, Emerald, Amethyst, Ruby, etc. After his death Emma transferred the gold heart to her own bracelet. Natty also framed, in a blue and gold surround, the pencil scrawl which he wrote out as a cladder for the press anouncement of their engagement. As for Emma, she kept every line he ever wrote to her, and she inscribed a little bible bound in red leather in her flowing script: "This Bible was used by Lord Rothschild when sworn a Privy Councillor on August 11th, 1902."[5] Then, on the opposite page she wrote, "Remember now thy creator in the days of thy youth. Ecl. XII.i." — for she was a sternly religious woman.

Rosebery noted years later in his diary[6] that when a letter arrived for Natty — he was staying at Mentmore for the weekend — containing the news that Emma, who was at Biarritz with her sister, did not feel well, he packed his bag and left immediately for the Basque coast.

The tragic death of their daughter Evelina and her prematurely born child following a railway accident, prevented Natty's distraught parents from journeying to Frankfurt for his marriage. This melancholy circumstance gave posterity several descriptions of the wedding week in the form of relatively detailed letters home. Natty, who was almost disconcertingly indifferent to the outward manifestations of wealth, was nevertheless delighted that his parents should lavish such wonderful gifts of jewelry on his bride and thus amaze his future father-in-law, the Baron Carl, with the generosity, purchasing power and industry of the English House — for a necklace such as this was evidently the reward of years of assiduous collecting of individual pearls. For the Rothschilds who fell in love with Rothschilds didn't care two hoots about what the world thought, or whom they might *épater* — except one another. Natty hated letter writing, but under the circumstances he yielded with as good a grace as possible.[7] "I have this instant given your beautiful jewels to Emmy. I need hardly tell you what wonder and amazement was depicted on the faces of all.... I think the pearl necklace excited the most astonishment — they all agreed the pearls were like pigeon eggs." He described the party which took place in the evening in considerable detail (for him) and remarked that for beauty and distinction and exquisite dress no one could hold a candle to Emma. Alfred, his younger brother, was also impressed by his future sister-in-law — who in days to come was to infuriate him beyond endurance by her stubborn refusal to invite his mistress, Mrs. Wombwell, to parties at 148 Piccadilly.* "At the civil ceremony", (on the 16th April, 1867) he wrote, "Emma looked truly beautiful where she wore [in deference to the half mourning for her sister-in-law] a lilac gown and a white bonnet and white Indian shawl."[8] Brother Leo wrote in similar vein: "As you may suppose they are radiant with joy and you have no idea what a handsome pair they made. Emma looked beautiful last night...."[9]

They were married the following day, and at five o'clock in the afternoon left for a honeymoon in Heidelberg — stopping at various places en route. For the next fifty years of their lives Natty and Emma remained a mutual admiration

---

* Alfred could never forgive Emma for her implacably straightlaced censorious attitude concerning his deviations from her severe code of sexual morality, i.e. his open liaison with Mrs. Wombwell. "All Emma's children must have been conceived under protest", he remarked sourly. At one dinner party he told a very *risqué* story, and then added innocently: "Emmie told me that". In his will he tried to annoy her by his bequest of the beautiful but naughty Greuze painting *Le baiser envoyé* — [10] a picture Emma immediately gave to her daughter-in-law Rozsika who in due course sold it to the Baron Maurice in aid of Jewish victims of Nazi persecution. Her son Charles also disapproved of Alfred de Rothschild, although he appreciated his peculiar sense of humour, and suggested that it was the greatest art of all to let others work for you throughout your life. Privately he told Rozsika that "to be rid of Alfred was cheap at any price." But in those days differences of opinion within the family were not publicised in the popular press. Walter, however, must have quailed at the thought of his mother's humiliation if it ever became known that *her* son shared certain weaknesses with Alfred.

society. One long black shadow fell between them — Natty's bitter, harsh, unyielding disapproval of his eldest son. Walter got under his skin. Eventually he gave up the unequal struggle of trying to understand him and, stepping aside, left Walter to Emma.... Yet Natty had always handled his two younger brothers with great skill and kindness, encouraging them, praising them, mothering them — like an old hen when they first came up to Cambridge — and in later life finding just the right niche for each of them at New Court, where they made maximum use of their respective talents. Sometimes at meetings in the partners' room he forgot himself and announced: "I have decided...." and Alfred would murmur discreetly, "*We*, Natty — *we* have decided...."

"Porchy" Carnarvon (who claimed to be Alfred's grandson)[11] remarked that one of the most attractive human attributes is shining good health, and Emma and Natty were both endowed with this particular quality. Neither of them experienced a serious illness before they were over 65 and both possessed enormous, almost ruthless vitality and drive. Curiously enough they also both manifested a natural but impeccable neatness about their clothes,* as if they enjoyed a negative charge for dust and disorder, and Natty's buttonhole — whether it consisted of his favourite *Dendrobium* or a pink carnation — always seemed to have been picked at that very moment, with dew still on the petals, and Emma's hair, by some mysterious circumstance, arranged by a skilled hairdresser only five minutes previously. Walter inherited this characteristic, and one of his ribald nephews, struck by the contrast between his uncle and his scruffy self, once remarked with awe: "Isn't Uncle Walter CLEAN."

The lines on Emma's hand would have delighted and surprised a soothsayer, for her fate line bisected her palm vertically in a uniquely deep, straight, unbroken, unwavering furrow — born a Rothschild, married a Rothschild, loved and lived a Rothschild, for ninety odd years. After her marriage she threw herself wholeheartedly into Natty's activities and immediately gave up her own hobbies, her painting and piano playing — she played Chopin with a delightfully light touch, dispensing with a score — were soon abandoned for there was no time to pursue them seriously. She also gave up swimming, although she always remained proud of the fact that in gaining her life-saving certificates she had plunged into the water fully dressed and had swum with a full length skirt billowing out behind her. She certainly quickly dismissed the quaint accomplishment of walking upstairs on stilts, but that, too, was sometimes recalled with evident satisfaction and amusement. The seven sisters were

---

* Natty, perhaps hoping to conceal the fact that he was not a tall man, deliberately wore trousers a shade too long for him, so that they "concertinaed" round his ankles.

a lively bunch of girls — but Adèle and Emma should both have been born men, for — although Rosebery was wrong in evoking a lion rather than lioness — both women had a masculine streak in their make-up, and Emma was far tougher than Natty. "Your grandmother is not the complaining sort," said Rozsika curtly to one of her daughters who marvelled that Emma made no mention of pain or discomfort after her operation for glaucoma, and only talked admiringly of her surgeon's skill. Although no more was said, Rozsika's tone implied that, alas, her own daughters were not cast in the same heroic mould.

Emma was a perfectionist and did everything well. A tiny but impressive skill which symbolised her life-style was the way she packed up a small parcel. Everything seemed to be miraculously at hand — the stiff white linen paper, the thin, dark blue string, the navy blue sealing wax and pink tipped wax vesta matches. The ends of the packet were always symmetrical, the seal proved a perfect oval and the imprint as sharp as the model itself, the address fitted easily into the available space, yet the writing was clear and free. No professional parcel from Garrards or Cartier looked as good.

Emma took a lot of trouble with her written English when she first married.* Conversation presented no hazards; she had learned the language in the nursery for her mother, Baron Lionel's sister, was English. Nevertheless since her early letters to Baroness Lionel presented a Germanic turn of phrase, she diligently worked at this problem. Her written English eventually became so flawless that her son Charles — returning from a round the world trip — sent her a brief article which he had written for the *Entomologist's Record*, to criticize and correct.[13] The subject was the work of a charming Japanese Lepidopterist, with whom he had fallen a little in love, and he felt that only his mother's impeccable grammar and style were worthy of the occasion.

Emma and Natty were both very sociable, outgoing people, and now, in the early days of their marriage, and later, when they moved to Tring, Emma's letters to her mother-in-law were all about the family weekend parties, fine dresses, drives and walks through beautiful parks and gardens, trees — especially trees which she loved — flowers, new acquaintances, the war in Germany, Royalty, the Dreyfus case, balloon post and well selected gossip. Thus, Lord Stanhope's son "informed us that Lord Rosebery who had also been to Balmoral 'on trial' caused Princess Louise to waver. Though he had not forgotten Miss Fox he was quite ready to become the husband of the charming

---

* Baroness Lionel was also interested in English. She annotated a letter from Natty at Cambridge. "What does avuncular mean? We cannot find it in any dictionary. I suppose you coined it."[12]

Princess", and again, "Although the Queen of Prussia had a long *tête-à-tête* with Papa, H.M. gave him no information, the fact is Augusta knows nothing.... it seems the dinner at the Royal Table was extremely bad and the only good things were the Sèvres plates and dishes of dessert."

She certainly did not inflict her good works or her considerable erudition on Baroness Lionel. Books were rarely mentioned, although on one occasion she advised her father-in-law to read Wilkie Collins. She and her cousin Constance together paid a 24 hour visit to Blenheim — which delighted Emma, who was always interested in everything new and unusual. On this occasion the glass beehives, which enabled you to watch the industrious insects at work, particularly took her fancy. There were also at least a dozen emus — very tame — in the park, and kangaroos, but the grapes in the hothouses were not as fine as those at Gunnersbury.

> *I was enraptured by the marvellous pictures. The magnificent Raphael, several beautiful Van Dykes and a whole room full of more or less indecent Rubens.... The Duchess devotes much of her time to schools and has a small Industrial Home of her own where she trains girls to become maid-servants. The Ladies Churchill teach there; they are very nice well mannered girls and the youngest who is not out is very pretty and an accomplished musician."*

In a reply to some remark by Charlotte about the marriage of the Prince's daughter to the Earl of Fife, she replied: "I pity poor Lady Ely for being compelled to witness the love making of Princess Louise and Lord Lorne. It must have been equally disagreeable for both parties."

It was more surprising that Emma referred only briefly to religious matters and education in her letters to Baroness Lionel, for these were subjects in which both took an interest. When Leopold, a freshman, wrote to his parents from Cambridge he was careful not to neglect this point:[14] "I was listening to Ruskin's lecture which was quite the most beautiful I have ever heard, it would have pleased you, dear Mama, as it partook more of the nature of a sermon than a lecture." There were rebukes to Dons "for at the University they paid more attention to science than the great moral truths of life."

What amazes the reader a hundred years later is the variety of activities Emma managed to cram into her days, and the number of visits she paid. She went frequently to Aston Clinton, for she dearly loved and admired Connie's mother who was an angel and wrote and painted with great skill. As for Connie, she was fat and jolly and high spirited, and at that time her deep-seated streak of spitefulness was not apparent. Both young women enjoyed the country and rural pursuits and they were extrovert by nature and liked to entertain a lively crowd of visitors. Both were born Rothschilds. This was a great boon — for explanations of so many awkward facts — religious, racial, social, familial — were then unnecessary! There was much to be said in favour of female as well as male cousins. Though surprised and deeply shocked by the announcement of Connie's forthcoming marriage, Emma stood loyally by her cousin. It was not so much that Cyril Flower was a Christian, but she considered him a man of dubious character and quite unworthy of Connie. He was far too beautiful — decked out in a wine coloured velvet coat — and several of his male friends were also endowed with an undesirable type of good looks....

After her husband's death Connie decided to leave Aston Clinton and move to Overstrand, abandoning for ever the Tring district, which as a girl she had loved so much. She left a strange directive for her heirs, which somehow belied her jolly insouciant exterior: bells were to be tied to her wrists and ankles before she was placed in her coffin. All her life she had suffered from a secret horror of being buried alive.

The other cousins at Mentmore were visited almost as often — Natty commenting freely on Aunt Juliana's love for the sound of her own voice and the joint weight of the couple, which much have been around 30 stone....[15] Even if interesting men were present, commented Natty — who firmly believed in the superiority of the male sex — Juliana insisted on talking rather than listening. On one occasion a Russian Princess, who was travelling with the house party on their way to watch the Derby at Epsom prevailed upon Hannah (Juliana's only daughter) to practise her powers of suggestion upon her. The Princess, at Hannah's bidding, fell into a profound sleep and could not be roused for several hours — long after the party drew up at the Durdens. Hannah was frantic with fear and swore that she would never again hypnotise a living soul. In her old age Emma would tell the story with great gusto, enjoining her grandchildren "on no account to tell Cousin Sybil" (Hannah's daughter) who, Emma commented dryly, was quite potty enough as it was....[16]

Hannah is rarely praised in contemporary memoirs, but Winston Churchill, like Emma, appreciated her, and saw beyond her faintly embarrassing, doting love for her husband. He described her[17] as "a remarkable woman on whom he [Rosebery] had leaned.... She was ever a pacifying and composing element in his life which he was never able to find again because he never could give full confidence to anyone else." Churchill realized just how much Rosebery depended on her, despite her husband's attempts at concealing this behind a not altogether attractive form of banter. Chaff was something, said Emma, a trifle censoriously, which could be overdone, and Archie had caught it from the Rothschilds themselves, who used it both defensively and as a form of wit, and it had become a tiresome habit.

Rosebery never quite recovered from Hannah's death. Winston, who had a unique genius for the *mot juste*, described him without her as "maimed".[18] What Emma really thought of the wisdom and propriety of Hannah's marriage outside her faith we shall never know,[19] but she and Rosebery conducted a life-long friendship, based on enduring mutual respect and admiration. Loulou Harcourt noted in his diary that on one occasion Emma and Evelina had been thrown out of a pony carriage — Emma "a good deal hurt". Rosebery, hearing of the accident on Sunday "came over to enquire for her expecting to find her alone and when he plunged into the dining room and found us all there a look of horror came over his face and he incontinently fled."[20] In her old age Emma paid visits only to a small circle of acquaintances, but Rosebery was among them.[21] Maybe she felt that, in their own ways, they had both suffered a similar sort of extinction when their respective spouses had left them. And now there were just the grandchildren.

Emma was not creative like Natty, and did not have the same knowledge of the world and the human race, but she was more skilled, more efficient and although so full of vitality and so earnestly resolute, was also cooler — nor did she grant favours easily. Her judgement was more detached. But like many cool people she inspired burning passions in others and — long after her death — a crystal clear memory of her personality remained in people's minds. One of her neighbours at Mentmore wrote to her in 1871: "I always think your perfect truthfulness of nature has not been sufficiently dwelt on. It is one of the greatest charms in my eyes that I believe you to be incapable of lying."

Another woman friend who evidently felt insecure and worried lest Emma did not return her feelings so ardently wrote in fine, flowery Victorian style: "I have.... an implicit trust in your affection. You cannot tell how highly I value it and how tenderly I return it. Some things can never be put into words.... There must be some mysterious link that will bind us to one another until our lives end and I do not believe anything could give me such exquisite pain as to think that we should ever drift apart. Never fail to remember that neither joy nor sorrow can touch you without reaching me.... How I shall yearn to pour out my heart to you when we are far apart. Where there is such utter trust the relief is sometimes unspeakable and you don't know how much good you have often done me." And again, "I believe you give me the only light in my existence." Emma copied this into her note book and it thus escaped destruction by her conscientious executor. There were others in the same vein — letters in German and also in French.

It is a curious fact that the Rothschild family, so *avant garde* in many areas, ranging from their agricultural settlements in Palestine to jam in tubes, were pitifully slow and unenterprising in recognising, let alone using, the business ability and political talents of their women folk. Their wives and daughters were assigned the traditional role of powerful matriarchs and gifted hostesses in the home, but in financial matters and party politics were kept in total ignorance. Natty, so go-ahead with milk recording and electrification, was a supporter of the anti-suffragette league and firmly opposed to women's franchise. Probably Mayer Amschel's testament (dated 1812)[22] aggravated this conservative attitude in the family. Article 1 includes this phrase:

*I will and ordain that my daughters and sons-in-law and their heirs have no share in the trading business existing under the firm of Mayer Amschel Rothschild & Sons.... and belongs to my sons exclusively. None of my daughters, sons-in-law and their heirs is therefore entitled to demand sight of business transactions.... I would never be able to forgive any of my children if contrary to these my paternal wishes it should be allowed to happen that my sons were upset in the peaceful possession and prosecution of their business interests.*

Clearly Mayer Amschel — founder of the family fortunes — had little confidence in his daughters' choice of spouses, and unmarried females were apparently not worthy of consideration. Even Nathan Mayer's Will failed to put matters right, for on his deathbed in Frankfurt[23] he enjoined his "Dear wife Hannah.... to co-operate with my sons on all important occasions and to have a vote upon all consultations. It is my express desire that they shall not embark on any transaction of importance without having previously demanded her motherly advice...."

From 1900 onwards the numbers of male Rothschilds available as partners for the business dwindled alarmingly both in numbers and talent, whereas there were several unusually intelligent and able women in the family. For instance Walter's cousin Alice (sister of Baron Ferdinand and nicknamed by Queen Victoria "the all powerful"[24]) who never married, was as practical and efficient as Emma, and although not such a blue stocking, was also a remarkably well informed perfectionist. Miriam* (Baron Edmond's daughter) had infinitely more brains and originality than either of her brothers, and for the greater part of her adult life had no ties, while her grandmother Mathilde, who, like Emma, was born a Rothschild and married a Rothschild, would have been a Golda Meir to-day, or maybe a concert pianist, for she was an able pupil of Chopin[25], whom her aunt had popularised in Paris when he was a virtually unknown piano teacher. She published a little book of her own pieces.[26] A fair proportion of the Rothschild women drew and painted well, played the piano

---

* She also contributed a very large fund to Israel in the shape of the Beni Israel Trust.

agreeably and wrote with a fluent pen. Had they been trained and orientated towards the city, there is little doubt that the dreary doldrums in which N.M. Rothschild & Sons found themselves in the twenties and thirties would have been avoided.

Thus Walter's sister-in-law Rozsika was a brilliant business woman, although unlike Emma, she had received no formal education and was perhaps in consequence, a trifle too impulsive. In the world recession of the thirties N.M. Rothschild & Sons was drawn into the maelstrom, and at the time of the Creditanstalt Bank crisis in Vienna (of which Walter's cousin Louis de Rothschild was president) they found themselves in very deep water. Anthony immediately rushed round to consult Rozsika. Behind the scenes she did an enormous amount of work on his behalf, and a year later I was present when he returned to Palace Green to thank her. "All this probably wouldn't have happened if Charles had been alive," said Tony ruefully, "but thanks entirely to your advice and your help we have weathered the storm. I — we — can never express our gratitude sufficiently." "All this" would certainly not have happened and the firm might have achieved great things between the two wars if the partners had invited Rozsika to fill Charles's vacant place in 1923. It would have taken her only a few years to get a thorough grasp of their business as she had learnt the broad outlines from Charles. Lord Newton, who knew Rozsika well as a shrewd and tireless worker for the rehabilitation of Hungary between the two World Wars, wrote in his diary when she died: "3 July 1940.... A dreadful loss. She was one of the cleverest and most charming women I ever met. A real loss to all sorts of people here and in Hungary."[27]

Emma's talents, like those of her future daughter-in-law, were allowed to lie idle. In 1867, within three months of her marriage she became pregnant, but since at this period Emma and Natty shared at least one parental roof there are no letters announcing the happy event. Emma gave up riding only temporarily, but after Walter's birth she stopped hunting. With astonishment she found that she — the skilled and daring horsewoman who loved a fast gallop across country — had lost her nerve. In old age she found the admission was irritating even in retrospect. This was the first tangible impact upon her of her fierce over-protective love for her eldest child. The invisible strands of the iron apron strings had suddenly materialized.

# CHAPTER 4

# Emma — the iron apronstrings

WHILE WALTER was in the nursery Emma and Natty laid the foundations of the mini-welfare state at Tring.[1] It became part of the background of the boy's daily life. Natty tackled the wider issues, slum clearance and the Chiltern Hills Spring Water Company, progressive agriculture, farm stock breeding, the support of local industries and later, following a typhoid epidemic, the building of the Isolation Hospital. Emma concentrated on a more personal angle, and took special interest in the welfare and health of the women and children and in their education. By 1879, when Walter was eleven years old, Emma had entries in her personal account book[2] of some 400 local charitable causes which she had espoused. These ranged from building repairs for several Christian churches, to paying the passage of a pet lamb to Canada for a poor Jewish emigrant family.* In addition to these various personal matters she subscribed to 177 worthy charitable causes in and around Tring,[3] among them convalescent homes, choral societies, lying-in societies, needlework guilds, organ funds, Church Girls Union, Young Women's Christian Association, Tring United Band of Hope, Literary societies and a score of Parish Church projects. At Tring she organised a carving class with a tutor, a dancing class run by Mrs. Minall, reading and club rooms, one of which she built at Hastoe, which, although outside the town, was centrally situated for the estate workers. Prizes were awarded for any piece of skilled work by children in any field of activity from handwriting to recitation.

On anniversaries like Royal birthdays or weddings, or on New Year's day, the Tring children would queue up to receive a commemorative mug filled with sweets and a bright new shilling, handed out by Walter standing dutifully at his father's side. Walter, with the keen eye of a budding morphologist, noticed that one or two boys paraded past him twice, thus receiving two shillings apiece; they had nipped back to the end of the line.... But Walter kept his own counsel.

---

* Natty paid the cost of emigration for 200 families who wished to leave England for Canada during the eighties.

At this point in time there were no old age pensions, no Workmen's Compensation Act, no educational grants, very little if any public transport in the country, and no health service. Emigration grants and county libraries were unheard of. Keeping warm in winter was a major problem. All this is reflected in the ledger notes alongside Emma's list of personal expenditure: for coal and clothing; coal club; coal delivery; for three men involved in an accident; for compensation for loss of fingers; for people suffering from lightning accident; for loss sustained from swine fever; "another horse for a man"; for apprenticeship premiums; for apprenticing young girl and boy; for music lessons for child; for tuition of voice. There were endless items for the benefit of medical cases: blankets, medicines, screens for hospital, necessities for the sick; advice re cancer of the shoulder; operation on child's neck; removal of growth from child's throat, and so on.

She and Natty organised a two tier health service for Tring. For a subscription of £1 per year anyone qualified for free medical attention, free nursing and a nursing home which Emma herself built and equipped. Any employee on the Tring Park estate had this same service free of all charges. In memory of her mother Louise (Lionel's sister) whom she dearly loved and greatly pitied, she built and endowed a row of eight small houses, the Louisa Cottages, for retired employees of Tring Park. It says much for the builders of that period that in the ninety years of their constant occupation no major repairs have been required.

Any man living in Tring who wanted a job was employed in the park woods. Such men were paid piece work and cut the glades and rides and dug up roots. Any wood retrieved in this way was available as a bonus. Everyone in Tring could also lease an allotment garden,[4] either 20×10 poles or one acre in extent — via the Estate Office — for a token sum. They recovered this at Christmas by the gift of a joint of beef and a brace of rabbits.[5] Prizes were awarded for the best garden and the best individual plants. Every child in Tring received a Christmas hamper with food and a choice of toys and a bright new shilling. New cottages were built for all Tring Park employees.[6] In this day and age it is difficult to believe that Emma found quite a number of their tenants objecting to chimneys on the grounds that they were draughty. "I stuff a bag of potatoes up mine," said one of the occupants cheerfully. "To work at Tring Park," said Mrs. Gutteridge, the horseman Arthur's widow, "was insurance from the cradle to the grave. When my mother was took ill Lady Rothschild sent us a message. 'She is to have everything she needs, spare no expense'. She was a good friend to us."

There was nothing paternalistic in Natty's and Emma's attitude. They respected and liked their employees. They considered it right and proper that in return for honest work a man was entitled to a good house, a fair salary, free milk, a pension, free medical care and a suitable place for retirement. And plenty of time off. That was important. The indoor staff at Tring got up at 5 am (Natty went riding himself at 6) but they had alternate afternoons and evenings free. Emma was interested in every department of the farm, the stables, the gardens, the carpenters' yard and so forth. She admired Bert Hinton, the dapper little groom who rode on the step of her carriage — among other things because of his great skill in polishing leather and ironing a top hat;[7] she thought Haystaff the carpenter was a genius — and he certainly could copy any piece of furniture in Tring Park like a practised forger — she appreciated Mrs. Gutteridge's green fingers in rearing a Shire horse foal by hand (the mare having died when the giant baby was born, which in due course grew up into a champion sire), and Mrs. Gutteridge for her part admired the way her Lady-ship in middle age nipped over the stile into the long meadow like a young girl. Emma was a genuine admirer of her marvellous old shepherd and his golden touch with the flock of half breeds. "You wears well, my lady," he remarked smilingly when they met in the Park — two eighty-year-olds enjoying the pale March sunshine. As for Miss Lawrence in the dairy, she was unique — working the first electric churn, but on the other hand modelling the most marvellous tea roses out of butter. She never revealed the secret of her skill which, alas, died with her....

Both her sons inherited from Emma her greatest of gifts, which remains the despair of the biographer. For how can you describe the opposite of a thrower of cold water? How can you explain the lively genuine interest and apprecia-tion which, coupled with knowledge and the rare ability to listen, boosts the other chap's ego, straight away increases his self esteem, and somehow, raises his performance? It is a quality as elusive as the end of the rainbow.

On one occasion Emma was leaving for London after her weekend visitors had departed and she went to the nursery to say goodbye to the children. Walter showed her what appeared to be a battered Tortoiseshell Butterfly which, with the nursemaid's help, he had boxed up while it was sitting on the Michaelmas daisies. "A nice tortoiseshell,"* said Emma perfunctorily. "No," said Walter firmly, "It's not. It's different." Who but Emma would have taken the trouble to sit down on the sofa and get down Morris's *British Butterflies*[8] and hunt through it with care and concentration? Most mothers would have said, "Is it, darling?" and kissed the child an affectionate goodbye. "You are quite right, Walter," said Emma a few minutes later, "It is different. It's a Comma Butterfly. It's really very, very clever of you. How did you know?" But

* Emma knew the commoner species well. In letters to me as a child she referred to them by name.

that was too much for Walter — who was 5½ years old — he merely searched his mother's face with his lively blue eyes and hugged the pill box to his chest in delighted silence. He just *knew*.

Emma was a curiously restrained woman where deep personal emotions were involved, and almost uncannily undemonstrative. Her daughter-in-law, who was herself rather reserved by nature, was taken aback by the lack of cosiness and casual affection in Natty's life. When alone in London he lived in austerity amounting to discomfort, for in winter while she was "in residence" at Tring, Emma refused to remove the dustsheets in which she swathed the furniture at 148 Piccadilly — except in his study. "I am here," Natty wrote to Rosebery in Emma's absence, "like a solitary hermit",[9] and thanked him for letting the children come to lunch. Cosiness was something he had missed in his mother's house, and subconsciously, he must have longed for it all his life, for like Charles he himself was demonstrative and affectionate, and yet both men had selected as their wives women who lacked just these qualities. Emma into the bargain was straightlaced and severely censorious. Even if she did not openly display her profound attachment to her eldest son — on the contrary she bullied him in a kindly fashion all his life, treating him like a child rather than a man — and they never even exchanged a formal embrace, nevertheless her protective relationship with Walter must have been a subconscious source of jealousy and frustration for Natty. A number of men seek their mothers in one side of their wives — which no doubt explains those unexpectedly happy and enduring childless marriages, where the husband basks in the attention and the brand of affection usually reserved for his family. This facet of wifely love was totally denied to Natty. Furthermore there was nothing in Walter's character, his anti-talent for finance, his stubborn silences, his slowness and blank spots and phobic interest in natural history which could console or interest his father. On the contrary Natty viewed his eldest son with increasing dismay and chagrin as time went on, and, with a ruthlessness born of profound and baffled disappointment and regret, he abandoned Walter and turned with relief to his second son Charles.

But even Charles was disappointing in some measure, for he lacked his father's fierce dedication to the Jewish people, and Natty could not see him as his heir in community affairs. Charles in his heart was an assimilationist with socialist inclinations, who believed in participation rather than state ownership, and Natty suspected that in addition to holding these progressive views he might well have stuck his toes in and married outside the faith if he had fallen in love with a Christian girl who was willing to embark on matrimony without parental support. But Charles himself was also subconsciously jealous of Walter's relationship with his mother, although, as Emma wrote to her future daughter-in-law, "he has been a wonderful son to me". The closeness of their relationship was evident to Rozsika for he sent twenty letters to his mother during his three week honeymoon! But, writing to his sister-in-law Sarolta in later life, he referred to Walter, in his fifties, as "Mammy's darling" — for nothing ever makes up for what you somehow feel you missed as a child. Hadn't he been sent to a boarding school while Walter stayed at home with his mother?

Emma and Queen Victoria were faced with certain similar problems in bringing up their eldest sons. Both mothers were straightlaced and censorious, although Emma considered that the Queen went too far. For instance, she thought it absurd that when Prince Leopold remarked to Princess Louise, "Lou, if you have so many eggs for breakfast, when you are married you will have chickens instead of babies", the Queen was so shocked that she had the child locked in his room for the rest of the day. Their eldest sons were both unfortunately attracted to the stage and to undesirable company — especially that of beautiful actresses and dubious characters like Mary Anderson. Both had extravagant tastes, especially in clothes .... "Don't me look beautiful?", asked Walter as a small child, resplendent in black velvet.* And the Prince of Wales, according to his father, at the age of 18 "takes no interest in anything but clothes and again clothes".[10] But the Prince overcame his stammer, his fits of rage and exhaustion,[11] and sawed through the iron apronstrings, married and won his freedom — though at considerable sacrifice — while Walter never seriously made the attempt. He chose to remain the precocious and wilful child, escaping from his father's baleful glare into the fields with his butterfly net and pill boxes, but — despite his speech problems and clumsiness — living severely and greatly beloved at his mother's side.

Karl Jordan — curator for 44 years at the Tring Museum, who had plenty of opportunity for a close study of the situation — laid the responsibility for Walter's eccentricities fairly and squarely on his upbringing. He was, curiously enough for a scientist, unwilling to admit the role of nature and heredity. He blamed the circumstances of Walter's almost regal background and his mother's coddling and over protectiveness, and his father's sky-high expectations for the man's extraordinary contradictions, insouciant irresponsibility, nervousness, baffling ambiguities and almost schizoid streak. In retrospect this seems to be a mistaken viewpoint for without the sheltered environment she afforded him, and the escape from heavy responsibilities, though not from

---

* Walter, despite his size, always wore marvellously well fitting suits; in adult life his interest in clothes was restrained, although he wore a pillar box red silk tie which delighted his ribald nephews. "Uncle Walter is ready for the revolution," they explained.

duty — of which the willing, conscientious, oversensitive Charles shouldered a double load with disastrous results — without the sternly protective shell she provided, without her firm if cool support of his "hobby", what would have happened to the eccentric genius? In fact Walter's ebullient optimism — closely linked to his love for animal life — and his attachment for his mother were together his salvation. His obituary notice in *Nature*[12] stated that "he was better known at home and abroad than any other contemporary zoologist". This was true, but only because he had had Emma for his mother.

# CHAPTER 5

# Walter's Father — the *Eminence Blanche*

NATTY'S MOTHER, Baroness Lionel (Baron Carl of Frankfurt's sister), did not have a very high opinion of her eldest son's abilities,[1] and it is evident that she loved her two little girls more than this "thin ugly baby". But that did not signify, since he was a boy and as such welcome to his father and the whole family. Like Walter, "during his infancy he remained most delicate" and caused his mother great uneasiness. As an eight-year-old he was considered tall and slight for his age — which is strange, since in due course both his sons towered above him. His mother described him as "mild, gentle, kind to all around him and most affectionately devoted to his father and brothers and sisters, but he lacks cordiality and frankness, he is reserved and shy and not generous .... he is constitutionally indolent" — here Charlotte was extraordinarily wide of the mark, a fact she later acknowledged,* for no Rothschild eventually developed more drive and purposeful energy and passionate generosity than Natty — "and does not learn with great rapidity and facility, but his memory is most excellent for names, facts, events and even dates". Baroness Lionel was impressed by the fact that as a child Natty was extremely fond of reading, for him this proved an amusement not an exertion. She agreed with everything his tutor said about him: "Master Nathaniel is very kind hearted, benevolent, fickle in his pursuits, venerates what is great and good, and is gentlemanly in his feelings .... He has not much self-esteem but is fond of praise .... He has a great deal of imagination and is fond of the sublime and beautiful in nature — fond of order. Fine talents for history and miscellaneous information. He has a good memory. Fine taste for the drama and astronomy and colours". Charlotte noted his lisp was improving, but deplored the fact he wrote English, German, French and Hebrew with great difficulty and very slowly indeed. "He must practice writing regularly and indefatigably", wrote his mother, "His slowness and inefficiency in the calligraphic art being calcu-

---

* Writing to Louise[2], Charlotte (in 1867) notes that Natty is very industrious and sets his alarm every day at 4 am. Israel Zangwill described him as indefatigable, "to be found at his desk in New Court at times when even clerks make holiday".[3] The *East Anglian News* in an obituary notice stressed "his amazing industry .... and unrivalled capacity for hard work".[4]

lated to impede his progress in other more important branches of knowledge." She sighed because the estimable Mr. Piper was not strict enough ....

One wonders how Walter would have developed in the care of the relentless Charlotte rather than the fiercely over-protective, sternly indulgent Emma, for it is obvious that he and his father had very many traits in common — even their appreciation of colour and flowers and a poor handwriting — and especially the fact that Natty as a boy was shy and nervous and, according to Charlotte, "neither appreciated in society nor calculated to shine in it, nor fond of it". But even Charlotte and the different emotional environment involved could not have turned Walter into a Natty, despite the surprising number of traits they shared, for nature as well as nurture separated father and son.

A distinguished Victorian writer once remarked that every country gets the Jews it deserves. Be that as it may, it is certainly true that anti-semitism not only varies from country to country, but also from period to period and from one section of the population to another. The relatively mild brand flourishing in England in Natty's day is well described by Max Egremont[5] in his life of Balfour. The author rather lamely seeks to excuse this uncomfortable trait in his hero, whose meeting with Weizmann[6] in 1906 marked "the beginning of one of Balfour's greatest and far reaching enthusiasms".* The key phrase begins: "In common with others of his *time and class* [my italics] Balfour could make scathing reference to Jewish influence and character, and in 1905 had spoken during the introduction of an aliens bill to curb the influx of immigrants mainly from Eastern Europe, of 'the undoubted evils that had fallen upon the country from an immigration that was largely Jewish'.[7] But he had no connection with the anti-semitic[8] feeling rife in Edwardian Britain exemplified by the polemics of Chesterton and Belloc."[9]

In 1859, when Natty went up to Cambridge, Jews were still regarded by the undergraduates' parents as second class citizens.[10] They were considered highly undesirable companions for their sons both at school and the University. Proust records a similar situation in France. Natty's childhood friendship with Randolph Churchill[11] — which endured until his death — was frowned on by the pedagogues at the latter's private school near Aylesbury, and the Prince of Wales's genuine liking for Natty (at Cambridge he went so far as to ask him for a photograph, and let it be known that Natty was to be among the guests at the large balls and parties which he himself attended)[12] was considered by the

---

* The use of the word "enthusiasm" rather unfortunately brackets Zionism with lawn tennis in the reader's mind, but it is probably in this instance, *le mot juste.*

Queen as most unfortunate. In England to-day it is difficult to envisage the degree of religious intolerance and the rigid type of class distinction which existed less than a hundred years ago. Natty drew his father's attention to the fact that at Cambridge, for instance, noblemen and their sons could take a degree after seven terms and did not have to pass Little Go.[13] "Both noblemen and fellow commoners should be done away with, but I'm afraid these things never will", he wrote. Much of the spontaneous emotional antagonism directed towards those of different persuasions was absorbed by the Catholics* just as to-day in England the full weight of racial hatred is diluted or diverted from the Jews by the presence of many black immigrants** (who in addition are more easily identified) while much pent up violence is dissipated at football matches. Professor John Vincent describes the anti-Catholicism of the Victorians as "the anti-semitism of that era".

At the turn of the century a physician visiting a titled patient automatically rang the bell at the back door — a fine distinction which seems rather curious to us, but appeared perfectly normal both to the butler who let him in, and to the doctor himself.[14] Presumably it was also accepted unquestioningly by the invalid concerned. This powerful sense of hierarchy, as well as the current type of anti-semitism found among the British aristocracy, was the result of conventional but vigorous early indoctrination, and seemed to carry with them feelings of immutability. Jewish people were not only un-Christian, but they were also tradesmen and members of the lower orders, like the doctor and lawyer. A wonderfully telling example of the acceptance of this conventional type of anti-semitism was provided by Richard Haldane (Viscount Haldane) a true friend of Natty and Emma (and also in due course of their son Charles). In his old age Natty had occasion to remonstrate with him concerning a speech reported to have been made by the Governor General of the Sudan, Sir Reginald Wingate, to the Sheiks at Khartoum in which he referred to "the syndicate of Jews, financiers and low born intriguers — like broken gamblers staking their last coin, and in deference to the urgent demands of Germany and our enemies", etc., etc. "I am sure if the Sirdar used the expression he did so

* During World War II Dr. Heinrich Fischer, who visited a dockyard for the purposes of broadcasting anti-German propaganda, remarked that anti-Catholicism was still a reality; there were no anti-Nazi slogans to be seen but a large banner across the entrance proclaimed in foot-high letters: "* — *! the Pope".

** There was also great prejudice against black people in Natty's day too, but there were too few of them to attract general attention. From Cambridge — shocked — he wrote to his father:[15] "Each college is allowed .... to refuse admittance to anyone they may object to. A negro was refused admittance a short time ago at Trinity on the plea that the odour which he transmitted to the atmosphere could not be tolerated in the lecture room."

only colloquially and did not mean it literally", wrote Haldane in his reply. "But it was a wrong expression to use and I am going to enquire about it and see whether the point can be set right. Affectionately ...."[16] Haldane (then Lord Chancellor, but in Grey's absence temporarily in charge of the Foreign Office) who was an essentially tolerant, highly idealistic man without a vestige of anti-semitism of thought or feeling, nevertheless believed that his own explanation was rational and acceptable, while admitting that the point could have been put more tactfully!

Natty's father, Lionel, achieved the parliamentary emancipation of the Jews after a protracted struggle which he had begun in 1829, but which developed in earnest in 1847 and which he pursued doggedly for thirty odd years, until in 1858 a modification of the Parliamentary Oath made it possible for him to take his seat at Westminster. He was, in fact, a standard bearer not only for his co-religionists but also for Liberal and Nonconformist middle class elements especially of the towns — a middle class not entirely emancipated from hierarchy. But Natty won a more subtle and symbolic battle for the Jews — namely the inalienable right to be admitted by the front, instead of the back door; he and Emma changed their image in the eyes of those Englishmen who wielded political power in the Empire.[17] Natty's advice and expertise (and not infrequently the financial support of his firm*) were needed in matters of national and international finance, but in the process of procuring his assistance they discovered a fine clear character, a man of integrity and sincerity, devoid of personal political ambitions but dedicated in an uncomplicated and effective manner to tireless public and private service. Furthermore *Homo sapiens*, however cynical, worldly, indifferent or hostile the mood of the moment may be, in the final analysis respects a modest man who is proud of his creed and is prepared to defend it and his co-religionists, vigorously and uncompromisingly.**

The *Pall Mall Gazette*[18] held a slightly different view — which was shared by many others — namely that Natty did not reduce current prejudice, but stemmed the rising tide of a virulent continental brand of anti-semitism. "It is owing to the life of Lord Rothschild that Great Britain has escaped those collections of race feeling .... with which so many other communities have been

---

* Queen Victoria found it comforting to know that Rosebery (who at that time had agreed to serve in Gladstone's Government) was closely connected with the Rothschilds. In her view it added a measure of "security" to his other desirable qualities as a Liberal politician.[19]

** Natty's advice and motives were impersonal in the sense that he did not seek office, but one feels that he thoroughly enjoyed manipulation, psychological machinations and his sardonic string-pulling among the politicians.

embarrassed during the last generation. He was at once a Prince in Israel and an Englishman of whom all England could be proud." This dual role also impressed Lord Swaythling[20] who remarked that as acknowledged head of the Jewish community in England, and as the leading Jew throughout the world, he occupied a position held by no other Jew since the exile from Palestine.

When Victoria turned down Gladstone's request[21] of a peerage for Lionel Rothschild she was genuinely horrified at the thought of a Jew peer,[22] but twenty years later not only had Natty taken his seat in the House of Lords but he and Emma were staying at Windsor Castle. The Queen ordered a special ham-less pie for their dinner. Emma was touched; she thought this was both considerate and gracious, since it showed that the Queen was anxious to demonstrate publicly her respect for the restrictions imposed by their Jewish religion. She could so much more easily have struck ham off the menu altogether!

Randolph Churchill leaned heavily on Natty, especially during the period 1885-87 and his tenure of the India Office. "The Rothschilds' knowledge is as great as that of the Bank of England is small",[23] he wrote to the Viceroy, while Brett in his diary remarked that "Churchill and Natty seem to conduct the business of this country *together* in consultation with Chamberlain."[24] And Rothschild was such a good listener!

This was one of the very rare occasions when the extent of Natty's influence was revealed. No one was more anxious than he to remain anonymous and to conceal the role he played, for he had good reason to believe that publicity would serve no useful purpose but was on the contrary counter productive. Those of his friends who wielded political power felt likewise. He was, on the whole, outstandingly successful in hiding his light under a bushel. A note from Hamilton to Rosebery in 1887 shows that his judgement in this respect was pretty sound. "There is one thing", wrote Hamilton, "which (between ourselves) might, I expect, have got him [Churchill] into trouble; and that was the excessive intimacy of a man occupying the post of Chancellor of the Exchequer[25] with a certain financial house."[26]

The same thought is inherent in Lady Salisbury's conversation at dinner with Herbert Bismarck, recorded by Rosebery in a memorandum.[27] "Many of our present Cabinet were dull, but they could all be trusted, she said, and then launched out about Randolph who communicated everything to Natty Rothschild and hinted that people did not give great financial houses political news for nothing."[28]

Despite Natty's intense discretion and secretiveness — and possibly because of it — it was whispered that Churchill had leaked prospective budget secrets to him in December 1886.

Churchill in those early days indulged in the conventional anti-semitism inculcated during a Victorian childhood and was not above playing a useful anti-semitic card in his campaign about Egypt in the early 1880s. Thus he referred (in Edinburgh)[29] to a "gang of Jewish usurers in London and Paris seducing Ismail Pasha into their net" while "Gladstone had delivered the Egyptians back into the toils of their Jewish task masters".

But that was before his friendship with Natty had ripened; later he had the reference in question expunged from the published version of his speeches.[30] Furthermore he advanced the cause of men like Lionel Cohen and L.L. Isaacs within the Tory party — a powerful advocate — and rather ostentatiously resigned from a club which black-balled Wertheimer. There is little doubt that this change of heart was directly attributable to Natty's influence.

Emma's assault on the front door was conducted along rather different lines, but proved no less effective. First of all she had a gift, according to Disraeli,[31] for bringing interesting people together for dinner, and her chef, the elder Grosstephen, was probably the best in Europe. (His son, who was born, and trained, at Tring Park and worked there all his life, cooked the lunch when St. Amant won the Derby in 1904, but, according to Emma, was far less talented. Those were the days when Grosstephen bought smoked salmon at 1/- per pound; nevertheless he ran up an annual bill of £5000 at the fishmonger.)

It was a time when people took a deep interest in religion and in books — before radio and television — and while there was still time to read voluminously. Books indeed played a more important part in men's lives — and Emma was keenly interested not only in ethics and moral issues but also in literature. Thus in '88 she sent Gladstone[32] two "remarkable books" for comparison, and suggested he might review *John Ward*; later on he reciprocated with *A Southern Planter* for her appraisal.[33] She also enclosed with comments, a review article on George Eliot which he had somehow missed, and a list of eight German novels "which may I hope afford you some relaxation".[34] Also at his request she sent him the source of the Goethe he had quoted at dinner,[35] and some information from Natty about the size of birch trees at Hawarden.[36] Gladstone wrote to her "endeavouring to draw on your stores of information …. as in other days I have successfully drawn on those of that admirable woman Baroness Lionel."[37] He continued: "I have read in some Jewish author (it might be Rabbi Kalisch) but I am not certain where, during the last ten years a popular but able account of the Mosaic law compared with other contemporary or ancient scripture in its moral and social aspect on a number of points the comparison being greatly in its favour. I have the belief that you can …. direct me to the source". Gladstone hoped "the Dante studies prosper" and sent his kind regards to Nathaniel. Emma, after some careful research, duly directed him. She was, as we have seen, a deeply religious woman and for her it was indeed "a source of deep satisfaction to think that, though our creeds differ on so many points, the Christian and the Jew agree in their fidelity to those Holy Scriptures of which you say 'They arm us with the means of neutralizing and repelling the assaults of evil in and from us.'"* Gladstone signed Emma's copy of his *Impregnable Rock of Holy Scripture* and the same evening came to dine with her and Natty and Balfour at 148 Piccadilly.[38]

Disraeli in his turn sent her first a copy of *Lothair*[39] inscribed "Emma de Rothschild from her friend Beaconsfield" and then *Endymion* (in three volumes) to read and criticize, not only because he had confidence in her intelligence, but also because he always required a female counsellor. Emma, in a letter to her mother-in-law, asked if she had read Disraeli's preface: "Though very eloquent I think the great man pays himself too many compliments and does not mince matters as regards his own merits".[40] She acknowledged Disraeli's gift with a tactfully fulsome, but less sincere, letter of praise,[41] but "took advantage of his offer". "If I might venture to allude to a shadow amongst so much light," she began, "I would ask why your hero, possessing every quality likely to ensure a success, should be partly indebted to the worldly and prosperous position of a sister?" Asquith, Morley, Crewe and Froude all sent her their own lectures and books to read and comment upon.

Buckle wrote to Natty in 1912[43] to tell him in advance that he was about to resign from the editorship of *The Times*, for "you are so old and kind a friend of mine and have helped me so often in my work". Then he added: "Pray tell the news to her ladyship who has been so friendly and hospitable to me and for whom I have so great a respect and (if she will let me say so) affection ...." Emma with her implacable integrity, her wide reading and genuine scholarly approach to literature, her gifts as a hostess, her lively interest in the human race, her wifely devotion, sincere modesty and unshakeable religious beliefs, passed with great dignity and not a little assurance through every man's front door.

Natty's friendship with Balfour is almost totally neglected by the latter's biographers, although in fact he exerted a considerable influence upon the younger man, for his advice was frequently sought and generally heeded. In Natty's characteristically forthright manner he often passed on interesting

---

* The faint wind of change can be felt if one compares the tone of Emma's letter with that of Baroness Lionel's to Gladstone on March 9, 1875. "In accordance with my promise I send Dr. Kalisch's lectures and as you do not object to reading books by persons of our persuasion I venture to add Dr. Kalisch's commentary on Leviticus, a work of great learning and a line of thought not considered strictly orthodox from the Jewish point of view....".[42]

information, for he was exceptionally well informed* owing in part to the family connections on the Continent, with built-in suggestions: "I think it is right that you should know how astounded I have been to find that an intense feeling has been raised in the city against any concessions being made here in London to the bi-metallist party ...."[46] The letter, which was sent by messenger, ended with the hope "in which the writer joins" that Balfour would come to Tring that evening and stay over the weekend. The scribe was Emma who, as we have seen, frequently wrote Natty's letters for him if they contained confidential information. A characteristic note to Balfour which began "I am more convinced every day that it is absolutely necessary [illegible word] at this critical period to drop, altogether, ideas of Tariff Reform ...."[47] was dictated to Charles. Walter only very occasionally was privileged to act as scribe — although at this period (1906) he was himself an M.P. — for now Natty was barely on speaking terms with his eldest son.

In 1901 Natty had occasion to write to Balfour and remonstrate with him concerning a derogatory, and as it happened, incorrect accusation, against De Beers,[48] made by him on the floor of the House, which contained undertones of anti-semitism. Balfour wrote apologising: "My dear Natty,[49] I am very sorry that any answer of mine should have given just cause of umbrage to the De Beers company" — explaining how the mistake arose and promising to make the necessary correction at Westminster. What Natty thought in private about "an alien immigration" of Jews (see p. 26) we do not know. He had had the difficult and thankless task of serving on the Royal Commission on Alien Immigration,[50] attending 49 public sittings and personally questioning most of the 179 witnesses. At the time of the appalling pogrom in Russia in 1905 he asked Balfour to be present at the protest meeting he was organising at the Queen's Hall in London. Balfour declined, but he sent a message of sympathy which was to be read aloud by Natty. "....I may be permitted to send through you a word of sympathy for the Jewish victims of law and lawlessness in Russia. The treatment of their Jewish citizens by European nations from medieval times onwards is certainly the darkest blot on the history of Christendom; and I earnestly trust that the outcome of this appalling Russian tragedy may be to make Security, Liberty and Equal Rights the inalienable birthright of every Russian Jew."[51] It is sad that the covering letter was not one of those preserved in Charles's collection of autographs. The only surviving letter

---

* Disraeli described Natty in a letter to Lady Bradford[44] as "N. Rothschild who knows everything". In 1906 Natty told Haldane "for what it is worth" that he had heard from Cairo that the Sultan of Turkey would evacuate Akabar. He added: "This does not seem to agree with Reuter's message. At the time of the Marchant crisis however our agent was better informed than Lord Salisbury."[45]

written on New Court notepaper by Walter dealing with matters other than zoology, was one concerning the plight of the Russian Jews in 1890:

*My dear Mr. von Bleichröder,* [52]

*Pray accept our best thanks for your very interesting letter which we received on Saturday. It is evident that the Russian Government has entirely misunderstood the accusations brought against them. It is not the question of new laws being promulgated against the Jews but the putting in force of the existing ones which have been held suspended for some time and which are so harsh and oppressive that they may be the cause of many Jews becoming violent Nihilists. We hope you can use your powerful influence at St. Petersburgh to prevent the Government from putting into force the old cruel and senseless laws. We hope soon to hear some better news from you on the subject.*

*I remain, Yours very sincerely,*
*WALTER ROTHSCHILD.*

A large book could be written concerning Natty's efforts on behalf of the Jewish victims of persecution and pogroms in Roumania, Russia, Morocco and elsewhere. One of the letters in Charles's autograph collection is from the Foreign Office to Natty in 1893.[53] "I am directed by the Earl of Rosebery to inform you that a Telegram has been received.... stating that the Sultan has reprimanded the Moorish Kaid accused of ill treating the Jews of Morocco city and ordered him to treat them as favourably as the Mohammedans. The Vizier has also informed the Jewish Community.... and conveyed to them the Sultan's good will."

In November 1905, Balfour sent Natty a copy of the telegram he had dispatched to Spring-Rice in St. Petersburg for the information of Count Witte, pointing out that the massacres and plundering of the Jewish community in various parts of Russia, unopposed by the police, were creating a most painful impression in the public mind, and that his country risked losing the sympathy of both the English press and the public unless the outrages were checked and punished. Balfour hoped that it "would have the effect of stimulating the Central Government to put an end to the atrocious attacks on the Jews."[54] He also told Natty that there would be no difficulty in obtaining consular assistance for sending help to the victims of the outrages. The family did indeed send large sums of money to Russia and arrange for their distribution by the Consul.

More publicity was attracted by Natty's refusing a loan to the Russian Government (£100,000,000) and thus forgoing a £2,000,000 profit for the firm.[55] In 1908 he prepared — with meticulous care — a memorandum on the question of the Jews in Russia,[56] signed by himself and his two brothers, which the King was prepared to convey to the Emperor, whom he visited in 1908.

*London, 3 June 1908.*

*To His Majesty The King*

*Sir,*

*We should not venture to address Your Majesty on a subject which preoccupies us more than any other, were it not that we know from long experience extending over nearly fifty years, the warmth of Your Majesty's heart, and the greatness of mind which always enlist Your Majesty's sympathy in the cause of the oppressed, and make Your Majesty endeavour by Your wise counsels to alleviate their position and to help them to obtain a happy and peaceful life, and thereby make them good and faithful subjects of their Sovereign, and industrious and law-abiding citizens of the country in which they live.*

*It is unfortunately true that since the conclusion of the Russo-Japanese War, there has been a considerable revolutionary and anarchical movement in Russia and, were it not for that movement, which was widespread, and affected all classes, the Emperor of Russia would not have instituted a certain form of Constitutional Government, and called together the Duma.*

*On the other hand it is equally true that a certain section of the Russian population, chiefly among the aristocracy, are very hostile to any Liberal movement, and would be only too glad to see the old state of affairs returned to and an autocratic régime again instituted; a large number of those who share the view that a repetition of the ancient régime would mean the salvation of Russia, belong to the Union of the Russian People and the Octobrists: these two Institutions or Associations were undoubtedly answerable for the outrages on the Jewish population and the horrors which ensued at Kishinev, Kiev, Odessa and elsewhere.*

*We need hardly remind Your Majesty that thousands of innocent people were killed, many women violated and the property of thousands destroyed: on that occasion, those of the Jewish Faith in different parts of the world sent money to relieve their coreligionists, and the large emigration from Russia was directed to countries where it was hoped the immigrants would not interfere with the existing Labour Market.*

*At that time, a certain number, a very small number of the Union of Russian People and the Octobrists were punished, but of late their punishments have been remitted, those punished have been restored to the favour of the Czar, and in many instances they have been decorated. — The result has been a recrudescence of the persecution of the Jewish population artificially hidden under legal devices. The Jewish population is again terrified, and naturally there are fears both in Russia and elsewhere that emigration may take place on a large and unprecedented scale, which would have the double effect of depriving Russia of industrious and sober workmen, and this extra influx of immigrants would certainly disorganise the position and condition of all workmen in many parts of the world.*

*We do not propose to trouble Your Majesty with the details which we have received of individual cases; they are heartrending, and it is incredible that they should take place in the Twentieth Century!*

*One of our informants, and one in whom implicit confidence can be placed could, if he dared, give all the details; but we prefer to put our case on more general and much higher grounds, and which likewise imply a much broader issue.*

*As we have already stated, it is regrettable, but unfortunately true that since the Russo-Japanese War there has been a widespread movement in Russia, not confined to one class, to one race, or to one religion: no doubt in this movement a certain number of Jews have participated, presumably the natural result of their oppression during hundreds of years, but the mass of the Jewish population are loyal to the Czar, and their only desire is to live and die in Russia and to serve the Sovereign of their country. Their present position may be summed up in a few words: — a few Jews, probably not so many as the proportion of Jews to the rest of the inhabitants in Russia, may have joined an Anarchical movement, and therefore they must expect to be dealt with accordingly, but let the remainder, the very large majority, be free from persecution, let them enjoy equal rights with their fellow subjects, and then they will remain faithful and devoted to their Sovereign and to their country.*

*We have ventured to write to Your Majesty on this subject, as we know the goodness of Your Majesty's heart, and that the paramount influence of Your Majesty has unceasingly and invariably been exercised with the most beneficial results.*

*We beg most respectfully to tender to Your Majesty our most grateful thanks in advance, and whilst presenting our most humble duty,*

*We have the honour to subscribe ourselves*

*Your Majesty's most faithful subjects and most obedient servants,*

> *ROTHSCHILD*
> *ALFRED DE ROTHSCHILD*
> *LEOPOLD DE ROTHSCHILD*

Lord Knollys, who one suspects disapproved of the King's friendship with Natty, wrote to him on June 3rd from Buckingham Palace:

*My dear Rothschild,*[57]

*The King desires me to let you know, in reply to the letter which you, Alfred and Leo have written to him, that he will speak to Sir Charles Hardinge and Sir Arthur Nicholson respecting the question which you have brought before him.*

*The subject would be a very delicate one for him to bring before the Emperor of Russia, and it is moreover one of considerable political importance.*

*His Majesty feels therefore that it would not be constitutionally correct or proper for*

> *him to speak to the Emperor, or to his advisers, unless he did so with the full*
> *concurrence of Sir Charles Hardinge and Sir Arthur Nicholson both of whom accom-*
> *pany him to Reval.*

*Yours sincerely*
*KNOLLYS*

This letter must have warned Natty that the Foreign Office were determined to suppress the document.[58] He was, as usual, outspoken on the subject and when told by Charles Hardinge[59] that Stolypin, the Russian Prime Minister, "contemplates legislation for the amelioration of the lot of the Jews in Russia" and therefore the King had decided not to say anything further on the subject at present, Natty replied that he had little confidence in Stolypin's assurances and was disinclined to express satisfaction which he could not feel.

In 1912-13 the charge of ritual murder against the Jews was again revived at a trial in Kiev. The ensuing correspondence between Natty and Cardinal Merry del Val,[60] in which the latter confirmed that an Encyclical of Pope Innocent IV declared the accusation to be baseless and wicked was duly published in the *Standard*. An appreciative, but intolerably cloying article appeared in the *Pall Mall Gazette*. "Not by their wealth but by their unswerving rectitude, public spirit and love of race the Rothschild family has become as it were the Judges of the dispersion". The article[61] was headed "As a King unto a King", and finished, "We, as a nation, may be proud that the head of this wonderful family dwells among us, an Englishman of Englishmen, as well as a Jew of Jews".

Natty also prepared a letter of protest regarding the charge of ritual murder, and among the signatories[62] were the Duke of Norfolk, Alfred Lyttleton, Montagu Butler, Lord Cromer, Lord Llandaff and Lord Rosebery. (Various members of the Government such as Lord Morley, Lord Lansdowne, Lord Crewe and Lord Loreburn felt that they were unable to sign, on account of their official position).

Rosebery, a Christian who married Natty's first cousin and was an intimate and sincere friend, had the infinite self-confidence which saves a man from anti-semitism or similar basically defensive prejudices.[63] He used to remonstrate with people who made derogatory remarks about the Jews, not by pointing out that he himself had married Hannah Rothschild, but by saying bluntly: "You should not say that to me. I am Jewish."[64] His acquaintances found this embarrassing! On one occasion Henry Ponsonby, describing an agreeable and amusing weekend at Mentmore, records the following conversation: "I attacked him [Rosebery] on his Secretary of Scotland Scheme and said the present Government were enlarging the powers of the office and would give the Secretary for Scotland the power of pardon in capital offences. 'I mean', I explained, 'that it will be for Lothian to advise the Queen as to the prerogative of mercy in the case of a MacLipsky'. Rosebery solemnly observed: 'You have wounded me in my two tenderest points — as a Scotsman and a Jew'."[65]

The Rosebery marriage was undoubtedly one of the most important factors in obtaining recognition in its widest sense, social as well as political, for the Rothschild family. Hitherto it was mainly their religious beliefs that had automatically relegated Jews to the status of second class citizens — the converted Disraeli had been accepted, even if his semitic origins had been the cause of some antagonisms. Hannah was not converted, yet Lord Rosebery married her.... Nevertheless, although on the one hand this established the Rothschilds in "society", it emphasised another and more pernicious aspect of anti-semitism, namely discrimination based on racial differences, an attitude rapidly gaining ground on the continent and which could not be eliminated by a change in religious beliefs or convictions. This is hinted at in a speech already referred to, in which Balfour suggests that the Jews constituted an alien element in the East End "who by their own action remained a people apart, and not merely held a religion differing from the vast majority of their fellow countrymen but only inter-married among themselves".[7] Balfour, however, had no reservations where Natty himself was concerned, but placed considerable confidence in him, as the following letter, written at Tring,[66] indicates:

*Tring Park*
*Tring*
*Sunday Night   Nov. 1911*

*My dear Natty,*

*I had hoped to have a word with you to-night; but I most stupidly forgot that on Sunday you keep very early hours.*

*I should be very unhappy if you should hear from anyone but me the news which I wished to discuss with you; namely, that I am determined on resigning without delay my place as leader of the Party — though* not *my seat in Parliament.*

*I greatly fear that this course will give you pain; and I cannot attempt to defend it within a brief compass of a note. But I shall see you on Tuesday morning; and shall then hope to convince you that I am right — both from a personal and a party point of view. In the meantime please say nothing about it.* Very *few people know it as yet; and it is most important that the news should not be prematurely divulged.*

*Yours affec.*
*ARTHUR JAMES BALFOUR*

*This is for* you *and no one else: till I see you.*

That same week Balfour had informed his "closest colleagues" of his intention to resign the leadership.[67] "There are many who think", wrote Lord James of Hereford some years earlier, when congratulating Natty on a speech on the Free Trade issue, "that you know Arthur Balfour's opinion better than anyone, and those many are thinking and hoping.... that what you have said so well, the Prime Minister will say shortly."[68] In fact the two men — so remarkably dissimilar — did know each other pretty well, but Natty never discussed his feelings for Balfour with anyone but Emma. And Emma kept her own counsel.

Someone once asked Dorothy Pinto (the wife of James de Rothschild) why she was so confident that A.J.B. was a staunch "Nattite", as Leo Maxse (editor of the *National Review*) characterised Natty's fans, and she replied: "As a child I thought Lord Rothschild *lived* at the Foreign Office, because from my schoolroom window I used to watch his carriage standing outside every afternoon — while in reality of course he was closeted with Lord Balfour."

After Natty's death in 1915 Balfour wrote to Lady Wemyss[69] that he had sustained a greater blow than most people would suppose. "I was really fond of him; really admired that self-contained somewhat joyless character. He had a high ideal of public duty and was utterly indifferent to worldly pomps and vanities. Moreover he was perfectly simple."

Natty was far from joyless, but he was a raspingly serious man. As the acknowledged lay head of world Jewry he could find little to laugh about in 1915. Furthermore he had none of those weaknesses which require humour as an acceptable crutch. The books all tell us that Weizmann, first in 1905, and again ten years later, evoked Balfour's genuine fervour for Zionism. This was true, but it was Natty who by his rectitude and sincerity had dissipated Balfour's streak of conventional anti-semitism and who first made him deeply aware of the real tragedy of the Jews.

Isaiah Berlin[70] has provided us with an illuminating and deeply penetrating vignette of Weizmann. This man — the very embodiment of Zionism — had an idealistic passion for England. "It was very central in him, and in his ideal, for he wanted the Jewish Society — the new state — to be a political child of English — almost exclusively English — experience."

Curiously enough Natty had also evolved a great love for the English *life-style* — he was intensely patriotic — although the two men had arrived at this similar viewpoint along such very different roads. Natty was not at the moment even considering the possibility of a Jewish state in Palestine, but he was deeply and emotionally concerned with the Diaspora. He believed, however, that the only way to solve the world problem of the persecuted Jew was along the lines he had himself experienced in England. The fact that he had achieved a unique position of respect and influence in a Christian community gave his predilection for the English way of life a superficial resemblance to assimilationism — a false accusation and one which could not be levelled at Weizmann! He also believed that the characteristics which they both admired and wished to propagate could only exist and flourish in a country whose army and navy, or geographical situation, rendered them safe from outside aggression.

Natty was so successful at hiding his light under a bushel that many of his activities have been completely obscured. One of the few mementoes that Charles kept — along with Nathan Mayer's silver piss-pot, Natty's war-time armband, and the truly amazing gold and diamond cigarette case "Presented by the Sultan of Turkey to Lord Rothschild for services rendered, August 1894"[71] — was a large ledger-like notebook with "Rothschild" scrawled in pencil on a white ticket on the cover. This contained the notes Natty made at meetings when, in 1889, Cardinal Manning invited him to join the committee endeavouring to settle the dock strike.[72] Manning himself was an eleventh hour choice of the Salisbury Government and it was in desperation that he appealed to Natty. The dock strike was in a sense a turning point in the labour movement in the U.K., and Natty's successful arrangement of the pay and the terms of settlement, together with the recognition he obtained for the Unions, was reckoned by Charles to be one of his greatest triumphs. Unfortunately his note book is totally illegible except for one page which is headed very clearly "Widows and orphans" together with the word "insist". There seems to be no record of the meetings (about six, judging from the ledger) in any contemporary book, memoir, biography, diary, library or record office. In this connection Natty seems to have engendered his most successful smoke screen. His cousins* in France had been following events with interest.[74] Baron Gustav wrote: "We are delighted to see that.... Manning has taken our Cousin Natty as an intermediary. It is most honourable and excellent that the workers should know the sentiments of our family in their favour." This was on September 7th. Five days later Baron Alphonse congratulated him on his role in achieving a happy result. Were it not for these two letters and the illegible note book the whole story could well be a dream.

---

* There was great rivalry as well as great co-operation between the English and French cousins, but "Nat" was acknowledged by Paris as the world lay head of Jewry. To Israel Zangwill Edmond said, "What Nat thinks is my opinion.[73] But he was not so appreciative when other people voiced "these sentiments". His son James was rapped over the knuckles for saying as much during a pro-Zionist lunch party in London.

One of Natty's great qualities was his helpfulness. No appeal was too trivial or too daunting to be laid aside and he never seemed to flag or grow weary of the endless requests he received whether they concerned music lessons for a Jewish student or the Chairmanship of the Red Cross. When he was lying mortally ill at 148 Piccadilly, Lord Haldane[75] asked if he could call on him. Upon entering the room the Minister suddenly realized that he would not see his dear old friend again, and hesitated in the doorway, deeply shocked.

"Come in, come in", said Natty, stretching out his hand. "I don't know why you have come except to see me, but I thought to myself if Haldane wants a cheque for £25,000 it's his — and no questions asked, of course...."

"No," said Haldane smiling sadly, "this time it's not a cheque. Grey is away and I'm temporarily in charge of the Foreign Office. There is a steamer which has started from South America — it is a neutral ship and we suspect it has supplies for Germany on board, particularly bullion. We really have nothing to act on...." He hesitated as he looked at the frail old man, "Could you stop the ship?"

"Of course", said Natty in a strong voice, and he propped himself up on his elbow, scribbled a note on the back of an envelope in his shaky handwriting and rang the bell. "That is a very simple matter...." He sank back on his pillows and Haldane, deeply moved, sat down at his friend's bedside.

The Rothschilds were basically doers not thinkers.

# CHAPTER 6

# Walter's Father — Nobody wanted to touch Lloyd George

ALTHOUGH NATTY was described by Lucien Wolf as an intense politician,[1] he was one by necessity only — by inclination he was dedicated to various aspects of social reform — believing implicitly in equality of opportunity, a higher standard of learning for all, as well as civil and religious liberties within a patriotic framework. His youthful admiration for Gladstone (which dwindled markedly over the Home Rule issue) was never obscured by the family's devoted friendship with Disraeli. From Cambridge he wrote to his father saying that he had studied Gladstone's budget speech carefully,[2] and both with regard to eloquence and context it compared favourably with Disraeli's last effort. On another occasion he wrote from Trinity: "I am not sorry that Dizzy has made so many mistakes.... it is the natural consequence of hypocrisy — and no party can hope to present a bold front to the enemy which does not represent some principles...."[3] and again, "I have always thought that one of the chief reasons of Dizzy's unstatesmanlike reputation was his great want of patriotism", and he proceeds to compare him unfavourably with Palmerston in this respect. Natty's forthright manner was already evident in his undergraduate days when he criticised his elders and betters frankly, with candour and self-assurance. "Bulwer can write such twaddle about ghosts and the like", he wrote, "He would like to be eternally young and is afraid of death."[4] In another letter he commented, "Lord Stanley was equally silent, sober or drunk."[5]

The brush he had with Lloyd George[6] in 1909 attracted so much attention that it has somewhat obscured Natty's twenty-five years of staunch Liberal allegiance. In his speech attacking him, the Chancellor, if he did nothing else, called attention to his influence on current opinion by denouncing Lord Rothschild as the Dictator of England. The East End of London was highly delighted.

*Really, in all these things[6], we are having too much Lord Rothschild. We are not to have temperance reform in this country. Why? Because Lord Rothschild has sent a circular to the peers to say so. We must have more Dreadnoughts. Why? Because Lord Rothschild has told us so at a meeting in the City. We must not pay for them when we have got them. Why? Because Lord Rothschild has told us so. You must not have an estate duty and a super-tax. Why? Because Lord Rothschild has sent a protest on behalf of the bankers to say he won't stand it. You must not have a tax on reversions. Why? Because Lord Rothschild, as chairman of an insurance company, said he wouldn't stand it. You must not have a tax on undeveloped land. Why? Because Lord Rothschild is chairman of an industrial housing company. You ought not to have old-age pensions. Why? Because Lord Rothschild was a member of a committee that said it could not be done. Well, I should like to know, is Lord Rothschild the Dictator of this country? Are we really to have all ways of social and financial reform blocked, "No thoroughfare, by order, Nathaniel Rothschild"?*

The historian, Cecil Roth, writing in 1939,[7] in a sarcastic "Limehouse" style, which he presumably thought appropriate for popular consumption, accuses Natty of urging an increase in naval armaments* while "strongly opposing expenditure on such unimportant trifles as old age pensions" and the social services, which were advocated in Lloyd George's budget. This was far from the case, but an impression which can easily be gained if — for want of other more personal sources of information — years after the event, they are based solely on the propaganda of newspaper reports of the day — in their turn making the most of the polemics of a splendid rabble-rousing political speaker. Perhaps this is the fate of all *eminences grises* who choose to remain discreetly silent, who burn letters and who do not keep a diary! Natty did not merely doggedly oppose Lloyd George as contemporary writers like to proclaim, he had various constructive suggestions to make. What he objected to strongly was an ill conceived old age pension scheme without a simultaneous reform of the Poor Law, and on March 22nd, 1909, Lloyd George had replied to a communication from him[11] on this very subject.

---

* Natty was already writing to his father about the cost of naval armaments in 1862, while he was at Cambridge, and he[8] retained a life-long vivid interest in the Navy, and a profound belief in its importance for Britain's security and power. But it is curious to find in Charles's autograph collection several letters from Winston Churchill marked "Secret" concerning the undesirable prospective sale of the battleship Rio to Italy by Brazil, which the Admiralty wanted to prevent. "Would you let me know whether you would be prepared to buy the ship from Brazil and hold her till her transference...." asked Churchill.[9] It is even stranger to find him arranging at Natty's suggestion to allow, in view of the special circumstances, Chilean officers to study on board British ships, and again a little later the same year, Asquith (then Prime Minister) was writing to Natty[10] thanking him for his letter — which he had shown to Churchill — and adding that two cruisers, the Hampshire and the Weymouth, were now at Beshika Bay and were being sent through the Dardanelles. Moreover the 3rd Battle Squadron — much the most powerful force in the Mediterranean and Aegean waters — was well on its way.

*11 Downing Street*
*Whitehall, SW*

*Dear Lord Rothschild*

*I am greatly obliged for your letter of the 15th instant and for your remarks upon the proposed scheme for a system of contributory insurance against sickness, invalidity [sic] etc. I fully recognise the weight of your contention that a scheme of this kind should be considered in conjunction with reform of the existing poor law and I hope that we may be able to take this in hand before very long.*

      *Again with many thanks*
        *Believe me*

*Ever yours sincerely,*
*D. LLOYD GEORGE*

On another occasion Natty attacked Lloyd George's housing scheme in one of his rare political speeches, delivered on this occasion at Tring.[12] Natty considered the Chancellor's promises to be without substance — so much lightweight political propaganda.

Lloyd George was annoyed. An exchange of letters followed.[13] He wrote to Natty on the day following the speech.

*Treasury Chambers*
*Whitehall, S.W.*
*18th November, 1913*

*Dear Lord Rothschild*

*I have read with considerable surprise your remarks as they appear in today's newspapers, with regard to the housing proposals of the Government. You are reported to have said that the Reserve funds accumulated under the National Insurance Scheme ought not to be invested in a "jerry building speculation". With that statement I entirely agree, and the fact that you made it in connection with the Government's proposals would seem to show that you are under a very serious misapprehension as to their exact significance. Stated shortly, it is proposed that money should be advanced to the State for Housing purposes from the Insurance Reserves, the advance to be made by way of loan on a State guarantee and at a rate of interest appropriate to State security. Thus any loss which might be incurred, — not that I admit that any loss is likely to be incurred — would fall not upon the Insurance funds but upon the Imperial Exchequer.*

*I would remind you that Friendly Societies in the past have with general approval lent enormous sums from similar funds on the security of house property owned by private individuals. In the present case the Government are proposing to borrow money for Housing purposes not on the security of privately owned property but of the State.*

*I also notice that you credit the Government with the intention of providing houses "which would keep out neither wind nor weather, and which would give no accommoda-*

43

*tion to those who lived in them". I need not say that the Government have no such intention, and I am not aware of any statement made on behalf of the Government which gives grounds for the suggestion.*

*Yours sincerely,*

*D. LLOYD GEORGE*

*The Right Hon.*
*Lord Rothschild, G.C.V.O.*

Natty was quite unmoved. He had had a great deal of experience of building cottages — for one thing he cleared the slums behind Akeman Street at Tring and replaced them with fifty model houses which he presented to the Town Council as a gift, on the condition that the tenants should only be charged a nominal rent during the first twelve months of occupancy. This was in addition to rebuilding all the cottages on the Tring Park estate which, according to Lady Battersea, numbered 400. Furthermore, in 1885 he had founded and then piloted as Chairman, the Four per cent Industrial Dwellings Company[14] — an astute combination of business and philanthropy — which had erected houses for 6500 persons. He replied:[13]

*New Court*
*St. Swithin's Lane*
*London, E.C. 19th November 1913*

*Dear Mr. Lloyd George,*

*I have the honour of receiving your letter of the 18th in which you allude to some remarks made by me at a meeting at Tring on Monday night. With regard to the first point mentioned by you, I am afraid I must adhere to the opinion I then expressed that the funds accumulated under the National Assurance Scheme ought not to be used to facilitate the building of cottages. This is a view I cannot alter, whether the money be invested directly in the cottages or, in another way, by the loan to the State of a sum sufficient to cover the cost of their erection.*

*With regard to your second point — the quality of the cottages — it has been stated by Mr. Runciman that they are to cost £150 each (including the cost of the land). If that be so, I am still of the opinion that either the accommodation will be insufficient or the cottages not as good as they ought to be.*

*You refer to the investments of Friendly Societies on the security of house property, but although I know little or nothing in detail of such investments, it will, in all probability, be found that when Friendly Societies have lent money on cottage property, they have only advanced what they supposed to be a moderate proportion of the value of the cottages, often with other security besides.*

*Believe me,*

*Yours sincerely,*

*ROTHSCHILD*

At the outbreak of hostilities in 1914, Lloyd George called a meeting of the leading bankers, business men and economists to discuss financing the war effort.[15] Natty was present and briefly stated his views. After the meeting was over, Lloyd George turned to his private secretary and remarked drily: "Only the old Jew made sense". He was delighted and flattered when in due course he received a congratulatory letter from "The Rothschilds" on his handling of the Stock Exchange crisis engendered by the outbreak of World War I.[16] In a sense this was history repeating itself, for Harcourt was highly delighted and flattered when Natty congratulated him on his budget proposals in 1894[17] and added that he [Natty] "does not seem to mind the prospect of Walter paying a quarter of a million" (in death duties).

In fact Lloyd George was a great admirer of the "old Jew". His son[18] once asked him whom he would select for the "ideal cabinet" if he could choose anyone he had known, from all walks of life, during his twenty years of political activity. The first on his list (which included Smuts, Winston Churchill, Philip Snowden, etc.) was Lord Rothschild as Chancellor of the Exchequer.

Natty died on March 31st 1915. A week later *Reynolds Weekly Newspaper*[19] published the following "warm tribute from Mr. Lloyd George":

> *Lord Rothschild had a high sense of duty to the State, and although his interpretation of what was best for his country did not always coincide with mine, when the war fell upon us he readily and cheerfully forgot all past differences and encounters, and gave me the benefit of his wide experience and knowledge of finance; but he never confined help merely to good advice. He was prepared to make sacrifices for what he genuinely believed in. It will therefore surprise no one who knew him to learn that he was one of those who recommended the double income tax, with a heavier super-tax, for the war expenditure. He was essentially public-spirited. We need such men in this crisis, especially when they are men who have won dominating influence. His death before our troubles are over is a sad loss to the nation.*

Many years later a wizened old trader in Leadenhall Central Meat Market — who somehow discovered my name — described the extraordinary veneration in which Natty was held in the East End of London — where he ruled the Jewish community as a golden hearted despot, thinly disguised as a fierce old man of granite. As he walked down the street towards the Jewish Free School or the Rothschild Buildings, people pressed forward and tried to touch him — much in the manner that pilgrims in Eastern countries seek to touch the garment of a wise man and thus receive a sort of blessing, or merely — at last! — a little luck. "You see," said the old trader, thinking back to his youthful days behind a street barrow, "he had the common touch — here was this man who was so rich that he could buy the Suez Canal for the country, but we felt *he was on our side*. We all felt we knew him. Everyone trusted him from the King to the sandwich man. The Rothschilds never pulled their horses. I was once a jockey in South Africa so I know what I'm talking about. At Christmas he gave all the cabbies a brace of pheasants — that was a real treat, *pheasants* — no wonder they tied yellow and blue ribbons to their whips. When he died we all put black crèpe on our barrows. Some of us couldn't find black boards so we put up another colour. We had lost a real friend, perhaps our only friend."*

But what about the Limehouse speech? I asked, What about Lloyd George's virulent invective? Didn't the left-wingers in the East End of London agree with the Chancellor and feel the great man was now a reactionary, who was only out to protect his own ill gotten gains, with merely a token sympathy for the poor and needy? Didn't they all agree that they had had just a bit too much of Lord Rothschild? A knowing look spread slowly across the old man's face and his eyes twinkled. "Believe you me," he said, tapping his nose thoughtfully, "Nobody much wanted to touch Lloyd George."

"Diplomaticus", a political writer of the day, who has been accused — rather unkindly — of a sycophantic attitude towards the rich and powerful, published an appreciation for the *Daily Chronicle* which displayed not a little insight into Natty's superficially rugged personality.[20]

> *It was not easy to know Lord Rothschild, for he was reserved to strangers, undemonstrative to mere acquaintances, and disconcertingly short and candid to fools and humbugs. But when you did get to know him you found you had added a singularly fine character to the circle of your friends. Lord Rothschild brought a really great personality to dignify the great name he inherited. It is a mistake to imagine him only as a leviathan of finance. He was richer in public spirit than in his millions, and it is no fulsome exaggeration to say that the public interest was always immeasurably more to him than any consideration of private profit. That, indeed, when you come to analyse it, was the real secret of the imposing position he won for himself in the public life of the country. He wielded a very great influence — an influence not accounted for by his great wealth and the peculiar magic of his name. It was due in much greater measure to his high personal character.*
>
> *If he was popular in the City it was not because the market-place knew much of him as a good fellow, but because of the high level at which he always sought to maintain the traditions of British credit and finance. This made him in a sense a sort of uncrowned king among the merchant princes of London, and there was nothing of which he was*

---

* Natty once sent £50 to the Costermongers and Street-sellers Union to help them in their fight against Lord Avebury's Bill. In their message of sympathy to Emma, they described him as the "revered and honoured Lord Rothschild.... a leader in every high and noble cause to better the poor and relieve suffering without regard to faith or denomination."

*more proud. It was the same in politics. There, too, his activities were not those which loom largely in the newspapers, but none the less he was an intense politician, and his sagacity, his vast experience of men and affairs, and his readiness to serve the public interests with all his tremendous resources constituted him a valuable national asset. Downing-street knew this well, and made ample use of it. The general public understood it tacitly.*

*But beyond finance and politics lay a larger field in which he achieved his greatest distinction, although its record would be difficult to write in detail. There he was known and loved as an inexhaustible source of charity, untiring in all works of benevolence and social progress, the owner of a heart which never failed to respond to a tale of human suffering.*

Rozsika, Walter's sister-in-law, was the most unsentimental of women, continually preaching the gospel of independence from material possessions; she kept no mementoes herself — for she was a true central European, quite capable of becoming, if necessary, the resourceful wandering Jewess overnight. Nevertheless she had a special feeling where Natty was concerned. She came into the nursery at Tring one rainy afternoon holding a packet of fifty odd letters tied up with pink linen ribbon — the sort used by government offices and usually designating red tape — and gave them to her six year old daughter. "These are grandpapa's letters — you must never lose them", she said. Who had retrieved them — envelopes and all — as they had arrived one by one on the nursery breakfast table? Presumably a nannie had received some explicit instructions on that score. But there was something in the manner in which Rozsika handed these letters to her daughter which so deeply impressed the little girl, that she saw to it that these almost illegible pencil scrawls, written with a shaking erratic hand, passed unscathed through all the bombings and all the moves and upheavals of World Wars I and II. They are among the very few of Natty's surviving handwritten letters*[21] — each with an enormous red seal

* Natty hated conventional letter writing. His parents complained about the brevity and "newslessness" of his communications, and to Rosebery, when he and Hannah were in India, he remarks that he is a bad hand at writing letters. He was not pleased when his future father-in-law asked him to pass on some information concerning Bismarck to his parents. "I take up my pen to send you his news (which I think he might have done himself)", he wrote. In later life he never used ink, but scrawled in an indelible mauve pencil which is the despair of the archivist. He gave one of these pencils to his granddaughter who recalls her tears and lamentations when her father confiscated it on grounds that it might be poisonous, due to some mysterious chemical ingredient. To add to the difficulties engendered by the mauve scrawl, Natty often wrote in what seems to us like riddles. "I have sown the seed, the corn is just beginning to grow it depends on yourself if you cultivate it", he says to Randolph Churchill (in 1887), and the reader, unless he is a specialist historian, is left wondering. Incidentally Walter was not the only member of the family who ignored punctuation. Natty rarely used a comma or a full stop, and these curious *lacunae* can be found in the current writings of the male Rothschilds.

on the envelope — sometimes with three seals, to make them more exciting. How much one longs to see those he had sent to Walter some fifty years earlier! He wrote:

*London*
*June 17th*

*My beloved and beautiful darling,*

  *A thousand thanks for your nice post card, I am so glad you like the sea. I find it very dull without you. Both Liberty and little Boy Blue are very good and talk a good deal about you.*

  *I did not hear any news at Tring. There were a quantity of young birds of all sorts and plenty of flowers, wild ones and in the hot houses, and young Kangaroos in the wood near Cholesbury Common. Daddy would have been delighted at some red crested grebe on the Reservoirs and bitterns in the reeds.*

  *I hope you will soon come home. The cows and pheasants at Tring are anxiously awaiting your arrival.*

*Thousands of kisses and my very best love to you dearest. Your*
*GRANDPAPA*

and again:

*Dearest little sweetheart, beloved beauty*

  *I hope that Mummy is much better and that you and little boy blue are quite well and happy. Granny and I went to the Royal Wedding at Chapel Royal St. James Palace it was a beautiful sight and the ladies in their smart dresses and beautiful jewels and all the gentlemen in their bright uniforms. Granny looked very well in grey with all her pearls and I was in my red uniform. The Prince and Princess Arthur of Connaught were all smiles — thanked everyone and when they drove through the Park there were thousands cheering them. I have sent you three.... palms [The rest of this sentence is illegible. The letter then becomes more precise.] Yesterday I saw all the cows. There were 2 Jersey cows, gold medal cows, whose milk made 3½ lbs of butter a day and a very big short-horn cow that gives 6 gallons or 60 lbs of milk a day\*. We saw quite a large covey of Goldfinches near Drayton and some herons at Puttenham and S.... and Orpington chickens at the farm at Buckden Common. You must ask Mr. Carr to take you there.*

---

\* This particular granddaughter was designated "the youngest milker" by *Country Life*.[22] She was four years old at the time and had allocated to her one Jersey cow, White Rose, who did not stamp or swish her tail and whom she milked most efficiently every afternoon, while the resigned Nannie waited alongside.

Palaces, jewels, scarlet uniforms, birds and flowers and baby kangaroos, gold medals.... the wonder was all there, yet a serious grown-up note was sounded, too, testifying to *"l'art d'être grandpère"* at which Natty was supreme. Evelina, a true daughter of Emma*, believed in schoolroom austerity and allowed her children only a single box of sweets at a time, an attitude which Natty simply could not understand. He found Evelina such a bore! Rozsika, who would not have deprived him of his sheer delight in being Father Christmas all the year round, was nevertheless sometimes taken aback by the piles of Marquis chocolates in different coloured boxes, all bedecked with rainbow ribbons, or the decorated desserts straight from his kitchen which arrived in a cab with a footman, unheralded, out of the blue. But the children never forgot the dazzling fairytale atmosphere he managed to create. If you blew on his gold watch it sprang open and showed its face.... and then it struck the hour and tinkled the minutes. It was magic. Nothing was ever to be quite like that again.

It was several decades later that I remembered how, once at the age of six, I saw the giant cuckoo clock in the nursery at Tring gain a quarter of an hour. My nurse, naturally, did not believe me, for — as she explained — clocks did not gain visibly, but *inside themselves*. She was wrong, for in this case the minute hand had suddenly dropped from the quarter to the half hour position. I have been an obsessional clock watcher all my life. I have watched them anxiously, waiting for the greatly loved to step out of delayed aircraft, or in hospital corridors, or day dreaming in delicious winter sunshine while the ski-lift came dangling down the slope, or butterflies dawdled capriciously over their egg-laying, or snow finches tinkled their dawn chorus — all day and every day — but I have never again witnessed the large hand drop those fifteen minutes of eternity. Strangely enough it took me almost as long to look back and realize that something equally improbable, but for me more significant, had occurred in the Tring nursery in the ambit of the cuckoo clock, with its loose screws and carved chamoix and the clink of pink rimmed tea cups and taste of Proust-like madeleines. I had fleetingly and unforgettably — linked to that childhood sense of happiness, yet tinged with terrible, inexplicable finality — known the most remarkable man I would ever know, and experienced one of the moments of understanding and insight that come suddenly out of the blue only in the presence of true excellence. Six years, seventy five years — it made no odds.

---

* Grosstephen recalled that Emma reprimanded him severely for putting *foie gras* in the game pie, as it was too rich for his lordship, but that Natty told him privately to slip it in, nevertheless, for he only enjoyed game pie with pâté. When the gift of a Christmas turkey was sent to a friend, Natty always had it stuffed with goose liver.

When Natty died at the age of 75, Emma sent one of his precious letters to Rozsika[23] for she had kept every line he had ever written to her since the first day they met. The envelope was addressed in Charles's handwriting.

*(Undated)*

*London — My Dearest*

   *It is sad to think all our grandchildren have gone particularly as one never knows if one will be there next year.*
   *I am coming down tomorrow by the 5. Little Alphonse\* from Vienna is here in search of a wife! I have no positive news but [illegible words follow] is in high spirits.*

*Yr.*
*R.*

It was not only his own grandchildren whom Natty loved, but children generally. All his letters to Rosebery contained references to his family, and he never forgot the birthdays of the "charming quartet". To him, children were not aliens but companions. Writing to his parents from Frankfurt[24] (before his marriage) he described how, after dinner at Uncle Willy's house, the three little girls played whist. "You would have laughed had you seen Minna standing on a chair discovering that her adversary had revoked, and remembering all the cards as if she had played ever since she had been born". Emma, in one of her letters[25] to her mother-in-law, mentions the delight with which Peggy and Sybil Primrose, then six and four years old, greeted Natty on a chance encounter in Brighton, and insisted on hearing his fairy stories, and claimed him as their uncle. An old woman in Tring described how, when she was a child, she and her sister were standing on the road verge one sunny afternoon, and, recognising the open carriage, waved their handkerchiefs as it approached. Suddenly a shower of glittering golden half sovereigns fell round them\*\*. "It was something I remembered all my life," she explained, "If golden coins could fall out of the blue sky, anything could happen." In the first photograph taken of Walter and his father, a touching expression of unselfconscious love and tenderness is evident in Natty's whole attitude to the child he is holding so carefully on his knee, and his subsequent bitterness, disillusionment and

---

\* Alphonse de Rothschild from Vienna, who married Clarice Sebag-Montefiore.

\*\* Emma was furious about this particular weakness of Natty's — which she considered both insensitive and insulting, but he insisted that children were children, and children were different, and told her not to be absurd.

hatred of his son's tastes and weaknesses can only be understood in this context.

But none of these scattered anecdotes or observations explains the aura which surrounded Natty during the first decade of the century. It may have been partly due to the glamour associated with an Aladdin-like wealth and the power which goes with it, or the fact that at this period there was neither cinema nor television to distract men's minds, who in consequence took a more lively interest in the doings of the great and powerful. This interest and curiosity could have been intensified by the slight sense of mystery which somehow surrounded all Natty's activities. Yet this is not enough to explain, for instance, the huge crowds which attended his funeral. There were no media apart from the newspapers to drum up public interest. There was a total lack of pomp and circumstance; the plain, unadorned hearse left 148 Piccadilly, drawn by two rather weary looking horses. Yet thousands of mourners crowded the roadway[26] and stopped the normal flow of traffic from Hyde Park Corner to Willesden, and five thousand jostled and pushed to gain admission to the cemetery[27] itself. Representatives of Jewish communities from all over England travelled to the simple ceremony. No other public figure[28] let alone a private individual, attracted such an enormous spontaneous manifestation of respect and sincere, deeply felt regret.

Natty remains a curiously isolated phenomenon, with no exact parallel in social history — he was there, at the centre of things, but it was something elusive, unique — a presence you could not describe but felt in the air. And now he had gone. The sense of bafflement is accentuated by the lack of letters and written records, and his determined and highly efficient secretiveness. Only one fact emerges with crystal clarity: for Walter, he proved an impossible father.

# CHAPTER 7

# Because I am so happy, Mama

THE PARAGRAPH concerning Walter in the *Encyclopaedia Britannica* is remarkably straightforward. The second Lord Rothschild did not marry; he was a zoologist who made the greatest collection of Natural History objects assembled by one man; the Balfour Declaration was addressed to him; his collections, including his library, were the largest single gift ever received by the British Museum. But this simple paragraph is deceptive, for Walter was an enigmatical man of colossal contradictions.Thus he worked at N.M. Rothschild & Sons in the partners' room for eighteen years, yet only one business letter out of the thousands he wrote from the Bank (New Court) and not one of those addressed to him in return has survived. There is no record anywhere of what he did or said during the hours he spent in the partners' room. On the other hand something like 80,000 letters and papers written to him and his Museum curators and relating to his collections, are now stored in the British Museum.

Walter never at any moment took anyone into his confidence — except perhaps his mother when he was a small boy. Paralysing shyness afflicted him all his adult life and he lived in a shell of silence, shattered at intervals by cracks of thunder. As I have indicated, it is impossible to write a conventional biography of Walter. All that one can do is to portray the people in the landscape in which he moved, and hope that the strange enigmatical figure, about which everything yet nothing is known, will somehow emerge out of his entourage.

As a baby, Walter was so beautiful that his nurse, with bated breath, likened him to the Infant Jesus.

She was not alone in making this comparison. His health, however gave rise to considerable anxiety although no specific details were ever given to the family. Years later a competent graphologist,[1] examining his mature handwriting, postulated minimal brain damage at birth — which might well have contributed to his speech problems — and, again, she discerned an aggravation of his lack of motor control in his thirty-fourth year. His mother makes no mention of any major anxiety in letters[2] to Baroness Lionel. Possibly she felt it unkind to distress her mother-in-law, for that particular branch of the Rothschild family was obsessively anxious to spare their nearest and dearest any form of worry. Nevertheless there are many references to the child's health in the sixty odd letters which survive: "Baby is quite well, thank God," is repeated several times. Then: "Baby is progressing quite favourably — the great irritability has subsided, though still very weak." There was never a mention of his specific problems, but Emma's relationship with Walter suggests that she was reacting to an exceedingly vulnerable, delicate little boy. Her attitude to her younger children was quite different, for although protective where they were concerned, she was not over-anxious. She also records Walter's frantic delight when zebras and camels in a small circus passed along the street. He "wished extremely to see them again". He was wild with excitement over the stag hounds and hugged and kissed them, and he was passionately fond of feeding the bantams on the hill. When he was ten his parents visited Günthersburg[3] and recorded that Walter was in high spirits and sent his grandmother his best love; he was "endeavouring to catch all manner of moths and insects which are not to be found in England". Once or twice his mother departed from her usual rather detached recording of daily events and confided a few of Walter's *bon mots* to her aunt/mother-in-law (for she and her husband were first cousins twice over, a brother and sister having married brother and sister). When she kissed the boy goodbye he startled her by saying "Enjoy yourself, Mama", and when she asked Walter how much he liked her, he answered with what proved to be penetrating insight, "Too much".

It was not surprising that this astonishing child who already, at the age of ten, recognised the differences between the insect fauna of Germany and England, dazzled his parents. He had grown from a chubby Leonardo-like baby with fair ringlets, into a frail, thin, beautiful little boy crowned with a halo of red gold hair, a pale sensitive face, lively blue eyes and gifted with an intense interest and love for nature and the animal world. His extremely retentive memory enabled him, already at this early age, to identify a whole string of birds and insects — the names apparently conjured out of the air — and this, coupled with his precocious interest, imparted a false impression of soaring intellectual ability.

His mother must have decided — perhaps rightly — that Walter's health was of paramount importance, and she embarked from the very first on a programme of intense and stern coddling which, at any rate from the purely physical standpoint, was an outstanding success. As a lad in his teens Walter appeared almost pathetically frail and Emma recalled that at his coming-of-age party a tenant took her aside and in a worried voice enjoined her to "fatten him up m'lady". "I succeeded," she remarked drily; the delicate little boy developed into a very large man, 6 foot 3 inches in height and weighing 22 stone. For his confirmation Walter received a pair of solid silver spurs of which he was inordinately proud and which he wore day in and day out — and one marvels how a lad of 13 could have had such slender feet and ankles. In fact all his life he had unusually small feet for his size and someone once remarked that "His Lordship bowled along the marble hall at Tring like a grand piano on castors". Walter took up boxing and regularly visited the Turkish baths in an attempt to control his weight, all in vain, but he nevertheless appeared to enjoy excellent health, prone to none of the minor indispositions which fall to the lot of the ordinary man. Certainly for the last thirty years[4] of his life he never had a day in bed, or missed a working day at the museum until at the age of 69 he broke his leg, but he nevertheless always retained a nagging anxiety about his own health.

His mother decided he was far too delicate to go to school: he must be educated at home by a governess and tutors. Dr. Heron, the family physician, was hopeful that as he grew stronger his vocal control would improve; he must do breathing exercises and also be shielded from infection, consequently his contact with other children was limited. His sister Evelina, three years his junior, was his only constant youthful companion. On the other hand there was no restriction on his passion for natural history and, once he had discovered Alfred Minall, the amateur taxidermist at Tring, he was always at his side. He began collecting stuffed animals at the age of seven, and setting butterflies at eight. Children with a time-consuming interest, especially one which "keeps them out of mischief" are very popular with their elders and betters. It must have been both agreeable and easy to spoil an eager, charming, friendly little boy — who was at the same time appealing, because so obviously shy and nervous — but also relatively simple to impose upon him a rigid discipline and iron restrictions. Nor was he likely to rebel, for none of these impositions interfered with his hobby. And spoilt he was. Walter accepted as a matter of course the unbounded admiration and adulation of the Tring employees who pandered to his slightest whim. And he took it for granted that he would get exactly what he wanted, whether it was a conducted tour of his father's ornamental pheasant pens or a set of pipes for blowing birds' eggs. In later years he looked upon wealth in much the same spirit: it was there; it was his right, and it was to be spent in the interest of his natural history. When he was a small boy his precocious knowledge and his successful collecting earned him the parental praise and approval which he craved all his life, but was never to find again. But in childhood he was listened to almost with awe — for when at the age of seven he described the specific characters of the Ruddy Shelduck[5] there was no-one in a position to contradict him. It was not a question of learning in the orthodox sense, he soaked up this type of information like a sponge — from picture books, from his own observations, from Minall's talk — with the air he breathed. He escaped the dulling and crushing effect of a formal education and all his life preserved the freshness of his boyish enthusiasm and a feeling of affinity with the animals he loved. He had to know everything about them. He wanted them all for ever. They were all his.

We have not a single clue concerning his relationship with Ellie Glünder, the German governess who took charge of the Tring Park schoolroom in 1875 when Walter was seven years old. Unlike his brother Charles, on the rare occasions when Walter spoke of the past, he never mentioned her. But he probably spent more time in her company than with anyone else. We do not even know if she liked insects! Two photographs have survived, and reveal that Ellie Glünder had a sinister face. In fact she possessed the legendary cast of feature of a Victorian murderess. But we shall never know what role, if any, she played in the development of Walter's plethora of idiosyncrasies. One has the feeling that it was his parents, not his governess, who overwhelmed his developing personality. It is, however, a fair rule of thumb that a mother should never, never employ someone to look after or control her children if she finds that person either attractive or outstanding. For in this eventuality she is apt to accept without question their version of nursery or schoolroom life and its little daily happenings. We know that Emma considered Ellie Glünder a "treasure", for she copied into the special leatherbound brass-locked notebook a letter defending herself against the charge of luring her away from an outraged lady to whom the governess had half promised her services, and who felt she "had been outbid". It is more than likely that Ellie Glünder did, in fact, wield a powerful influence over Emma, rather than over her eldest son. For nervous mothers anxious to do the very best for their first born are apt to listen attentively to nurses and governesses who they feel have had a vast experience of children which they themselves unfortunately lack. They are also afraid of "spoiling" their own offspring and feel it is almost a sacred duty to support the pedagogues and gently brush aside their children's halting and often obscure efforts to "explain". Had Ellie Glünder, as Charles suggested in later years, hypnotised his mother, who thereafter enveloped her in an aura of perfection, or was his obsessional dislike of the German governess and her influence, that curious and irrational hatred which neurotically or neurologically afflicted people develop for those they once greatly loved? Certainly we know Charles sent Ellie one of the few bound copies of his butterfly paper[6] which he published while still at Harrow — but was this a spontaneous gesture or one made in response to a gentle reminder from home? If she did nothing else for Walter she certainly taught him fluent German — although his turn of phrase remained strange — and in later years he frequently corresponded with both his curators[7] in this language, and appeared to read and write it almost as easily as his native tongue. In the photograph of the schoolroom group, which includes Ellie Glünder, Walter is in his late teens and has acquired a beard and a German tutor, Mr. Althaus[8]*. For the first time he displays his accursed lack of self confidence — the overwhelming necessity to keep his eyes fixed on the ground — which he strove unsuccessfully to overcome in the years ahead. But the day Ellie Glünder arrived at Tring Park Walter was out in the garden with his butterfly net — almost as big as himself — and his pill boxes and his enthralling specimens, for he possessed a marvellous capacity for living joyously in the present without a thought for the morrow. The impending arrival of a new governess did not so much as cross his mind. Furthermore, in the garden, believe it or not, he had caught a Chalkhill Blue. After tea his mother looked at him attentively. Something troubled her. "Walter," she asked at last, "Why do you sigh so often?" Walter looked up, mildly surprised at this interruption, for he was arranging his butterflies in a cork bottomed box. "Because I am so happy, Mama."

*your very loving*

*Emma.*

---

* He was first engaged as a holiday tutor but proved so popular that he was invited to remain on as a permanent addition to the household. Althaus's mother objected on the grounds that it was high time her son began to prepare for a future career. She was a determined character and actually bearded Walter's father at New Court to put her case. Natty promised to secure a good post for her son if he stayed on at Tring for a few years — a promise he kept faithfully for he installed Althaus in a city firm of stockbrokers of which his grandson is to-day the senior partner.

# CHAPTER 8

# I have nearly driven Althaus out of his senses

LEOPOLD DE ROTHSCHILD visited his brother Natty at Tring in 1878 and '79 and on both occasions he was greatly impressed by Walter's collection of birds and butterflies which the boy — with great eagerness — showed to his uncle. Leopold records this in letters to his own mother.[1] Walter had, in fact, already achieved the ambition he had voiced as a seven-year-old, and had founded his museum with Alfred Minall[2] as "curator". The Museum at this moment in time consisted of a shed at the end of the garden in Albert Street, the walls whitewashed and lined with shelves. We know only one fact about Walter in his early teens: accompanied by his governess, Ellie Glünder, he paid regular visits to the Natural History Museum in London. At the age of 13 this blond, nervous, yet rather imperious boy, with his precocious grasp of zoological classification and passionate interest in animals and plants, attracted the attention of the Keeper of Zoology.

Albert Günther[3] was captivated by Walter's wild enthusiasm, and avid thirst for information, and was soon escorting him round the galleries, instructing him as they went. A friendship was born, a very fortunate one for the boy, since Günther provided him with a taste of the formal education he lacked at home, for Walter was entirely self taught in all matters zoological until at the age of 19 he went to Magdalene College to work for the Cambridge Natural Science Tripos. Very soon,[4] however, the pupil was imparting information and correcting, for example, some misconceptions about the breeding records of the Bittern in the U.K. Günther, during the next few years, also played a not inconsiderable role in guiding Walter in the development of his museum. More important still, his advice was sought regarding a choice of curators, and it was through Günther's good offices that the original contact was made with Ernst Hartert. When the Keeper died in 1914, Walter wrote a distressed note to Günther's son:[5] "It is as if I had lost a parent." At the age of 16 he was scrawling exuberant letters to his mentor, urging him to come to Tring, offering him excellent shooting of waterfowl on the reservoirs ("we still have several thousand left"), and interesting plants in the greenhouses, and sending him live specimens, ranging from badgers and newts to orchids. Walter was intensely proud of everything at Tring.

In one letter he suggested that Günther come down for an overnight visit to look at his ever expanding collections, adding: "My father will not be at home".[6] This was a hint of the growing paternal disapproval of his obsessive concern with Natural History — an attitude which contrasted with that of his French great uncle, Baron Nathaniel, who was faced with a somewhat similar problem in relation to his own son, then a lad of only twelve. For James, Walter's cousin/uncle, was a passionate bibliophile[7] who insisted on spending every spare moment of his time visiting the famous libraries of France, accompanied by his tutor. He had already begun collecting books, specialising in the classics. His parents and grandparents, however, encouraged the boy in every way and laid the foundations of James's great library[7] by gifts of rare old books. We do not really know what Emma thought about Walter's overwhelming interest in natural history. She certainly did not disapprove, although she never discussed the matter in later years. One also senses that Natty's objections were based on other less tangible factors — the *lacunae* he perceived in Walter's personality and character, and the fanatical overprotection meted out by his mother, which he certainly disliked and of which he was probably subconsciously jealous. Both parents themselves took a great interest in animals, and they must, to some degree, have understood their son's predilections. Dogs played a big role in their lives and, apart from an interest in farming stock, Natty bred ornamental pheasants and albino peacocks, and had installed a small aviary for tropical finches at the Home Farm. Constance Battersea, in an anonymous novel,[8] gives a description of the Tring birds fed by Walter and Evelina. "Two children were feeding a large flock of pretty birds. Brilliant golden pheasants, peacocks with sapphire breasts and magnificent fan-like tails, turtle doves of tender grey, creamy white pigeons with rosy chests, splendid parrots robed like oriental monarchs — all flocked round the children fluttering, hopping, dancing.... perching and forming a circle of magic multicoloured wings. The oblique rays of the sun burnished the birds' feathers and formed a luminous halo round the childrens' heads."

Baroness Lionel recorded in her diary[9] that at the age of sixteen Natty "superintends the arrangement of the stables and the farm, the kitchen garden, the orchard, the green and the hothouses claim his attention and he understands all these matters perfectly well. He continues to take great interest in Botany, Agriculture and the Natural History of domestic animals." But Natty lumped all such matters together as delectable and not unimportant sidelines or attractive hobbies — but not a central theme which a serious man could pursue. Emma, up to the time of her marriage, studied physics and optics with a lecturer from the university. In Frankfurt in the 1860s this was considered an unsuitable subject for a young woman to master, and even the tutor himself, much to her amusement (for, as an old woman, she often repeated this tale), asked her how this knowledge would assist her in after life. Furthermore Charles's letters to his mother from his round-the-world trips were full of descriptions of the flora and fauna, which he evidently knew would interest her, and we find him, ten years later, inviting her to the Natural History Museum to see an entire skeleton of a Dinosaur which they had recently acquired. Be that as it may, later on, in the twenties and thirties, when Walter and his widowed mother were living at Tring together in relative seclusion, more like an old couple than parent and child, and Walter returned every day from the Museum to lunch and dine with his mother, they never exchanged a single word during the meal about either the museum, the new specimens acquired, or additions to the library, let alone his own publications. Very occasionally she asked after his curator Jordan*, and regularly on his return from the quarterly meetings of the Trustees of the British Museum, she would enquire about the health of the Archbishop of Canterbury, one of his fellow Trustees: "And how was the dear Archbishop?" And Walter, after the usual hesitation imposed by his speech problems — at their most acute when he was suddenly faced with a direct question, however familiar — would reply perfunctorily: "Quite well, Mama."

In 1932, when Walter was forced by blackmail and other financial liabilities to sell his bird collection, his mother, together with the world at large, learnt the facts from the newspapers. It remained a mystery — especially to his sister-in-law — why she never offered, even at this late stage, to rescue the birds herself. Incredible though it seems, it is unlikely that mother and son ever discussed the matter. It is certain that she had no inkling of what the loss meant to him, nor of the significance of the ornithological collection itself. In fact she had preserved a copy of his presidential address[11] to the Zoological section of the British Association among her last personal papers. In letters to her granddaughter in her old age she records that Walter had taken her to the Zoo, shown her a photograph of a magnificent stuffed gorilla and a new species of butterfly that he was describing, etc. Every Christmas she wrote to Miss Thomas, the Museum librarian, to enquire what book her "dear son" might appreciate as a gift. Yet she never visited the library. To his mother, Walter must have remained a boy in the schoolroom, and the museum, along with the collections and *Novitates* (did she ever look at the bound annual addition to the

---

* I once pressed my grandmother, then a formidable old lady of over ninety, for her impression of Dr. Jordan when he first arrived at Tring. "Ah, old Jordan," she sighed, for despite the fact that he was twenty years her junior, she persisted in referring to him thus: "Old Jordan.... he was so extraordinarily good looking. I can't recall anything else!"[10]

series housed in the billiard room at Tring Park?) a commendable and harmless hobby — an antidote to that unfortunate *penchant* for the stage; imperceptibly, without her realizing it, Jordan and Hartert had, in her mind taken the place of the faithful Miss Glünder and the tutor Mr. Althaus.

At the age of 16 Walter appeared to have no friends at '.ll of his own age. In any case his contemporaries were mere babes in arms when it came to matters which interested him. His photographic albums, unlike those of his brother Charles, who made several life-long friends both at school and at the university, contained only pictures of the family, his dogs and other animals, his tutor, the family doctor and the Rabbi who taught him Hebrew. This isolation was mainly due to his education at home in the country, his mother's fear of contagious disease, and worries, real or imaginary, about the precarious state of his health. On the one hand these anxieties seem to have been carried to absurd lengths, with an obsessional fear of draughts and damp feet, yet on the other hand we find his uncle writing to his own parents "trusting that Walter would soon be well enough to be riding again", when the child was only three years old![12]

Presumably Walter discovered at a very early age that his only effective defence against his mother's merciless and restrictive coddling was silent deception. This negative attitude became a deeply ingrained phobic habit, developed in conjunction with the impediment in his speech, which in any case made communication difficult and explanations impossible. In fact at no time in his life did he feel competent to discuss ideas or emotions. At best he could comment on current affairs, or describe the specific characters and habits of his animals.

A good illustration of his mother's unreasonable fears for his safety, and of their galling and embarrassing consequences, concerned a prospective visit to Dr. Günther[13] at his own home one Sunday afternoon — a visit planned long in advance, and to which Walter was looking forward with great impatience. The carriage was at the door and he was running down the steps of 148 Piccadilly when his mother decided *it was so hot that he might risk a sunstroke* and forbade him to go.

Walter's favourite amusement at this period was shooting. It fulfilled a dual role: it was a fashionable sport in Victorian times and he was thus able to participate in the Tring scene. Furthermore it gave him the opportunity to shine not only in society, but in his father's eyes, and thus bolster his ever wilting ego. More important still, it was the means by which he could add to his museum and satisfy the collecting urge which gripped him from childhood to old age. He practised assiduously to improve his skill, and Minall designed and constructed a trap for releasing rabbits and pigeons so that Walter could learn the turns and twists of a wild quarry appearing suddenly from behind bushes and out of long grass. In later years, when serving with the Bucks Yeomanry, he became a crack shot with a pistol, and marbles were then released from a trap to demonstrate his quick eye. But something always seemed to go wrong with Walter's attempts to win approval from his father, for on one occasion, while Minall was operating his machine, Walter peppered him with shot. He was never allowed to forget this terrible accident. In later years one of his zebras seriously injured a groom. His father gave the man a pension for life,* and chalked up another black mark against his son.

It was at this stage in Walter's life that Frederick Theodore Althaus was introduced into the Tring household as a tutor to replace Miss Glünder, although this lady remained to care for his sister Evelina. He was of German origin, a quiet, gentle bearded young man in his middle twenties, and a graduate of Oxford University. There was no doubt Althaus developed a profound and lasting admiration for Lady Rothschild during his sojourn at Tring. Up to the time of her death he occasionally dined with her at 148 Piccadilly, and faithfully sent her a book for Christmas and on her birthday, and began to plan the design of the luxurious leather binding about mid-summer. When Natty died, his widow sent Althaus a Georgian silver cigar box with an inscription on the bottom stating that it was "from his writing desk".

In 1884 Walter probably took Althaus very much for granted, looked on him as yet another Tring employee, and was quite happy in his company. No doubt because his tutor was familiar with his home background, admired his mother profoundly and appreciated his pupil's natural good humoured courtesy, the two eventually became life long friends. In later years they went on several journeys together and Walter was quite a frequent visitor to their house, and in due course, attended his son's wedding. A strange and sad episode suggests that perhaps Walter preserved a closer relationship with the Althaus family than anyone suspected. In 1917 their only son was reported killed in action, and the shock was so great that Mrs. Althaus experienced a stroke from which she never wholly recovered. In fact the report was false, and within a few days it was known that her son had been wounded, not killed. Only the immediate family knew of this tragic error — it was never published — yet Charles, at the time desperately ill in Switzerland, wrote to his wife deeply shocked by the news that young Althaus had been killed. Only Walter could have told him of this tragedy, and according to the "dead" man's widow this seemed uncanny. How could he have known unless Mrs. Althaus had told

---

* The man eventually died from his injuries, for these were pre-penicillin days.

him herself? But at the age of 16 Walter was unconcerned and not greatly interested in Althaus's arrival. Minall's company was what he really sought, for every moment of his time, when he was not engaged in formal lessons, was spent bird watching, and collecting specimens with immense energy and concentration, with both gun and net. This activity filled him with a tremendous sense of excitement and expectancy, and made every day a holiday. There followed stuffing[14] and setting of the material concerned, and reading books on the fauna and flora of the world. Caring for his small collection of live animals also proved time consuming. The successful mating of the Dingo to another Dingo, for instance, involved him in a lot of correspondence, and no little trouble, and then to his dismay she proved dangerous when her litter was born, and bit the horses in the stables, a minor disaster which was carefully concealed from his father. Already, at the early age of 14, he was an adept at delegation — a skill he failed to impart in later time to his curators. Without issuing any obvious or precise orders, but by a sort of unspoken princely expectation, he had the full weight of the Tring Park estate working away for him — the office ordering his printed forms and labels, the secretary obtaining pipes for blowing eggs, the keepers collecting small mammals or shooting rare birds, or crossing pheasants, the footman taking his material, both plants and animals, live or dead, to London or elsewhere by carriage or train. There is little doubt that both governess and tutor were meekly writing letters to dealers and book-shops on his behalf. Minall, however, was his personal assistant, and had charge of his collections and livestock. In fact at this period the Minall family saw more of him than anyone else, for Mrs. Minall simply adored and hero-worshipped Walter — first as a charming, lively little boy, and then, as what she considered a God-like young man. All his obsessive embarrassment vanished in her company and, in the warmth of her unconcealed ardent admiration, he felt delightfully at ease. He never caught his breath in painfully long drawn out gasps at tea with Mrs. Minall. At the age of eight, when he arrived triumphantly clutching his boxes of freshly trapped mice and the latest insect capture for Minall to set and mount, he assured her earnestly that not only would Minall, in the future, look after the museum, but he, Walter, would build her a lovely cottage to live in. Later on, after a long session in the museum-shed in Albert Street, he often dropped in to talk to her, and many years after Minall's death, when the old lady was in her eighties, she used to stand in the doorway of her Park Street house (which had been allocated to her for life at a rent of $1/-$ a year) for the pleasure of seeing Walter walk past on his way home to family lunch. Mrs. Minall was a great personality — and in fact she outlived both Walter and his mother by ten years. It is the lack of details about this early period of the collection and the way in which his interest developed that makes one so deeply regret the destruction of the Minall diaries by the British Museum.

During his seventeenth year Walter suddenly produced a "tiresome cough" and he was sent with Althaus to Princes Hotel at Brighton for six weeks of sea air. He took with him his live Australian Opossum and his tame Dingo, which he walked along the esplanade on a chain. She proved very troublesome but — not suprisingly — attracted universal attention. Despite the splendid litter of eight "even" pups which she had eventually produced, the Dingo had always been a bit "unreliable". The opossum, on the other hand, he described as "the most delightful creature". It slept under his desk during the day and between 6 to 8 o'clock in the evening raced around and played like a dog, but, alas! during the night ran amok and had to be removed from his bedroom. He collected assiduously round Brighton and among the birds he shot, recorded a peregrine falcon, Dartford warbler, great grey shrike, kestrels, widgeon, gannets, pintails and a bird which he first mistook for a sheldrake, but which turned out to be an escaped New Zealand variegated sheldrake. He also collected 105 fish — 35 different species — and purchased locally a very fine skin of a three-year-old goosander, the breast being "of the deepest salmon" colour. He took nine young gannets home with him — eight of which were well and happy, but the ninth was unfortunately set upon and murdered by the others.[4] The beautifully mounted albino specimens to-day on exhibit in the Tring Museum were mainly shot by Walter while in his teens. He was particularly interested in albinos at the time, and noted that the year 1886 was remarkable for the number seen round Tring. He again described his successes to Günther: "I shot a redwing thrush which was pale cream coloured all over, and now I am after a rook which has white wings. It is very singular but white and coloured varieties seem on the increase for since last October I have seen no less than 5 albino robins besides the one I got my[self]." This love of white varieties was enduring, for in 1890 he refused "point blank" to renounce his claim to the albino jays and nightjars which the Keeper himself wanted for the Natural History Museum![4]

At this period Walter's letters to Dr. Günther were very endearing. They demonstrated that, despite his speech problems, embarrassing shyness, and curous inhibitions, he could be extremely talkative at times — childishly eager to share all his marvellous animals with his friend, generous and exuberant, and delighted with the endless wonders of nature, and eager for the appreciation of his mentor — of his collections and the latest prize specimen he had shot, and the long lists of species he had spotted in the field. He was for ever inviting Günther to participate in the Tring shoots, for he was not only anxious to provide good sport for his one and only friend — like his father he was determined that his guests should enjoy themselves — but he took a boyish pride in the big bags and felt personally identified with the unique excellence of the Tring scene. It was a feather in his cap if they all had a marvellous time — a Tring time. His cousin Harry Rosebery displayed exactly the same characteristics — the same personal identifications — at Mentmore some thirty years later. Walter was not only a good shot for his age, but he was quite knowledgeable about guns[15] in general, and we find him advising his mentor about the use of E.C. powder which had a greater lateral expansion than black powder, and the advantages and disadvantages of a shrapnel shell with a ½ inch spindle, if he wanted to kill birds from 80-130 yards.

Walter always seemed to be pouring out words in a spontaneous rather confused torrent. But we know this was an illusion since in reality he wrote, as he spoke, with painful slowness. He never indented any of his paragraphs and frequently omitted punctuation — which proved a trifle disconcerting until one became accustomed to it, and began to enjoy his own inimitable style. Perhaps the lack of formal education, linked to the repressive atmosphere of the Victorian English public school, preserved in him a certain *naiveté* and verve which he would otherwise have lost. It is a strange fact that many great men are endowed with a certain simplicity of mind — a characteristic of Churchill and Weizmann as well as of many lesser lights — which proves distinctly disconcerting to the man in the street. This feature was carried to extreme lengths in the case of Walter, whose simplicity, even childishness, verged not infrequently on a form of mental aberration — baffling if not embarrassing to those engaged in conversation with him. All his life — to everyone except perhaps Mrs. Minall — Walter remained a profound enigma. Furthermore most of his contemporaries, and not least his nearest and dearest — even his remarkably sane father — responded to his eccentricity by themselves behaving very oddly indeed. Had Natty been blessed with a family of four or five sons he might have been well satisfied and amused by Walter's niche in the Tring Museum and accepted him as a more distinguished if more peculiar version of Uncle Muffy*, but as it was, he grew to regard his eldest son's disposition as a major disaster. Furthermore Walter got under his skin. We do not know precisely at what stage of his life Natty, figuratively speaking, threw up his hands in despair and wrote him off; but his mother, his brother, his curators, his sister-in-law, his mistresses, his niece and his valet, who bullied him, sustained him, helped him, loved him, nannied him and regarded him with humorous respect or bewildered affection, all eventually abandoned

---

* Muffy was a family nickname for Mayer Amschel.

any serious attempt at understanding Walter — a wayward genius, or rogue elephant, overgrown schoolboy or world famous zoologist, walking encyclopaedia or half baked eccentric, romantic lover or figure of fun, into whatever category they mentally placed him. Moreover, no-one had previously known anyone remotely like Walter. They were all baffled, including to some extent Albert Günther himself, the only individual outside the family who had won his confidence — albeit in a very restricted area. In fact Walter never confided in anyone, nor was he intimate with a living soul, for he had erected an impregnable defensive wall of awkward unintelligible silence around himself.

The Keeper of Entomology at the British Museum, N.D. Riley, himself a butterfly specialist, was once asked for his impression of the great man. "When I was quite a new boy in the department I was sitting at my table, bending over a series of swallowtails," he said, "when I heard a heavy footstep approaching from the back of the gallery. There was no mistaking the deliberate, ponderous tread of Lord Rothschild. The footsteps — echoing on the boards — came nearer and nearer and finally stopped immediately behind my chair. I did not dare look round — I sat frozen, listening to his stertorous breathing, and wondered nervously what impossible question he was about to put to me. But no word was spoken. After what seemed to me an age there was the renewed sound of ponderous footsteps, but this time going in the opposite direction, and eventually they thumped away at the end of the gallery."

This episode, in a sense, epitomises Walter's relationship with the world at large. But in his teens his gifts — his euphoric, infectious enthusiasm, his precocious knowledge of zoology, his kindliness, his spontaneous generosity, classical good looks, and those nameless qualities which combine to make a natural gentleman, overshadowed and concealed these curious and baffling contradictions and *lacunae*.

In the autumn of 1886 a momentous decision was taken. The silver cord was to be severed at last, and Walter, with Althaus in attendance as a sort of watchdog-cum-crammer, was to spend six months in Germany at the University of Bonn.

In later life his mother displayed no great affection for Germany or the German life-style. In fact she never failed to point out that she was born and brought up in the Free City of Frankfurt and was not a German in the accepted sense of the word, and certainly not a Prussian, even though her father became a friend of Bismarck. Yet the governess and tutor she engaged for Walter were both Germans, and she selected a German university for his first venture beyond the maternal roof. Surely it was no coincidence that Walter's one personal friend in the zoological world was the German born Albert Günther, with the result that the curators he eventually chose for his Museum were also both German nationals.* It was simply another subconscious demonstration of his maternal allegiance. This foreign entourage must, inevitably, have added to his isolation when he was a boy at Tring, although it is extremely unlikely that he himself ever gave the matter a thought, for his nervousness emanated from more profound causes. Unlike some of his Jewish contemporaries, he was supremely unselfconscious about such matters as religion, questions of nationality or material wealth. He treated all these problems in a princely manner. Half German? An alien? A Jew? A multimillionaire? These were the facts of life which he took completely for granted and accepted as a matter of course. What interested him were not such fundamental issues, but the birds in the Chiltern beech woods.

It was strange that Walter's father, as well as his brother Charles, both of whom made a mark later in life, were such poor scholars. Natty had a good brain, an even temperament and wisdom beyond his years, together with a capacity for long hours of work, yet he was no good at his books. In a letter to his parents from Trinity College Natty writes:[17]

> *I am sorry to say that I am fully aware of Walton's estimate of my capabilities. I told him to-day what you had written about me and he wishes me to inform you that on receipt of a letter from Cox* (illegible word) *enquiring what chance I had of getting through, he wrote to that worthy and told him that if I remained up here and read VERY hard I might just SCRAPE (scrape was the word) through. As a rule one ought always believe 1/100th part of what one hears and 1/50 of what one sees.*

Natty's parents pressured him on the one hand to work harder at his books and on the other to write more regular and longer letters to them! Natty had his own views about how best to spend his time, and the tone of his letters to his father was often more like that of a parent to a son than the reverse! Already as a very young man, he saw reality in a harsh clear light, which in no sense conflicted with the sentimental and tender streak in his nature, which in its turn never obscured his judgement — except, sometimes, concerning the quality of art treasures. In due course Walter also had his own views on how to spend his time at Cambridge, and they also conflicted with those of his parents. But if, like Natty, he put his ideas on paper they are unfortunately lost to us.

Charles, despite the fact he enjoyed the advantage of conventional preparation at school, which both his father and brother lacked, was equally inept at

---

* Albert Günther, like the two future curators, always preserved a heavy German accent. Lord Lilford, in a letter to Alfred Newton concerning a visit from the Keeper, twice refers mockingly to Günther's son as "de Poy".[16]

examinations, which completely baffled Stanley Gardiner, then Professor of Zoology at Cambridge, who maintained that Charles was far and away the best student he had ever had — yet he only took a poor third in his finals. Walter, despite his fantastic memory where zoology was concerned, was the most ineffectual of them all when it came to the more conventional subjects. He was certain that he would "be plucked" for the Cambridge entrance examination, for quite apart from his failure to grapple with the Greek language, his knowledge was obscured by the fact that he wrote so slowly and laboriously and in such a curiously immature style. We do not know what his mother, who wrote with a fluent pen and was herself no mean scholar, thought of this family failing. But one senses that both her sons were dismally conscious that they had not lived up to her expectations. One letter to Günther outlines a rather dreary picture of the first term at the German University, but Walter's usual good humour nevertheless seeps through:[4]

> My dear Dr. Günther,
>
> I see by your many kind letters that you agree with that abominable English proverb "out of sight, out of mind" because you think it necessary for me to be at Tring in order to remember to send my friends a Xmas hamper. But my only holiday was 5 days at Xmas in Frankfurt with my father and grandmother. I also have only been able to attend lectures very little because I have had to devote all my time to preparation for the Cambridge Trinity College examination in March which is very hard Sallust's Catiliana and Aristophanes' Plutus the latter causes me great trouble and I have nearly driven Mr. Althaus out of his senses by my stupidity. As you told me, there has been no game shooting here for me but I have shot 40 birds of other descriptions of which I prepared the skins.... I was extremely sorry to hear of your illness and also I hope you will soon be all right again. We breed a large number of pied pheasants every year because they make the days bag look more interesting but we shoot all down and do not leave any to breed wild in the woods as they first of all weaken the breed and then have the extremely bad habit of wandering far which in our exposed situation is not desirable; but I can always supply you with birds to stuff.

Walter, as usual, added a careful list of all the specimens he had shot and skinned, and mentioned that he had — despite the limited time at his disposal — managed to acquire another 14,000 Lepidoptera (4000 species) for his collection, of which several hundred were not represented in the British Museum, and a number were "new". Only a systematist would appreciate the significance of this remark and marvel at the fact that Walter, who was essentially an ornithologist, could, at the age of 19, have the knowledge to sort out several hundred species new to the British Museum from a collection of this size. He also discussed various pheasant crosses and concluded that the finest pheasant known is a half bred *P. colchicus* × *P. versicolour*, or else possibly *P. ellioti*. And he showed his keen interest in zoological books by adding 130 volumes to his library. This was good going for someone permitted only "1½ days per week for natural history"! His mother apparently had not been able to make the trip to Frankfurt to visit him at Christmas, and father and son stayed briefly with his grandmother Baroness Carl. It would be fascinating to know what Natty thought of Walter's progress, but the daily letters written to his wife are lost. Was he still under the influence of his son's nervous charm and his encyclopaedic knowledge of the animal kingdom, or was he already disillusioned, apprehensive and coldly critical? Certainly he was not pleased with Walter's choice of subject, for he wrote to Rosebery,[18] "Walter likes Bonn to my disgust he wants to study anatomy". It was decided that his son would remain another six months in Germany, returning briefly in March 1887 to take the Cambridge College entrance examination. After that he planned to spend a fortnight in London with his mother, followed by a collecting trip wandering round Europe until October — providing there was neither "a war nor an outbreak of cholera".[3]

But events occurred which curtailed his sojourn in Bonn. A cousin from Frankfurt paid Walter a totally unexpected visit. What he found surprised him, and he reported his reactions to Natty and Emma. They were not surprised by the contents of his letter: they were astounded. A peremptory telegram — obeyed immediately — ordered Walter and Althaus to pack their bags and return forthwith to Tring.

A casual comparison of the photographs of master and pupil as young men will provide convincing evidence that, of the two, Walter had the stronger personality, and no doubt it was he who had persuaded Althaus to stray from the narrow path of virtue. But perhaps the tutor felt with justification that it was high time that Walter, now a man of 18, cast off the intolerable restrictive coddling of his mother, and enjoy himself. And Walter, all his life, retained an immense capacity for spontaneous and huge enjoyment. In a trice he could forget his worries and doubts, however traumatic and menacing, and find pure delight in the immediate present — whether it concerned a beautiful woman or a cross between two charming kangaroos. In this instance Walter had thrown off his worries about Aristophanes, and his slow projection of Plutus onto paper: he had fallen head over heels in love with his landlord's daughter, and was having a whale of a time. It is a curious facet of this brief interlude, described by the shocked cousin from Frankfurt as "riotous living" on the part of both mentor and pupil, that, a few years later, Althaus married Walter's girl. Walter had early learned the dire necessity for silent deception, and we do not know if, during his travels in the long vacation, he re-visited Bonn. He was writing to Dr. Günther from Germany that summer, and describing with great enthusiasm a kite, so tame, that it was hopping about on his balcony. It would be astonishing indeed if, on his wanderings through the Rhineland, he had not contrived such a romantic interlude. Walter never did anything by halves. Although in 1887 photography was a time consuming and tedious business, he amassed no less than 540 separate pictures of his love. Later he had some of the originals re-photographed in Aylesbury, the background obliterated and the head enlarged. At the time of his death he had in his possession a locked black leather trunk, containing these photographs. Had he forgotten its contents when, at the age of 69, after his mother's death in 1934, he left Tring Park for the first and last time, and moved to Home Farm? Unknown to him was it put on the removal van by some dutiful but ignorant manservant? We do not know. One can see from the pictures, miraculously unfaded, that the girl was beautiful, serious and unsophisticated. And she bore a startling resemblance to the youthful portraits of Emma. No doubt Althaus had also noticed this strange likeness.

*Lady Rothschild*
*with T. J. Althaus'*
*best wishes.*

# CHAPTER 9

# Who is the pink and gold boy in the corner?

WALTER, now twenty years old, still with auburn tinted hair and beard, sat the College entrance examination in March 1887. We do not know if Trinity had rejected his application and Magdalene was his second choice, or whether, for some other reason, Magdalene College was preferred. In any case he was inscribed there as a pensioner by July, passed the University Entrance in the autumn, and went up to Cambridge in towering spirits accompanied by a small flock of kiwis, for he could not bear to leave them at Tring. A delightful vista of four years of uninterrupted natural history and delicious freedom lay ahead, for he had "done with classics for ever".[1] Of the scores of letters he wrote hurriedly to his parents and brother from the University, only a single one, sent during his first term, has survived.[2] It shows that Walter, despite the year of independence spent travelling around Europe and the romantic interlude in Bonn, was still very much under his mother's severely puritanical influence. As usual, punctuation was minimal.

> *26 Jesus Lane*
> *Cambridge*
> *25 November 1887*

*My dear Mama*

*I send you Ridleys full name and address and also the name and address of his lawyer in his own handwriting On my return here on Friday I attended the first big dinner I have attended (55 people) and I am quite sure it is the last for I was nearly killed because I would not drink and tried to get out before the toasts began it ended in my flying home through the streets as hard as I could run without cap or coat or anything except my evening dress Mr. Hugh Smith is here on a visit to his sons I have no news of any kind*

> *Your affec*
> *WALTER*

Walter, in fact, soon discovered that he had an exceptionally strong head — a quality deplored by his mother — and could easily drink any of his contemporaries under the table. In the end they succeeded in making him drunk by doctoring his wine with spirit — one of the few stories he liked repeating to his nephews, but he remained very abstemious all his life and generally refused the half bottle of port offered him before breakfast by his father when they returned from an early ride. Unlike Natty he never smoked.

In December he told Günther that his first term had been extremely fruitful and he had expanded his embryo museum which now contained 5000 birds, 38,000 Lepidoptera and about 3000 "other creatures".[3] One of Walter's truly peculiar traits was his meticulous recording in his letters of the number of specimens he had acquired and a list of the birds he had seen during his walks. In future he even included such records in telegrams to his director Hartert. He was like a schoolboy recording the runs he had scored in house matches. But, he noted, more important than the thousands of new skins, his collection of Rallidae, Gallinulidae, Screamers and Jacanas and related species was almost complete, and he planned to publish a monograph on the group, and if all went well he would include all the fossil and extinct birds such as *Aptornis*. At this period Walter had not yet acquired a curator for his museum — he was still in his teens — and only the faithful taxidermist Minall was in charge. Yet the idea of big revisions — for which Tring eventually became justly famous — was already in his mind, conceived in fact long before either of his future curators Hartert and Jordan came on the scene. Since those far off days the views regarding authorship or joint authorship have undergone considerable change, but looking back on the fantastic amount of time and effort and skill Walter put into the amassing of the bird material, it is astonishing that he was not included as joint author of Hartert's papers which were based almost entirely on the collections in question. Such generosity and modesty would be unthinkable to-day, and also historically misleading.

It is very frustrating that the letters to his parents are lacking — they were all burnt following his mother's instructions — for it would have been of considerable interest to compare them with those of his father and brother written from Cambridge. The characteristics he shared with them were wide horizons, curiosity, zest, interest in nature and animals and boundless energy, but he lacked a cool brain, organising ability — all his life he allowed things to get on top of him — common sense and intellectual power. But he outdistanced them both in single mindedness and breadth of vision. His memory was superior even to that of his father, about whom Disraeli remarked that if he wanted a date in history he asked Natty.[4] Charles, in addition, was a master of detail — like an artist of the still-life Dutch School — whereas Walter, who loved colour and form, could be compared to an impressionist painter with a photographic memory. He liked big revisions encompassing large groups of animals and insects — there never could be enough of them — and he could not conform in later years to his curators' more conventional, painstaking, critical, logical methodology, and scientifically acceptable criteria. He swept on in his own peculiar fashion, sloshing around on a huge canvas with cans of paint and brooms and brushes — rather than pursuing the straight and narrow path of the careful well trained systematist;[5] he left the creation of miniatures in ivory with a camel hair brush and a mapping pen, to his curators and his successors. Jordan, when he wrote his biographical essay about Walter, was still too near to the scene. He could not step back and survey their joint achievement at Tring objectively. Had he been able to do so he would have recognised Walter's immense ability to provide a delectable landscape, and — with his usual well directed, silent and awkward, yet implacable expectancy — to inspire his well chosen specialists to fill in details of the foreground with skill and insight. In his first year at Cambridge it was already evident that Walter was the architect, although he had not yet found his incomparably gifted builders, Hartert and Jordan. At this stage he may well have been slightly disillusioned with his mentor, Albert Günther. In his autobiography the Keeper made a point of stressing that he had "never given him [Walter] advice or taken part and advised him in his purchases of collections, building a museum, sending out collectors...."[6]

It is interesting to quote Günther on the subject of his first visit to Tring — at Walter's invitation — for it shows how even he immediately fell under Natty's spell and, in a sense, abandoned Walter in deference to his father's wishes. After describing the initial meeting in the Natural History Museum, Günther continued: "He sought to cultivate my friendship which developed so far that I received invitations to Tring. There I met with a most friendly reception on the part of Lady Rothschild but less so from the father. I fancied that the latter did not look with satisfaction on the son's devotion to the study of Natural History. In this I was not mistaken and *I determined not to assist the youngster in any of his big undertakings for forming a large collection* [my italics] — much as I rejoiced in them — but merely to maintain a very friendly intercourse with Walter and his family.[7] Walter, however, never lost an opportunity of showing me personal attachment and I returned those feelings of hearty regard." Günther then naively remarks that thereafter his pupil "followed other counsels in his scientific efforts", and later comments adversely on his secrecy. Although Walter never bore Günther the slightest grudge and accepted as inevitable that he — like everyone else — bowed to his father's wishes and would under no circumstances risk the great man's displeasure, he must have suffered deeply from this unspoken but obvious reservation, and he never again took Günther completely into his confidence. During his first year at Cambridge he wrote to him only twice — once, as we have seen, in the

Christmas vacation, and again eight months later, during long leave, on a collecting trip in the Swiss Alps.[8] After he left the University and came of age they began to correspond once again, and he took Günther to Paris and various European museums to examine types and collections and went to immense trouble to make the trips enjoyable for his friend. Still later, as Trustee of the British Museum, he was instrumental in obtaining for Günther's son the Keepership of Zoology. But the relationship had undergone a subtle change.

Furthermore he had found a new pole star in Cambridge in the person of Alfred Newton, the professor of zoology, who was himself a distinguished ornithologist,[9] and although not a museum man like Günther, a far more inspiring teacher. Walter himself described how some specimens from the South Seas shown to him by Alfred Newton so fired his imagination[10] that he immediately began to dream of collecting the island faunas of the world.[11] (See map first edition) Walter was never an idle dreamer. On the contrary he was apt to attempt, on the spot in a flurry of furious activity, the translation of the latest castle in the air into reality. He immediately found himself a sailor/collector who could make a good skin, and sent him off to Chatham Island to collect every bird which occurred there. It thus transpired that his first published paper was "The description of a new Pigeon of the genus *Carpophaga*"[12] which Palmer collected and despatched to Tring from Chatham. From there Walter sent him on to Hawaii with instructions to collect the whole group of the Sandwich Islands, with the object of studying the variation which occurs from island to island. Palmer's amazing sojourn in this area (1890-93) is described in *The Avifauna of Laysan*, Walter's first major work, published in three volumes (1893-1900), and beautifully illustrated by the Keulemans. These two artists were very closely supervised by Walter, who again and again asked for slight alteration in the colour of this or that part of a species. He had an incredible memory for finer colour shades — a gift which proved very useful to a systematist/ornithologist concerned with birds and butterflies. Was this one of the secrets of his ability for flash identifications in the field? His companions would hardly be aware of the presence of any birds at all, and Walter in great excitement would announce that he had just seen an Isabelline Warbler.... In a letter written by Lord Lilford[13] to Professor Alfred Newton he remarks, "That young Rothschild must have sharp eyes if he picked his Chilean pintail out of a lot of fowl." There is more than a hint of incredulity in the phrase. It is always difficult to imagine fellow creatures with greatly superior eyesight and accommodation, whether they be man or bird. Furthermore Walter's hesitant manner in announcing his extraordinary feats of rapid identification always roused doubts in people's minds. Whenever a check with a gun was possible he was proven right.

In old age Walter still possessed this unusual gift. In 1934 he was driving rapidly across Hyde Park in the back of his Daimler, when he noticed out of the corner of his eye, a uniformed chauffeur, with a rug slung over his arm, holding open the door of a stationary car.

"Stop! Stop, Christopher!" he shouted to his driver in great excitement, "that rug is made out of the pelts of Tree Kangaroos".

Impossible, Lord Lilford would have said. But it was true. Curiously enough Walter and Dollman were at this time in the process of writing a paper on the genus *Dendrolagus*. Eventually the owner of the rug parted with it for £30 and it was placed in the Tring Museum.[14] Like so many of Walter's clients she tried at the eleventh hour, to squeeze another £20 out of him, but by this time he had reached the mid-sixties and had learned the hard facts of life. She failed.

Apart from the organisation of his collections we have no knowledge at all of how Walter passed his time in Cambridge. These two years are shrouded in mystery like so much of his life. Did he hunt with the Drag like his father? Did he regularly attend the tripos course lectures as he had expected? Did he make any friends? He did not once write to Günther during term time — possibly because letters to the family proved time consuming — but also maybe because he felt disillusioned by Günther's sycophantic attitude to his father. Why did he remain only two years instead of four at Cambridge? Apart from the fact that he failed the second part of the Previous Examination in October 1887, retook it in June 1888 achieving a third class,[15] and at the same time obtained a first class in additional German, we know nothing.[16]

There is a traditional tale that his mother disapproved of his involvement with the A.D.C.* Surprisingly enough, he was a gifted actor and, like many nervous people, his speech impediment and paralyzing shyness disappeared on the stage. His scrap album certainly contains various photos of Cambridge plays, but he does not appear to figure in any of the groups of undergraduates. The story goes that on one evening he received a staggering ovation — flowers, palm leaves, turnips and oranges, and who knows what else besides, were hurled on to the stage and the evening ended in an uproarious wildly enthusiastic brawl. This episode was reported to his parents, and his mother ordered him home. In view of his father's own involvement with the stage while he was himself an undergraduate in Cambridge, the tale seems hardly credible. Natty in his day not only organised the finances of the A.D.C.,[17] but selected plays on their behalf and was writing home to his father to say that acting was remark-

---

* University of Cambridge Amateur Dramatic Club.

ably good for his brother Alfred who was in this way overcoming much of his nervousness by public speaking.

Something much more serious may have occurred, linked in the minds of those who were not well informed, with his stage ovation. But not a whisper of such a scandal has drifted down over the years. A less dramatic explanation seems likely. Walter probably failed his examinations and decided he was too busy to remain at Cambridge and waste his time sitting the Natural Science Tripos. He wanted to organise his museum and his collecting, and to travel, and he persuaded his father to let him leave. His parents may have had additional reasons for agreeing to this suggestion and may have decided that an early entry into New Court would prove salutory.

We do know, however, that during his first year at Magdalene he spent a great measure of his time sitting silently day after day in Alfred Newton's room, reading books about ornithology and listening attentively to every word of "shop" that he could assimilate. But he never joined in discussions.

"Who," asked one of the puzzled visitors in a whisper as he left the room, "WHO in the world is the pink and gold boy in the corner?"

Alfred Newton was initially — like everybody else — completely captivated by his new pupil — but Walter soon managed to rub him up the wrong way, and the Professor's feelings rather rapidly turned to active dislike. Furthermore he had a low opinion of his work. It took several years for this disagreeable home truth to seep through Walter's barrier of wishful thinking — erected by his hero worship. He wrote Newton friendly and enthusiastic letters long after he had left the University, but by the time he was 28 he was enjoining Günther not "to tell the Newtons" for they were "bound to perpetrate some mischief".[8] Both as a schoolboy and a freshman — despite the spell at Bonn — Walter was a naive, immature extrovert who must tell everyone everything about the marvels of the animal kingdom and his own equally marvellous discoveries. One of his most curious contradictions was revealed by these bursts of extrovert-like behaviour set in the general framework of his secrecy and introvert traits. His father denied him the approval and praise that he so ardently desired and all his life he somehow felt betrayed and deprived of recognition and struggled in this fashion to overcome a nagging sense of inferiority. It was as if, denied something vital at a critical moment in his childhood, his craving and need for that particular thing became insatiable. It was Newton himself who roused Walter's wild enthusiasm for the fauna of the South Sea Islands, but the Professor was nonplussed and exceedingly put out when this raw pupil — who could not spell, let alone write the King's English, pass the simplest written test nor speak coherently — suddenly, after a bare twelve months at Cambridge took off and soared into the blue like some boisterous and unruly young eagle, outpacing his mentor and heading wilfully without restraint for what Newton considered his very own territory and preserve.

The root of the trouble with the Professor and various other contemporary scientists was sheer envy of the young man's independent means, for financial support at that period in time, when travel was difficult and hazardous, yet still yielded such rich zoological rewards, was even more important than it is to-day. Furthermore Walter had big ideas and was both headstrong and yet secretive about his collecting expeditions. His secrecy arose from the simple fact that matters must be concealed from his uncannily well informed father, but incidentally it served another purpose, for he could thereby sometimes steal a march on other collectors and obtain the coveted new species and undescribed faunas ahead of them. Undeniably, it was a feather in his cap to put on display in the Tring Museum one of the first entire stuffed Okapis to reach the western world from the forests of the Congo. The animal — one way and another — cost Walter £300, which at that time was more than his future curator's annual salary. It should be realized that, in those days, the successful investigation and description of unknown animals and plants was the equivalent to the modern laboratory's discoveries in cell science or the retrieval of rock samples from the moon. It seems to us quite extraordinary that Walter should have been expected by other ornithologists to give a full detailed account of all his plans in advance and the fact that he chose not to do so was considered a heinous offence. Walter soon discovered that his naive confidence in his fellow scientists was misplaced, and that they blandly took advantage of his childish eagerness to tell all. Wounded and disillusioned he felt with some justification that the professionals regarded him as fair game. From that moment on, his attitude underwent a subtle change, and by the time he sent Palmer to Laysan (he was then 22) he was far, far less forthcoming. Although he could never resist sharing part of his rose-coloured plans and the latest discoveries with his acquaintances, he invariably held something back. Originally he had offered Newton participation in an expedition to the Chatham and Sandwich Islands, but later he refused the Professor's offer to co-operate with the committee when they eventually formed one for the same purpose. He knew perfectly well that Newton was, naturally enough, after his financial support, but by now he realized that being in a minority of one, Tring would not have been allocated a fair share of rarities.

The fact that Walter — a mere undergraduate — could afford to send his own man forthwith to the Islands, and that he — through the committee's inefficiency and procrastination — reached the Pacific ahead of them really infuriated Newton. Palmer, the *Avifauna of Laysan* and even *Novitates Zoologi-*

*cae*, were damned in Newton's eyes. His letters[18] and his subsequent behaviour do him no credit. Walter, in some unfortunate manner, had laid bare a totally unsuspected and unworthy streak in the man's nature.

One of the Professor's letters to Walter, sent two years after he had left Cambridge, contained, as subsequent events proved, a quite unjustified attack on Palmer, and also insinuated that Walter was prevaricating, or even lying, about information he had previously supplied about the expedition.

*Magdalene College,*
*Cambridge*
*16 December 1891*

*My dear Rothschild,*

*Your long letter tells me of course several things I did not know before — among others of your application to Mr. Wilson for a portion of his son's collection. With his refusal of course I have nothing to do but I have understood from Scott Wilson that he has since sold you all he could, so that I do not see your position is any the worse — I certainly don't remember your having proposed to me a joint venture to the Chatham and Sandwich Islands nor have I any recollection of Mr. Palmer's name being mentioned to me — but I should no doubt have declined any venture with a man of the kind he appears to be for the making of [illegible word] collections has never been my object and to help in extirpating an expiring fauna is a notion I can't abide. Until nearly 18 months ago I was still in hopes I could prevail on Scott Wilson to return and finalize the job he had begun. It was only when I found there was no chance of this that I bethought me of applying to the British Association and the Royal Society as being the one way left. I think you are mistaken in supposing that you had before that communicated the details of your plans to me — you subsequently told me what they had been and I have your letters — but this was after you wrote saying that you had counter-ordered your men.*

*Unfortunately, as I thought at the time, and as it has since proved, the Sandwich Island Committee delayed taking action. I had a man ready to go and as I still think a very fit one. If they had been provided by me, he would have been out there at the beginning of this year, and though undoubtedly he would not have killed birds by the score, we should have done better, for we should have known a great deal more about the ornithology of the group than we are now ever likely to learn — for he is a trained Zoologist and no mean collector — However I have reason to think that he possesses perseverance and powers of endurance.*

*However there is no need of controversy between us though I can't agree with you in thinking that Zoology is best advanced by collectors of the kind you employ. No doubt they answer admirably the purpose of stocking a Museum; but they unstock the world — and that is a terrible consideration.*

*Believe me to be, Yours very truly,*
*ALFRED NEWTON*

But Newton made a grave error in challenging Walter's memory. He should have known better. His pupil's hesitant manner, halting speech and downcast eyes were misleading; Newton might annihilate him in an argument, but to try and "put it across him" showed a singular lack of judgement. Walter had no secretary, nor did he keep copies of his own letters. He had no need to, for just as he remembered all conversations verbatim, so he apparently remembered all the letters he had written. He had no difficulty in citing — by return of post — the one containing the relevant information which he had indeed sent Newton ten months previously, showing clearly that his accusations were, in the main, unfounded. Again replying by return, Newton lamely acknowledged his mistake: "My memory gets bad as I grow older and I had quite forgotten writing to you last March — when, as you say you replied telling me about your men [It was only one man in fact] having been to Kanai before taking ship, etc., etc." Then Newton spitefully turned to attack Walter's prospective book:

> *You have, I think, only twice before mentioned your book, and then you did not give it a title. It seems a pity that you should bring in the Sandwich Islands only by way of "supplement" to one on a much less important group. I write only in the interest of ornithology and ornithologists — but I see it would be far better for the information your men have imparted to you to be contained in Wilson's book than elsewhere — to say nothing of the fact that courtesy expects a certain amount of deference to be awarded to the man who is first in the field and is doing his work to the best of his ability. There is another thing in which it is well to consult the convenience of naturalists, and that is by describing new species in some recognized journal like the P.Z.S. [Proceedings of the Zoological Society] rather than in a separate book which cannot fail to be an expensive one. I don't know whether you believe other people's opinion, but they will like you all the better for this concession to almost universal custom which I know from experience does not hinder the sale of such a book as yours would be.*

It is evident today that this criticism was groundless. It is perfectly true that most authors published a description of new species in current journals — they still do to-day — but this was because they wished to establish "priority" and not risk the long wait usually associated with the publication of a review or a monograph. As for the suggestion of incorporating in Wilson's book information obtained during Palmer's three year sojourn in the Pacific, this was quite unrealistic. As it was, Walter published three separate parts of his *Avifauna of Laysan*, the last one after a gap of nine years. But the really revealing phrase is the gibe concerning the cost of Walter's book, and the impoverished zoologists who could not afford to buy it, but would, it seems, subscribe to the P.Z.S.

Newton added what must have been a sarcastic postscript to his letter: "I am glad to hear your museum is making such good progress."

It was a sad ending to Walter's brief but ardent spell of hero worship — for what is more sincerely flattering than the pupil who follows in the master's footsteps? But Newton could not conceal his dislike of Walter, and Walter saw all too clearly through his one time-idol. The uneasy situation could not even be accorded the dignity of a love/hate relationship.

# CHAPTER 10

# I am not a Lieutenant, I am a Captain

IT WAS ASSUMED that Walter would follow closely in his father's footsteps — obediently and enthusiastically: Cambridge University; N.M. Rothschild & Sons; the management of the mini welfare state at Tring; the promotion of agriculture and horticulture; politics as a side line and a duty; responsibility in connection with the affairs of the Jewish community; support of local activities. There was probably a brief moment when the prospect appealed to Walter's spark of megalomania, and when he saw himself in his day dreams as the natural successor to his father — a great lay leader of the Jewish people.

At least one small facet of the paternal programme pleased Walter. He enjoyed soldiering, and took his association with the Bucks Yeomanry seriously. He had received his commission as a 2nd Lieutenant in the Regiment in 1889 and had risen to the rank of a major by July 1903,[1] proving himself rather a good officer for he always had high expectations: furthermore he knew, for once, that everyone would obey his orders however crazy they seemed to be — and they did.

Walter was a good horseman, an excellent driver — although inclined to drive his four-in-hand much too fast — and a crack shot with a rifle. It was one of the curious contradictions in his make-up that a man whom Lord Balcarres (a fellow trustee of the British Museum) could describe as the clumsiest fellow he had ever met[2] was, despite this handicap, such an excellent marksman and a skilled driver. Walter knew a great deal about firearms and in 1894 presented a Maxim gun to the Company — now the first Yeomanry Corps to possess a Regimental machine gun. He thoroughly enjoyed pig sticking* and tent pegging, and shone at target practice. Equestrian drilling was not disagreeable either. While in camp he insisted on taking a full-sized hot bath in the lines — set up in front of his tent. A batman was instructed to install a huge oblong tub and fill it with kettles of boiling water, and when the temperature was to his

---

* Pig sticking consisted of mounted bayonet or sword practice. At full gallop the rider had to spear a sack of potatoes or some other suitable object lying on the ground. It had nothing to do with pigs — a fact which grievously disappointed Walter's eight-year-old nephew when it came to light.

80

liking he stepped in stark naked. It never entered Walter's mind that this might seem a trifle eccentric to the men under his command. He liked a steaming hot bath in cold weather, so he had one. Why not? It was as simple as that.

Needless to say he wrote regularly to Hartert[3] giving him instructions about every facet of the museum and the livestock, and inflicting upon his curator the inevitable list of the local birds he had seen. He was disappointed to find that, compared with Tring, Buckingham was, ornithologically speaking, such a poor area. After drill he had been for a twelve mile circular ride and could add only five species to the list he had sent Jordan yesterday. One hour round Tring had produced twenty. Another letter to Hartert recorded, with evident satisfaction, that their inspection was a great success "because we had an Inspection Officer home from Africa who had seen the Yeomanry at work there and who had realized what their use was that is for outpost and irregular work where each man has to think and not be used as a regiment for show purposes and parade drills. I have seen a great tit nesting since you left...." There followed a list of nesting birds observed, with each site described. Hartert visited him in camp and wrote every day. If five officers were present at the weekend Walter returned home to work in the museum from Saturday afternoon until Monday morning.

It may well be that the army was the only arena in which — despite his speech impediment — Walter outdistanced his illustrious father, but nevertheless in his presence — discussing the Yeomanry — he felt for some reason like a private facing a court martial.

Natty was commissioned a Cornet in 1863, a lieutenant in 1871, and finally in 1884, a captain. During this period Emma's letters to her mother-in-law make no mention of camp but before his marriage Natty wrote to his father from Buckingham and said that all the officers were assembled at Stowe for the inspection and had a miserable time drilling in the rain.[4] Thirty years later — not surprisingly — Walter wrote an almost identical letter to Hartert deploring a howling gale and driving rain for their inspection. Natty, like his son, was a competent and fearless rider across country, and on several occasions out hunting it was noted he was among the few who jumped the brook, but pig sticking was not exactly his line. He was deeply interested in the organization of the armed forces, which is evident even in the letters he wrote to his parents from Cambridge. All his life Natty was deeply concerned with the welfare and morale of the troops on active service. He believed that a luxury parcel or two made all the difference, and during the Boer War and Egyptian campaigns arranged their delivery on a massive scale. In Charles's collection of autograph letters there is one from Arthur Balfour from Downing Street reassuring him that "Kitchener will get all he wants...." Natty was not satisfied. He wrote again from Tring[5] on September 8th 1901:

*Dear Arthur,*

*Many thanks for kindly answering my last letter. I am sure Kitchener gets all he asks for but a good War Minister, according to Sir Walter Scott, gives his generals twice as much as they ask for.*

*There was a clever article in the "Daily News" the other day which ended by saying that, unable as His Majesty's Ministers were to make peace, they were still more unable to carry on war. I have been informed, and believe it to be correct, that Kitchener wants not only mounted men but some good Cavalry Officers sent out at once.*

*It will be far cheaper in the long run to make a big effort now, than to run the risk of the war dragging on another year. I am leaving London tomorrow and only send you these few lines as I think it right you should know both what the public feeling is on the subject and also the anxiety which is felt by some out in Africa that there is a desire to save money and thus in the end be forced to incur a much larger expenditure.*

Arthur Balfour did not reply — indeed there is little one can reply to a sharp rap over the knuckles.

A week later Natty again put pen to paper:[6]

*I returned home on Thursday night after spending ten days in the North of England. I think it right to tell you that I heard a good deal of hostile criticism directed against the Government about the conduct of the war in its later stages before the news arrived of our slight reverses. Everyone I met was asking why.... [illegible word] of reinforcements had not been sent out to the Cape and Natal and why Generals had not been sent out to replace those who were coming home and to replace Kitchener. I suppose you will be back in London soon to inculcate into Broderick's mind that a good War Minister is the man who gives his Generals three times what they ask for. Do not think I wish to bore you. I think it right that you should know what the public feeling is.*

Kitchener meanwhile sent him a grateful telegram thanking him for all the "comforts" and followed it up with a long letter[7] from Pretoria in which he assured Lord Rothschild that his generosity and kindness would never be forgotten, for, thanks to the lavish scale of the consignment, there would be but few of those engaged actively with the enemy who would not participate....

From Egypt Colonel Ewart (2nd Life Guards) forwarded Natty a poem celebrating the charge at Kasassin, written by Trooper Tom Froude,[8] and privately printed, in which his gift of cigars to the regiment is mentioned.

*And Colonel Ewart in the camp at roll call said: "Men*
*They'll never say in England that you cant fight again"*
*And afterwards when small clouds rose from Rothschild's soothing weed*
*We fought again the glories of many a daring deed.*

1 Walter's great great grandmother, Gutle Schnapper, born in 1755, was undoubtedly the most remarkable member of the family. She bore 19 children, of whom 10 survived to maturity. Her protruding lower lip can be traced in a large proportion of her Rothschild descendants, including Nathan Mayer. (Oil painting by M. Oppenheim.)

2 The Rothschild home in Jew Street (Judengasse) in the Free City of Frankfurt. (Painting by Karl Goebel. Collection of Earl Rosebery.) (Photograph by Tom Scott.)

3 Walter's mother (aged 18) in her riding habit. She was an intrepid horse woman, but abandoned hunting when her eldest son was born.

4 Walter as a baby, with his father Natty.

5 Walter aged 3 to 4.

6 Walter aged 5 – the year the family moved to Tring.

7 Walter, aged 3½, on his first pony. Before he could walk he was taken riding in a pannier-type saddle.

8 Tring Park, rebuilt, with the Lord Lieutenant's flag flying from the roof. Walter's emus and rheas are beyond the ha-ha. This divided the Park from the mossy lawn on which the grandchildren's new-laid eggs bounced unbroken.

9 Alfred de Rothschild, Walter's uncle. He never forgave Emma for her refusal to invite his mistress to her house.

10 Sir Anthony, Nathan Mayer's second son, who suggested his brother should buy Tring Park for Natty. He was a great spender and gambler.

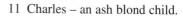

11 Charles – an ash blond child.

12 Walter's sister Evelina, aged 7, with her 'circle of magic multicoloured wings'. (Portrait by Louise Jopling.)

13 The schoolroom group at Tring Park. Charles, Evelina, Walter, Miss Glünder, Mr. Althaus and two dogs. (Blackie in the foreground.) Walter is struggling not to look at the ground. The strong family likeness between Evelina and Charles is apparent, although in character they were poles apart. They both resembled their mother when young, but lacked her impressive Roman nose.

14 A corner of the shed in Albert Street – Walter's first museum.

15 Alfred Minall working at his bench in 1890, ten years before the Museum opened. Walter and Minall together arranged all the exhibits. Living as well as stuffed animals were at that time housed in the shed in Albert Street.

16 and 17  Walter and his tutor Mr. Althaus before their departure for Bonn. These photographs suggest that the pupil had the stronger personality.

18  Elizabeth Lipschitz – Walter had over 500 photos of his landlord's daughter.

19  Walter's mother. The resemblance between her and Miss Lipschitz is striking. Both have fine, sensitive Jewish faces.

20  Walter, a Lieutenant in the Bucks Hussars.

21 (*top*) Rozsika with
her sisters Aranka and
Charlotte in 1899. Forty
years later Aranka was
beaten to death by guards
wielding meat hooks on
arrival at the Nazi extermination camp Auschwitz.

22 (*left*) Rozsika aged
20. Her chief delights
were skating, dancing, and
lawn tennis.

23 Walter watching his most gifted pupil, Charles, astride one of the giant tortoises at Tring.

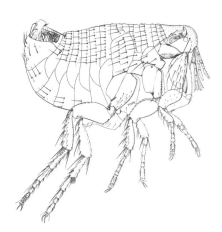

24 *Xenopsylla cheopis* Rosthchild – the vector *par excellence* of the Black Death. Discovered by Charles Rosthchild on an expedition to Eygpt and the Sudan – probably the most important discovery of all Walter's collectors. Drawing by Karl Jordan.

25 Charles, the year of his marriage to Rozsika.

26 The opening of the Museum (1892) to the public.

27 Kangaroos at Tring. Walter had various species breeding in the Park.

28 Early Museum cases of marsupials and antlers, arranged by Walter with Minall's help. The crowded arrangement proved very attractive to the public. It gave them a rapid survey of the different groups of animals.

29 A corner of the Zebra cases in "My Museum".

30 One of the flock of kiwis which accompanied Walter to the University – subsequently stuffed in the Museum.

31 Male *Gorilla gorilla matschei*, whose size and stature thrilled Walter. He took immense trouble about the preparation of this specimen. (Collected for him in the Cameroons.)

In 1902 Lord Kitchener declined further gifts, the troops now had all the comforts they required, and Lord Rothschild's generosity should perhaps be diverted to their dependants at home[9].... Broderick, it seems, had taken the hint.*

Curiously enough, Walter's first cousin, Henri de Rothschild, was also deeply concerned with the welfare of troops on active service. He was a qualified pediatrician, interested in child nutrition, and — in World War I — his mind turned to the monotonous diet of soldiers in the trenches. In order that they should have strawberry jam, preserved milk, mustard and other nourishing and "appetizing additions" to their dreary fare, he designed plastic tube containers which could be sterilized and easily transported. This invention caught on in later years, and is now a familiar sight in every grocer's shop on the continent, but at the time the Baron Henri was ridiculed for his toothpaste tubes filled with strawberry jam. Had he marketed the containers he would have made a fortune, but Henri and Walter had much in common — they were both creative people, but they lacked business acumen.

Whatever Balfour may have thought about Natty's insistence with regard to Kitchener's supplies, he respected his wide knowledge of army organization and frequently sought his advice on financial matters. We thus find Natty sending him a brief memo on the "Danger of entirely abolishing deferred pay", which he had apparently concocted at Tring over the weekend.[11] The first paragraphs ran thus:

1. *Every man discharged after long foreign service takes a holiday on going to Reserve. He must have something besides Reserve Pay on wh. to keep himself.*
2. *We want more men of 8-12 years service. But if men were induced to re-engage by seeing no certainty of employment probably ¼ would do so. It is our experience that ¼ of Reserve men who serve 12 years go on to 21 years. The result of this would be a*
   *1. Long Service Army*
   *2. Diminution of Reserve to 20,000 or 10,000 (It ought to be at least 60,000)*
   *3. A pension list of at least £1,000,000.*
   *We may relieve our present estimates but to do so at the expense of a future pension list — beginning say 14 years from now — and at the expense of the Reserve wh. is absolutely necessary, would I think be a serious error.*
   *I incline to a bonus of say £10 on going — capable of a diminution according to*

---

* Natty's misgivings about the welfare of the troops were amply justified according to Thomas Pakenham's "The Boer War" in which[10] he quotes from the talk of the ordinary soldiers, ".... It was a cruel war, it was.... we were half starved all the time."

> *conduct. £250,000 a year is the outside expense (if you deduct ½ deferred pay*
> *and all good conduct pay) and anything below £12 [illegible word] as a bonus is a*
> *further gain.*

Natty then proceeded to suggest and outline a reorganization, which would strengthen both Militia and Army.

Father and son were both fond of detail, but Walter's involvement with the armed forces was altogether on a different level. In December 1899, when the Yeomanry asked for volunteers for the South African War, Walter immediately offered his services — perhaps he considered that he qualified as one of the good cavalry officers his father had had in mind when he wrote to Balfour, but he thought it advisable — as usual — to keep the matter to himself. Besides the obvious reasons for volunteering — the whole troop had done so — he had an additional motive, for the terrain they were destined to fight over was, zoologically, of the greatest interest and he had already planned some ambitious collecting expeditions behind the lines. The secret eventually leaked out. His father, on hearing the news, did not conceal his displeasure. He looked like a thundercloud — furious at what he considered an act of unparalleled, irresponsible frivolity. An M.P's duty, however dull and prosaic in comparison to that of an Army officer, lay in the House of Commons and similar areas — not gallivanting about in the Veldt and probably getting himself killed *bird watching*. Ostriches.... Yes, watching ostriches.... To Walter's great chagrin, at the eleventh hour it was decided he could not go, although he had already been selected. The official reason given was that it now appeared that he was on the heavy side for the rather small African cavalry horse operating over rugged terrain. (At the end of the campaign the Regiment was recommended in any case to go in for a cobbier class of horse than formerly "for their staying powers were superior in the hardships incidental on active service".) But Walter suspected that his father's views on the subject may well have leaked out.... Curiously enough, *The Citizen Soldiers of Bucks 1795-1926*[12] suggested that he did in fact embark for Cape Town on the *Norman* on the 10th February with the 37th and 38th Companies. But this was not true, as their records confirm.[13]

When the 37th and 38th Companies of the 10th Battalion of the Bucks Imperial Yeomanry returned from South Africa — 53 men and 5 officers* — a

---

* The Yeomanry were severely mauled during the campaign. Among those killed was Mrs. Patrick Campbell's husband. Lieutenant Arbuthnot fell dead with the cry: "Let's rush them, boys!"[14]

county reception and Welcome Home was arranged for them.[15] They disembarked from the Tintagel Castle at Southampton and arrived on June 17th, 1901 at Tring Station, where they were met by H.R.H. the Duke of Cambridge and Lord Rothschild (the Lord Lieutenant, heading the reception committee).

Natty laid on a hearty lunch in their honour.

*The Menu*
Cold Salmon, Sauce Tartare
Surrey Chicken & Tongue
York Ham & Jelly
Pigeon Pie
Sirloin of Beef
Veal & Ham Pie
Hind Quarter of Lamb and Mint Sauce
Salad
Fruit Tartlets
Champagne Jelly
Charlotte Russe
Cheese & Biscuits
Strawberries & Cream
Hock Cup Claret Cup Beer
Aerated water

The band of H.M. Scots Guards and 1st Bucks Volunteer Band played "Selections of Music" during the meal.

Afterwards Natty gave every man a silver watch.

Although Walter must have been one of the officers[16] directing the guard of honour lined up in the station yard to meet the regiment, his name had, somehow, been omitted from the lengthy account in the local newspaper.

When Hartert next received a letter from the Camp at Buckingham Walter made no mention of Africa and seemed in good spirits. He was elated because in a large flight of swifts he had spotted an Alpine (*Apus melba*) — one of the earliest British records of this rare vagrant — thus confirming his suspicions of last summer. Scrawled along the side of the paper was a postscript: "I am not a lieutenant I am a captain".

# CHAPTER 11

# Please ask Walter

CHARLES was Walter's star pupil. A gap of ten years separated the two brothers, and Walter was already an ardent and knowledgeable butterfly hunter and a skilled "setter" when the new baby arrived. At the age of five the child was doing him credit and writing descriptions of birds in careful script between double lines.

> *Dear Mama*
>
> *I hope you are quite well. One of the fishermen shot a velvet scoter it is a black duck with red legs it is as big as a goose and a red throated diver. there is a Albatross 14 feet from one wing to the other.*
>
> *I remain Your loving C.*

By the time Charles was eight, he was hero-worshipping Walter. His elder brother's immense knowledge about all matters natural historical, and his good looks and kindly nature, coupled with the obvious delight he took in sharing his love of animals with a small, gifted disciple, made him the idol any boy would admire and emulate.

"Please ask Walter," he wrote to his mother in his first letter from his private school, "if he can spare me an emu's egg. I would like to give it to a boy here."

Much later, in a letter to Dr. Hartert, Charles now 17, wrote thanking him for the "Kittiwake Gulls and the most interesting article on the flight of the Raptores" and added that "zoological news from here is at a discount and lectures so far have been dry and uninteresting but I have no doubt Walter's will be most interesting; the announcement that he was going to lecture was received with a burst of applause I have rarely seen equalled and never beaten." A newspaper reporter once recorded that Walter was the slowest speaker he had ever listened to — it was the only way he could cope successfully with his impediment. In due course Charles noted in one of his last letters from school, that he was "never so interested in a lecture as in Walter's especially in the Corals. Everyone who was there was unanimous in his praises."

As the years went by the brothers exchanged roles. Walter fell heavily from his pedestal and turned into the irresponsible *enfant terrible*, and the quiet, self-effacing, modest, reliable Charles became the son on whom Natty pinned all his hopes[1], and to whom everyone turned for advice, information and support. Walter accepted this switching of roles without a moment of bitterness, and perhaps in one sense with secret relief, for he had a curious and rather engaging insight into his own limitations and he shied away from responsibilities. He also acknowledged that he richly deserved his father's disapproval, and he gradually withdrew into his ivory tower — which consisted of his museum and his uneventful and secure life under his mother's roof, enlivened by an occasional spree in London. When his father's Will was read and it became apparent that his younger brother had completely supplanted him — that his name was barely mentioned, and that Tring, which was his home, had been left to his mother for life and then to Charles — Walter drew his embarrassed sister-in-law aside, and struggling with his vocal impediment, said in a sepulchral whisper, "My father was absolutely right — I can't be trusted with money."

This spontaneous and generous gesture was never forgotten by Rozsika.

Fate had, in fact, been very unkind to the brothers. For if they could have exchanged only a few genes, a very happy and highly creative partnership could have resulted. For one brother had received a crushing overdose of the sense of responsibility and over-concern and over-sensitivity for the ills to which human flesh is heir, while the other had retained an almost infantile insouciance regarding other people's problems. But Karl Jordan, Walter's curator, who was a highly intelligent, although not a very objective onlooker, took the view that it was nurture rather than nature which was at the root of the matter. The brothers had, according to their make-up, grappled with the difficult problem of sky-high expectations, rigid Victorian moral codes, an impeccable parental example, and a famous autocratic father, coupled with excessive love, care, and over-protection, the possession of great wealth and the enormous additional stresses and strains imposed by World War I.

"I wonder what the dear boy will do next?" was the maximum criticism that Charles ever allowed himself, and that only in the privacy of his own home.

One winter afternoon in 1918, Walter came back from the Museum to learn that Charles had arrived on a visit after his two years absence convalescing in Switzerland. He immediately rushed upstairs and found his brother lying on his bed with his eyes closed. He sat down hesitantly on a small cane chair by the open door and stared in silence at the floor. Walter at this time was 54 years of age and weighed 22 stone, and he completely hid the chair from view, and appeared to be balanced on an invisible column of air. No word was spoken. After waiting for about 40 minutes he rose to his feet like a wounded elephant and blundered out of the room. One bewildered ten-year-old, who witnessed this scene, remembered ever after the sense of Kafka-like desolation which hung in the air.

But that was thirty years on.

Walter had been considered too fragile to go to boarding school and even too delicate and vulnerable with his speech problems to be "Barmitzvahed" in Synagogue. A private service was arranged for him at Tring. A different approach was tried with Charles. At the age of eleven he was sent to a private boarding school at Farnborough with the object of rubbing shoulders with other boys, gaining self-confidence and losing some of his oversensitivity and nervousness. He seems to have been rather happy at this little school and wrote some engagingly naive, unaffected Victorian letters to his mother.... He *never* forgot to say his prayers and he usually remembered to sponge himself and clean his teeth and he would always prize her letters "if I live, and perhaps [they will] stop one from doing an action which you would not like, by seeing them." He seems to have been worried by the thought of death and ill-health for he prayed every night that "you and all the family may keep well and see me again," and even his birthday greetings to his mother contained an anxious note: "I hope you and I will live to see many more happy returns of the day". On the other hand he was delighted with the food, had caught a very rare moth *Hyria murinata* and a large mouse, Sims had told him he was *universally* popular, he told his anecdotes and made masters and boys roar, and thought it delightful that the school had eight W.C's in a row and boys were not allowed to shut the doors so that "when you were sitting there someone else comes and talks to you. Isn't that fun?"

Just before leaving he wrote a letter to his mother in a rather different vein, since he wanted to stay on for another year.

> *I hope you dont think I like school better than "Home", but I find it very difficult to tear myself away from people and boys who I really like and who have been so kind to me: perhaps you dont understand this but I think one does get to like strangers like Mr & Mrs. Carter and Mr. Cumberbatch and one gets used to surroundings where one has spent some unhappy and many happy hours. But you must not for a moment think that I do not like all you and my home 10,000,000,000,000 times better than anything here. I am at present quite well and I hope I shall keep so.*

> *Your loving son*
> *CHARLES*

The Headmaster seemed to share the boy's regret. He was mildly reproving about the parental desire to see his sensitivity toned down and suggested it was in fact a useful quality, which enabled him to enter into the standpoint and sympathise with the view of others. "His departure will be a grief to me," he wrote to Natty, "Charles is a boy of the highest moral integrity. He is modest, courteous, scrupulously honourable and true."[2]

Unlike Walter, Charles was confirmed in the Synagogue. He was, apparently, a stalwart young chap, tall for his age, very blue-eyed, very fair. The *Jewish Chronicle*[3] described the ceremony in a way which must have weighed him down with a sense of responsibility.

### Barmitzvah of the HON. CHARLES ROTHSCHILD

*The welfare of the Jewish community in London is so closely bound up with the great house of Rothschild that any event, joyous or mournful, in that family partakes of a public character.... The Rev. S. Singer felt sure that from the impressive manner in which the Hon. N.C. Rothschild had recited his portion of the law, and from his general demeanour that day that he would grow up to shed additional lustre on the race to which he belonged, and to be a credit to the illustrious family from which he sprang.*

Among the numerous presents he received, the place of honour was assigned to the gift from the pupils of the Jewish Free School, of which his father was President. Charles must have suffered when he read that "the subscription was limited to a penny, but in the majority of cases the extreme poverty of the parents did not permit of their children contributing more than a farthing". The following Monday 3300 toys, suitable, the *Chronicle* noted, to each child's age, were distributed among the pupils of the Jewish Free School, and a piece of cake and an orange — a greater treat in 1890 than to-day — given to each child from Lord and Lady Rothschild.

Charles then went on to a public school. His mother still preserved all his letters, and when he died she sent them, with photographs and locks of ash blond curls, to his widow to keep in his memory. But there were not above twenty from Harrow School and they were the emptiest, flattest most colourless communications ever put through a letter box. Did they never stir a sense of unease in his mother? Did she not compare them with the lively letters from Farnborough and puzzle over this startling falling off in quality and content? Did his father not ask him about his school life or did Charles merely give non-committal, evasive replies, feeling that interference from parents would just make matters worse?

The lack of communication between loving — in this case doting — parents, and loving devoted children is one of the strangest phenomena of all time. Not so extraordinary, but nevertheless astonishing, is the average school master's stupidity and lack of insight concerning his pupils. When Charles left Harrow, E. Bowen, his housemaster (whose handwriting was a disgrace to the school) wrote:[4]

89

*On all the more serious aspects of school life I have heartily to congratulate you on his career. He has been steady, willing and friendly, has kept the respect of his companions; without any of the special helps that bring some boys forward he has managed to be liked as well as valued; he has been right-minded and dutiful, and what always wins regard he has been modest and simple in tone and temper.*

*Intellectually he is an unusual mixture of strength and weakness. In most of the regular subjects of lessons he has been rather below than above average and except in his favourite subjects his.... [illegible word] is strangely bad. And yet in general intelligence he is, I think, distinctly ahead of the boys who beat him in school, and in the things for which he has a pronounced taste seems to us surprising. It will be most interesting to see how it turns out.... we feel very grateful to boys who have helped keep the tone of the house high and his steady and willing example has tended week after week to guide them right. I hope that Charlie will have a good deal to look back to at Harrow with pleasure. That the feeling that he has secured and.... our gratitude will perhaps be the best recollection that he has.*

Charles looked back on Harrow with something akin to horror. The historian, George Trevelyan, a fellow pupil admitted that what cemented their friendship as boys was their mutual unhappiness at the school — the extent of which Trevelyan glossed over in his autobiography. He remembered most vividly his reserved friend's skill in mounting small mammal skins and setting butterflies, for by the age of ten Charles set butterflies well nigh perfectly. Fifty years later, as an old man, Trevelyan wrote,[5] "I often think of him". In a letter to Hugh Birrell soon after his wife became pregnant, Charles remarked, "If I ever have a son he will be instructed in boxing and jiu-jitsu before he enters school, as Jew hunts such as I experienced are a very one sided amusement, and there is apt to be a lack of sympathy between the hunters and hunted."

Shortly before he died, when in the grip of encephalitis, Charles was the victim of prolonged bouts of obsessional thinking, and over and over again he would return to the unfair treatment he had suffered at Bowen's hands, and the fact that his mother was invariably hoodwinked by the governess and schoolmasters — long since dead — in whom she placed such unjustified and misguided faith and trust. He felt so bitter about this aspect of his childhood and the baleful influence of his German governess Miss Ellie Glünder, that he never allowed a resident teacher in his own home. A pony and trap removed the pedagogues after tea and returned with them at 9 am.

But not a sign of his pupil's unhappiness or his tendency to depression seems to have impinged on Bowen's mind, and we do not know if Charles's letters worried his mother by their emptiness, or if she merely thought he was — like herself — rather a poor correspondent. Nor did she seem to have had an inkling of Ellie Glünder's past misdeeds, real or imaginary. After he left school

Charles returned at least on one occasion: "I hope to go to Harrow for the House Supper. I am not sure if I shall enjoy it very much." He was then 19. From Cambridge he wrote to Dr. Jordan (by now Walter's entomological curator) that his brother had kindly promised to give him a *new* microscope which was shortly coming from Zeiss, "a very nice present as my old one is not a good instrument". The old microscope had a quaint history. In 1885 during Randolph Churchill's tenure of the India Office, a Rothschild-backed company acquired the Burma Ruby Mines when the country was annexed (Natty was at the time advising Randolph on financial matters). The shares in the Mines were almost valueless, but with an eye to their appreciation in the dim and distant future Natty put a few in each of his children's portfolios. Suddenly and unexpectedly the prospects of the mines improved and the shares soared in value. Jestingly Natty asked Charles, who was aged 12, what he would like to do with his new-found wealth. "Buy a microscope", said Charles unhesitatingly. Obviously his father liked the reply, for years later he told the story to both Professor Nuttall[6] (from the Molteno Institute) and Dr. Günther[7] at Tring — but as a judge of microscopes he does not appear to have shone. Walter, it seems, had to come to the rescue. Nuttall on this occasion was, like everyone else, captivated and flattered by Natty's "charming hospitality" and noted in his diary that he had "an interesting conversation alone with him". Nuttall described the bird skin and butterfly collection in Walter's museum as "stupendous".

Four years later, when Charles departed for his collecting expedition along the Nile and his round the world trip — which proved a period of great happiness and delight — Walter still loomed very large in the foreground: "Please thank Walter for his numerous epistles". And again: "I have been trying very hard to get a turtle for Walter". "The cattle in Shendi are very interesting beasts. I'm sure Walter can tell you the whole history of the breed". "I expect you heard about my rare bird from my wire to Walter". "Tell Walter I found a curious assortment of bats at the Pyramids of Mevae, i.e. the long tailed bat, and *Rhinolophus*, bright *orange* red in colour, and a most lovely creature". To Birrell he declared: "Walter's book on extinct birds is a great book and the reviewers have but done it justice".

The year after Charles's marriage a tremendous family row was engendered. Walter, who was then 36 years of age, had decided he could no longer bear to read his personal correspondence.* The information it contained was too

---

* One or two of his cousins, Louis de Rothschild for instance, after his forced exile in the U.S.A. as a refugee from Nazi oppression, likewise refused to open letters.

painful and aroused unendurable *angst*,* He obtained two large wicker baskets (5′×3′) from the Tring Park estate laundry, and when the mail was brought to him with his early morning tea he sorted the letters into two piles, one large and one very small, the latter with a well known dreaded handwriting on the envelope, and dropped them unopened into their respective baskets. When the baskets were full, he shot the iron bar through the two pendulous metal loops, secured the padlock and turned the key. Over the months which followed a large number of similar baskets accumulated in his room, one piled on top of the other. It was Walter's own, inimitable, highly original fashion of disposing of his worries. The ruse succeeded very well for over two years — an astonishingly long period. There is no record of the incident that eventually sparked off the conflagration. A vague apocryphal tale describes the arrival of one of the writers of the incarcerated letters, demanding to see Walter's father. Eventually Charles was called in, and with the assistance of four clerks working all day and most of the night for six weeks, he opened, read and sorted the mail, and unravelled the tangle of confused threads.

It was not until thirty years later, after Walter's own death, that the last of the linen baskets divulged its ugly secrets. His greatest act of folly was the decision to set aside, securely locked, this one basket into which he had dropped the smaller packet of letters. It fell to the lot of his horrified sister-in-law to discover the existence of a charming, witty, aristocratic, ruthless black-mailer[8] who at one time had been Walter's mistress, and, aided and abetted by her husband, had ruined him financially, destroyed his peace of mind for forty years and eventually forced him to sell his bird collection.

Meanwhile, in 1908, Walter left his desk at New Court for ever, the debts were paid, and the Museum's finances were reorganised on a sound but limited budget. Charles, worried about the future, decided he would himself endow a trust to secure pensions for the whole museum staff, and he arranged for the estate to build a house in Tring for Karl Jordan. But Walter's father could never forgive his eldest son for his irresponsibility and lack of rectitude.

At this juncture, why did not Walter leave the parental roof? His father stubbornly refused to speak to him, and even the simplest remark had to be relayed by his brother if he happened to be present. "What does he say, Charles?" was the stock answer. But Walter could never face separation from his mother. In the long run it was unthinkable.

But for the moment he must leave. He had no wish to discuss the contents of

---

* The graphologists notice considerable lack of co-ordination and increased signs of stress in his handwriting during this period (1902-7).

the linen baskets with Charles; moreover, once the keys had been handed over, an immense sense of relief swept over him. Apparently in high spirits and with considerable excitement, he packed his bags and with Dr. Hartert left for a collecting trip in Algeria. From a scientific point of view it proved a huge success. Soon after his arrival he booked an entire floor in the hotel and threw a tremendous party. But before leaving England he sat down at his desk in the museum and secretly wrote several letters to various collectors and importers of animals. For there was one consequence of his *débâcle* and disgrace that was unendurable, and so long as one penny of his capital remained safely in trust, he was not prepared to accept this punishment. He would not, and could not, give up his live cassowaries.

*Dear Sir,*

*In the course of the next few weeks you will receive a letter from my father forbidding you to grant me any further credit. But this is not to apply to my cassowaries. When these arrive please do not dispatch them to Tring but keep them safely for me until you get further instructions.*[9]

Thus in various parts of the world secret caches of live cassowaries began to build up. Although he had published his monograph on the genus in 1900,[10] he had not yet got them out of his system.[11]

Walter's affection and respect for his younger brother never faltered, despite the well nigh impossible role thrust upon him by his irate and chagrined father. When, in 1916, Charles fell desperately ill, it was decided that only one person could accompany him on his journey to Switzerland in search of medical advice and treatment. That was Walter's entomological curator. Karl Jordan was a German by birth — once out of the country his chance of returning before the end of hostilities was remote. What would happen to the department of entomology at the Tring Museum, already depleted of staff through the war, in his absence? What would Walter himself do, deprived of the assistance of his curator, his mentor and co-author[12] — and simultaneously battling with the problems of the impending Balfour Declaration? The question was put to him by Rozsika, for Charles was too ill to ask for himself. Could Jordan leave? Walter never discussed the problem or any of the implications, because he never discussed anything. He merely said: "Go".

# CHAPTER 12

# Rozsika, Walter's sister-in-law

BARON EDMOND'S daughter Miriam (a cousin of Walter) sat next to Charles Rothschild at dinner for the first time when he was 30 and she was an intelligent and beautiful young woman of 23. She was highly amused, and a trifle rueful, since from the moment the soup was served until the dessert had been cleared away, Charles talked of one thing only — the charm, beauty and brilliance of his own wife. "Never before or since," said Miriam, laughing, "have I known a man so ridiculously and obsessively in love." In his letters to Hugh Birrell, Charles advised his friend who was serving in Burma and in the throes of a nervous breakdown, to get married. Marriage to Rozsika[1] had solved all his problems including his tendency to depressions. "I am so glad, my dear Hugh, that you are really better and that the 'blues' are getting less. Marry as I have done and you won't have any at all.... My wife is a real treasure. I wish you knew her better." Lady Battersea wrote in her diary,[2] "Charles has.... married Rozsika, a Hungarian who will, I think, play a very prominent part in the family history as she is a very remarkable woman with much nobility and strength of character".

There followed for Charles eight supremely happy and productive years. He met Rozsika* while holidaying in the Carpathian mountains — accompanied by his friend Vaughan Williams — trapping mice for the sake of their fleas, and catching butterflies. He fell in love with her at first sight. But when he accepted an invitation to visit Rozsika's home at Cséhtelek[3] (in a part of Hungary since annexed by Roumania), he also fell a little in love with the whole family. Her father had been a professional soldier serving in the Austro-Hungarian Army — a tall, lean aristocratic looking man — who retired with the rank of Colonel to farm a small property in the depth of the country. There he earned a most precarious livelihood growing maize and breeding pigs. The pigs enchanted Charles too — for he was, as usual, delighted by unfamiliar animals,

---

* The meeting was not completely accidental. When Countess Teleki (an inveterate matchmaker) heard who was accompanying her friend Vaughan Williams on his weekend visit, she invited Rozsika to stay, as a suitable companion and possibly future wife for the unknown Mr. Rothschild. She never let him forget it either!

domestic or otherwise — and unlike their English counterpart, these Hungarian swine had long curly hair.

The Wertheimsteins were as poor as the proverbial church mice, but enjoyed life enormously. They were the first Jewish family in Europe to be "ennobled" (in 1791) and hence to put a "von" in front of their name. This followed Joseph Wertheimer's[4] appointment as *Kaiserlicher hof Factor* to the Emperor, but a legend persisted down the years that he owed his good fortune more to the charm and aristocratic beauty of his wife than to his honesty and business acumen.* Be that as it may, until Rozsika came upon the scene no Wertheimer had shown anything but an anti-talent for financial affairs. They were always in debt.... Two of Alfred's daughters were very tall and slender and resembled Joseph's lady, but Rozsika inherited her mother's type of good looks — her eyes were dark brown with violet edges to the iris, an extraordinary and unique characteristic. Her athletic ability also stemmed from the Grosz family. Professor Emil Grosz, her mother's brother, boasted of three consecutive generations of surgeon opthalmologists in his family[5] — one of whom had built and directed the first eye hospital in Europe. He was incensed because his second son played tennis and table tennis for Hungary instead of concentrating on the more serious problem of diseases of the eye — he had once felt certain of three, not merely two, surgeon sons.

The Wertheimstein family — the father and his seven sons and daughters, for his wife had died two years previously — were all living together at "Cśe" when Charles arrived in 1906, although two of the girls were married. They were a delightfully informal and gay Jewish family, with a serious, dedicated streak in their make-up, the girls full of *joie de vivre* and plotting ceaselessly with the devoted factor to keep their two brothers' hair-brained escapades secret from their father. It was the greatest possible contrast with the rich, heavy, constrained, almost regal atmosphere which prevailed at Tring, where everyone defensively kept their own counsel, and Charles decided instantly — although his own bachelor home at Ashton some 70 miles distant from Tring had only just been completed — that he would build a holiday house in the form of a one storey colonial type dwelling in the adjoining garden at "Cśe" for Rozsika and himself, complete with a bug collecting outhouse for his nets and traps and breeding cages. He immediately made a botanical survey of the garden and recorded a surprisingly large number of species — omitting the grasses — and re-discovered the so-called extinct broom *Cytisus horniflorus*.

---

* This inspired one of Walter's nieces to say with pride: "The most interesting and vicious streams of blood in Europe mingle in my veins — the Rothschilds and the Hapsburgs."

He asked the Colonel if he might propose to his daughter. "Do you think I can make her happy in England?" he asked earnestly.

By way of an answer the Colonel leaned back in his chair and smiled.

Rozsika was self-educated. A diminutive German governess, a sort of beloved maid-of-all-work, cum *dame-de-compagnie*, Fraulein Kecque, had taught all the children to read and write, but beyond that the girls received no conventional lessons of any sort. But Rozsika was a quite exceptionally rapid and voracious reader with a good memory. Furthermore she and her two erudite friends, the Ritok sisters, lent one another books and discussed literature and philosophy. In later years she subscribed to daily newspapers in four different languages which, to her daughters' amazement she read dutifully from cover to cover. "You are like the lady in Hans Andersen, Mum — you have read all the newspapers in the world, only you have *not* forgotten them again...." She turned the pages over with peculiar suddenness and her dry cough — a smoker's detached critical aside — associated with the multilingual newspapers, has come down the years as a sound remembered affectionately for its unique and endearing quality. She was a passionate lover of Proust — the *Recherches* were the most wonderful books ever written.

Rozsika used to say that had she not met Charles on that auspicious flea hunt, she would eventually have become the post mistress in Nagyvarad, the local market town where she was born. At the time all her father could afford to give her was an allowance of £50 per annum. Yet 17 years later, after World War I, the Hungarian Government relied on her to obtain Rothschild loans[6] to bolster their sagging economy, and, more or less single-handed, she was managing Tring Park and whatever the war, and several deaths in rapid succession, had left of Natty's fortune. But in 1906 Rozsika was mercifully unaware of the responsibilities and the tragedies which lay ahead. What she really enjoyed was tournament tennis in the summer, skating — especially waltzing on the ice — in the winter, and dancing all the year round. She was considered rather "fast" by her contemporaries: not only did she openly smoke cigarettes, but it was no secret that she jumped over barrels on skates — very unladylike — and did not conceal the fact that in learning the loop-change-loop she had had 70 falls in public on the rink. Nevertheless when she appeared with her skates the ice was instantly cleared and an admiring circle formed. When — in the days of the Doherty brothers — Rozsika became Hungarian ladies lawn tennis champion, she introduced overhand service into women's tennis, and this was considered most improper, for lifting the arm to serve exposed the area of the bosom to full view. An Archduchess of the day nevertheless demanded an exhibition match in Vienna which was duly arranged.

When Rozsika married in 1907, her passport recorded her age as 34, but she always insisted this was a false entry and that she was really seven years older than Charles, who was then 29. In any case she decided to bear him four children in the first five years of her marriage, and in consequence abandoned both athletics and dancing. She told her children over and over and over again — long before the advent of Hitler or the Nazis — that the only country in the world to *live* in was England. She sincerely believed it.

After Charles's death she had several opportunities to return permanently to Hungary. Two of her greatest admirers had never married and now pressed their suits again after a lapse of twenty years. But they were unsuccessful. Nevertheless had Rozsika been aware of the transformation which took place in her during her annual visit to Hungary, she might have entertained some slight misgivings about the benefits of a change of country relatively late in life. The fact was that, despite the metamorphosis which had taken place through the influence of her husband and her formidable mother-in-law, despite the fact that her interests had switched from tennis and skating to politics, the Gold Standard, and children — her own and those less fortunate — she remained at heart all her life a vivacious central European with a dire need for a continental climate, a sentimental love for the *Csárdás*, and a taste for paprika.

Rozsika met her new family for the first time on the day before her marriage in Vienna,[7] but a letter from her future mother-in-law had already touched her heart. She and Emma never wavered in their mutual respect and admiration; both were passionate but unaffectionate women, and during their 25 years of staunch friendship and unwavering devotion they never exchanged even a perfunctory kiss. But they understood each other from the very first. *En bloc*, however, the Rothschild family amazed Rozsika. They were all so isolated and remote — and gruff old Lord R. seemed so remarkably unmoved by the threats of abduction and robbery he had received, and which so greatly agitated the Viennese police. Walter was, as usual, in his father's presence, subdued, silent, embarrassed, with downcast eyes, and soon faded away unobtrusively to the Natural History Museum, where he had planned an exchange of entomological specimens with the director. As for Charles's sister Evelina and the complexes which drove her to wear a grey coat and skirt, and dress like Fraulein Kecque, all that was totally beyond one's understanding. But on her wedding day Rozsika was too occupied with her own worries to puzzle over her future sister-in-law's idiosyncrasies. These anxieties were not concerned with the elaborate arrangements for the service, nor the hotel bookings — which Charles had delegated to her — which were on a hitherto unimaginable scale of luxury and extravagance, for the party had brought seven servants with them,[8] nor her impending separation from her family, but due entirely to her incorrigible brother Victor who, on that very day, was fighting a duel — with rapiers of all weapons — and looked like getting himself killed or seriously wounded. At any price her father must be kept in ignorance of the facts.... When would they notice Victor's absence? Should the sisters take Charles into their confidence, or would it upset him too? Immediately before she left for the Synagogue her brother Heinrich — who had acted as second — rushed in with the news that Victor had, rather unfortunately, sliced off his opponent's ear, but had not himself received so much as a flesh wound. He would arrive in time to act as an usher. The sisters sighed with relief and their spirits rose in unison. Surely this was a marvellous omen.... At eleven a.m., on a bitter cold, dreary February morning, mused sister Sarolta, here was Rozsika dressed all in white satin and net, with a wreath and a veil on her head, any woman should look *like hell*, yet here she was, a radiant dream of beauty.... but — weren't all the Rothschilds *extraordinary*?

Did Emma wonder if the graceful Sarolta — with her wasp waist and gentle quizzical expression — might solve Walter's matrimonial problems — for he was still unmarried at 39? Walter had once toyed with the idea of proposing to Nellie Goldsmith — an aunt of the financier James Goldsmith. He could scarcely have made a better choice — but his mother had decided that owing to insanity in their respective families the union was unwise — a decision he accepted without question. In private the Wertheimstein girls had laughingly considered a possible union between Walter and Sarolta, who unfortunately harboured an unrequited passion for a Hungarian hussar — a Roman Catholic to boot — for Charles's elder brother was a handsome fellow, even if he was a trifle stout and pathologically shy — and how wonderful it would be for Rozsika to have her sister in England! Maybe if he had been a little more like Charles, thought Sarolta....

Walter on his side was also preoccupied — and with good reason. A few weeks previously his mistress Marie Walters had borne him a daughter, and the blackmailing peeress was again threatening him with exposure. Moreover his intimate friend Lizzie Ritchie was showing sinister signs of mental imbalance. He had mortgaged his estate in Bucks up to the hilt — and began to sell it piecemeal — and now he had raised the loan on the Museum building itself to £25,000 — at 5% interest. On his return home, he started putting his unopened mail into the third — or was it already the fourth? — linen basket. He was a profoundly worried man. But if some Hungarian gipsy had drawn him aside, gazed into the crystal ball and suggested that twenty years hence the bride, *yes the bride*, would be sharing the nursery floor at Tring Park with him, sleeping in an adjoining, though not communicating room — managing his affairs, organising his riding stable and paying his annual subscriptions and taxes for him — for Walter never paid taxes — he would have burst into a peel of thunderous laughter. Walter could not control his merriment any more than his speech. On the rare occasions when he laughed aloud, he raised the rafters with his tremendous roar.

# My Museum

WALTER BEGAN his career as a banker at the age of 21, and, hating every moment of it, he remained at New Court for 18 years. Incredible though it may seem, there is not one single piece of evidence apart from hearsay to prove that he ever worked there for one day, let alone nearly two decades. A study of his accounts suggests that he was earning a salary of £5000 p.a., but this is by no means certain. There are, in addition, various sporadic entries of large sums debited to his father and placed to his credit, but there is no indication why these transfers were made. All the outgoings recorded were of payments made to artists, taxidermists and collectors, or cheques to "self". His entire income at this period was apparently spent on the Museum and his live animals. Between 1908 and 1935, when he and I shared his mother's hospitable roof, I cannot recall that he ever mentioned New Court nor did he tell us a single anecdote concerning N.M. Rothschild & Sons and the partners. Sad to relate, I never questioned him either. Occasionally he ordered his museum curators to meet him at New Court. For instance when the row with Sir William Buller, a New Zealand ornithologist, was brewing, he dispatched the following characteristic telegram to Hartert:

> *Sir William Bullers letter is quite a conundrum to me. I years ago gave him a list that is three or four years but he must know I cannot always want the same. Come to New Court on Wednesday at or about 2 o'clock to talk it over.*

How did Walter react when his father announced it was time for him to begin his city career? Did he realize there was no alternative for him — although family precedents had been provided by his uncles Mayer Amschel and Baron Ferdinand who had both opted out of banking? He probably did not even dare beg for Friday off, so that he could enjoy a long three-day weekend working with Minall. Apparently he accepted this despotic father's decree without question — natural history was an interesting hobby, like shire horses and the cultivation of orchids, but not serious work. It is probable that Walter decided not once, but a hundred times, to broach the subject, but when it came to the point failed to find his voice. Did he dare talk to Emma about his hatred of City life which he voiced openly to his mistress Marie Walters? Did he ever suggest there might be some other way of supplementing his income or earning enough to endow his collection? We do not know. Strangely enough, in his long, chatty yet discreet letters to Hugh Birrell, which contained a lot of City news, Charles never once mentions Walter at New Court. He must already have taken a back seat when his younger brother arrived on the scene. Charles himself had made it a rule never to work on his fleas or on nature conservation during his 9-5 day at the Bank, but his elder brother had no such scruples and in fact he could not possibly have achieved what he did unless he had conducted a lot of his own business from New Court. For despite the severe rationing of his free time he had installed the public galleries at Tring, so that two years later, on the day of the Tring Agricultural Show in 1892, the doors of the museum were thrown open to the public — and 5000 people trooped in to admire and exclaim.

Walter invariably referred to the Tring Museum as My Museum as though it was part of himself — like a woman might talk of her hair. When he stepped over the lintel he apparently metamorphosed into a happy man. Minall or Young would take his coat and hat and he seemed to shed his worries and emerge from the dark grey serge genial and relaxed, and, as Phyllis Thomas (his librarian) expressed it in later years: "One of the staff. We all enjoyed our job." Walter was really meant to be happy. It was sad that fate and circumstances decreed otherwise. Sam Behrmann, with melancholy truth, once said that if a Jew was happy you could write him off.... But Walter's *joie de vivre* was always ready to burst forth — for whether it was swifts on migration or a wild pig, or a petite young woman — how interesting they all were! He dumbfounded a rather nervous visitor who arrived at Tring on a hot sunny afternoon by receiving him swinging unselfconsciously in a hammock, 22 stone in weight and stark naked.

Walter had inherited a gift from his father — which proves the most intangible and the most elusive and the most difficult for the biographer to assess. Somehow he divined what the man in the street was after — what the old man behind the barrow had called "the common touch". The Tring Museum was tucked away on the outskirts of the town and, after the opening day, was never advertised, and even lacked a sign indicating its existence, let alone its whereabouts. Nevertheless in the days when private cars were a rarity and there was no public transport, it attracted 30,000 visitors a year. A Trustee of the British Museum once made a study of people leaving the premises and remarked to Dr. Jordan that they looked happy and their step was buoyant, whereas the public trailing listlessly out of the Natural History Museum in South Kensington resembled nothing more than a group of resigned patients leaving a dentist's waiting-room — a harsh judgement which contained a fair measure of truth. Perhaps Walter's secret lay in the fact that in one sense both he and the man in the street wanted the same, rather simple thing, and he was prepared to fulfil their joint aspirations. At the turn of the century, when the museum was first opened to the public, there was no air travel, no organised motorbus safari, no cinema and no television screen to bring the jungle or the sea bottom into the living room. Walter found the animal kingdom the most exciting happening of the day and he wanted to share it not only with his fellow zoologists but with Tom, Dick and Harry. And they liked the way he did the sharing. Long before the advent of the Guinness Book of Records Walter knew the man in the street wanted to see the biggest sea elephant and the largest bath sponge in the world — looking as life-like as possible. He did too. Furthermore they both appreciated the great wealth and variety of the animal kingdom, but he considered it must be displayed in a relatively small space, so that the visitor left feeling excited and stimulated, not mentally and physically exhausted. Walter actually mounted 2,000 different species of birds for public display — something no museum in the world could equal.

In 1892 the only way to achieve this was to go out and collect animals, to bring them home and either house them alive or stuff them in a truly life-like manner, but in any case display them to, and share them with a gaping public. The sort of letter Walter liked to receive (this particular one was sent in 1895) was as follows: — "Would you please except [sic] of this two specimens of *Elathea strogona* (?strogosa) for your museum as I should like to have the pleasure of knowing that I had specimens in your museum. I would have one myself if I could, but as I am a working man I cannot...." What is more the specimens arrived in good condition.

From the very first Walter wanted to combine his museum with living animals free to roam in the Park. While he was still in his teens and later as an undergraduate, he achieved this on a small scale. He had live kangaroos — a number of different species which he attempted to cross with some success — zebras, wild horses, a tame wolf, wild asses, emus, rheas, cassowaries, wild turkeys, a maraboo stork, cranes, a dingo and her pups, a capybara, pangolins, several species of deer, a flock of kiwis, a spiny anteater, giant tortoises, a monkey (belonging originally to Charles) and a number of less exotic species free in the Park, or in enclosures in Dawes Meadow. The modest shed in Albert Street, the disused chapel, several outhouses and huts and of course his bedroom, housed the growing collections of stuffed and preserved material. The combination of a zoo and museum, also containing the experimental crosses which especially interested him — he tried successfully to mate a pony and a zebra — was a highly imaginative idea for his day, and one of the best he ever had, but unfortunately for science and future zoologists it was not considered practical. Few, if any, have really appreciated the fact that in preparation for his monograph on the Cassowaries Walter collected all the species he could muster *alive** — "something no one will ever try to do again," he rightly pointed out to Hartert. When he was aiming at a joint monograph with Günther on the giant tortoises he did likewise and based part of his specific determinations on differences he observed in their behaviour — which he pointed out in letters to both Günther and Hartert but omitted in his published descriptions. In this area he was far ahead of his contemporaries. He studied all his captive animals carefully: "Its the zoo's mistake to coddle cassowaries. My laying female has lived through 6 English winters without heat...."

Fifty years later (in 1938) my brother and I tried to revive the scheme, and the Trustees of the British Museum were offered not only the Tring Park house itself but the whole of the Park as an annex of the Tring Museum. This would have made the Natural History Museum the most interesting and attractive museum of its kind in the world, but Victor Rothschild's gift was turned down (by a subcommittee appointed by the Trustees shortly after the outbreak of war) — apparently for fear of high running costs, but also because several of the senior members of the staff were far from enthusiastic.[2] This lack of insight and understanding has already proved very costly for the taxpayer.[3]

But in 1888 the fate of the zoo section at Tring depended on events of a different category. First of all some children — the public were all allowed access to one half of the Park — teased the Black Walleroo (*Macropus bernardus*) which, not unnaturally, turned upon them. Although the animal was immediately caught and incarcerated in a secluded enclosure it earned Walter another black mark. His giant lizards escaped from the greenhouse into the garden and ate all the young arum lily shoots, which was another minor disaster. At this period Tring Park employed a groom with the improbable name of Jeeves. Without the owner's knowledge, the man used to take Walter's tame wolf down to "The Green Man" on a lead, and provoke a fight with any pugnacious local dog — with disastrous results for the dog. When the matter came to light Jeeves was dismissed, for this was a misdemeanour Walter could not tolerate — wilful cruelty to animals — but somehow, and most

---

* Even to-day the Natural History Museum fail to understand the significance of a large series of live specimens subsequently mounted in natural positions, and comment: "for some reason Lord Rothschild decided to have no less than 65 of these large cassowaries mounted as if for future exhibition, and as such they make a unique collection and something of a headache for the curator".[1]

unjustly, he himself was held to be partly to blame. But worse was to follow. One of the cassowaries chased Natty, who was hacking peacefully across the Park, and although the aggressive bird failed in its objective, it aimed some dangerous, slicing, sideways kicks at his horse. Suddenly his father realized the proportions of this livestock venture — for he was an extremely busy man and had not really noticed how the numbers had been discreetly increased — and put a stop to it. Walter was forbidden to add further living animals to his collection, and the birds, except the Emus and Rheas, had to be confined to cages or paddocks. But this row — for Natty was infuriated by his brush with the cassowary — crystallised Emma's belief that Walter's love of zoology was not a passing boyish fancy and that, moreover, he was highly gifted and deserved some encouragement — although not in the Park. She persuaded Natty, despite his disapproval of this all-absorbing passion for natural history, to recognise this fact, and put the matter on a more reasonable footing. Emma was also convinced that work in any form was a splendid antidote to frivolity — especially to Walter's unfortunate *penchant* for the stage which, to her great dismay, had been revealed at Cambridge, simultaneously with his understandable but disastrous fascination for beautiful women. Did she somehow hear of his love affair with Kitty — a classical Victorian "brief interlude" with a charming still-room maid, who subsequently named her only son Walter? We do not know. Natty, who usually listened to Emma, and respected her streak of masculine commonsense, and moreover wanted to give Walter a memorable coming-of-age, decided to build his eldest son a museum, on the perimeter of the Park, as a 21st birthday present. Walter was wildly excited. This, at last, was what he had dreamed of. He asked only one favour: a first class cottage must be provided for the Minall family. It was.

Natty and Emma and their three children were all great believers in local talent and local artisans: a Tring architect, William Huckvale (who also built Ashton Wold and the village of Ashton for Charles), was invited to submit a design and an estimate. On the 20th of March 1889, J. Honour & Sons, Builders and Brickmakers, agreed to erect, according to specification "a museum together with rooms and a cottage for the Hon. Walter Rothschild at Three thousand three hundred pounds." The opening of the museum was advertised in two London daily newspapers. As we have noted a large crowd of people streamed in to see the animals which Minall (who had stuffed a large portion) and Walter had arranged between them. A tremendous effort. Walter himself had set to and mopped out the cases with a bucket of water and a swab — he did not merely issue instructions.

In those early days Minall was, of course, the caretaker/taxidermist, and Mrs. Minall — a perfectionist, everything had to shine, including the floors

— was appointed cleaner, and nicknamed Queen Victoria. William Barber took charge of the live animals and the heating. There were by now far too many butterflies pouring in for Walter to set himself, so these were distributed among a group of ladies of which, alas! no photographs and only a few names — the Misses Britton and Miss Perrin — have come down to us. Payments for these services, and later, for handcolouring of plates illustrating the monographs, seem pitifully small, but compared with Karl Jordan's salary when he first came to Tring — £200 for the first five years and £250 for the next five! — were considered perfectly adequate. The cost of procuring and stuffing animals and equipping expeditions, was in comparison very large indeed.

The size of the new premises and the scope they afforded was the signal for Walter to embark on the expansion of his collection and his first big spending spree. He was soon heavily in debt.

Richard Meinertzhagen maintained that Walter would stroll into any taxidermist's shop such as Rowland Ward in Piccadilly and casually buy anything which took his fancy, without bothering to ask the price. He was certainly very careless when his sights were fixed on some particularly desirable object, for his enthusiasm and almost fanatical determination then overwhelmed him. Thus, when he finally decided to spend a large sum of money on an egg of the extinct Great Auk, he wrote to Hartert like a guilty schoolboy. "I at last succumbed to the temptation [after eight months vacillation] though I was not going to tell you until my return". He had hidden the egg at Tring in the box with the Great Auk's skin! Eventually he bought a second egg.... By this time he had amassed the biggest collection in the world, and among his treasures were some specimens collected by Audubon. All his life he retained a passionate love of birds' eggs and on his death bed he was discussing new clutches from Birds of Paradise which had just arrived and his tired eyes lit up at the sight of the pale spheres lying in their bed of cotton wool. But in the more routine purchase of specimens and the drawing up of contracts with naturalists or dealers, he was careful and very conscious of cost. Specimens from albino bullfinches to Okapis were frequently refused because they were too expensive. (£300 was offered for the first specimen of a captured Okapi, but on hearing Walter's name the vendor raised the price!) Walter was relentlessly victimised, bamboozled and deceived by dealers and collectors, who felt they were poor fellows scraping around for a living, whereas he was a man backed by unlimited wealth and affluence, and by hook or by crook they were going to profit from his fortunate involvement in their scene. Furthermore every museum in the world, whether it was Cambridge University or some modest county collection, expected the multimillionaire to present them with specimens free gratis — there was no understanding whatsoever of a Rothschild who wanted to *sell* or *exchange*. What a mean fellow he must be!

The deceptions with which Walter had to contend ranged from faked locality records (which increased the interest and hence the value of the specimens in question) to the "loss" of material which in reality had never been despatched, down to the presentation of accounts long since settled, or blatant overcharging. One Australian collector, for example, asked for and obtained a substantial advance which he subsequently added to the bill! On another occasion Walter dropped in unexpectedly on one of his dealers, on the Riviera, and caught the man redhanded, calmly putting imaginary locality labels onto a collection of skins destined for Tring. As late as 1910 — by which time he should have grown accustomed to this angle of his enterprise — Walter was writing in a disillusioned vein to Hartert from Valloir, saying that he had in his pocket the receipt for the account that had been presented once again in his absence — stamped and signed by the man himself — and added gloomily: "I think somehow animals, birds and insects seem to have the power of sucking the honesty out of people like the Vampire sucks blood." Under this rather relentless form of petty but infuriating persecution Walter seems to have become a careless and dilatory payer, or else the Tring Park Estate Office were slow in settling up their accounts, for the letters addressed to the curators contain a never ending stream of complaints to the effect that the collectors' money hadn't arrived, or their parcels or letters had not been acknowledged.

Walter never at any time of his life kept accounts, nor did he employ a secretary; and when the Museum was first opened he continued to pass all his bills to the Tring Park Estate Office where they attracted relatively little attention — for a clerk had automatically paid Minall's wages, his cartridges and the cost of his working material from the very beginning — when Walter was in fact still a child. These sums were engulfed in the much larger amounts expended on the livestock enterprises at Tring Park managed by Richardson Carr — the Jersey, Redpoll and Shorthorn herds, the Southdown sheep, the Shire horse stud, the riding and driving stables, the ornamental pheasant and peacock pens, and the duck breeding enterprise on the three Reservoirs and Dundale. The coke bill for the greenhouses was around £2,000 a year. Furthermore a number of people at Tring Park were running very considerable rackets of their own, a fact which only came to light years later when Rozsika, after Natty's death* — which roughly coincided with the onset of Charles's eventu-

---

* Natty was exceedingly lenient over such matters, although he objected when his head keeper bought a valuable racehorse. One of his head gardeners took advantage of the Tring unemployment scheme and entered a dozen fictitious "temporary hands" on his payroll, and pocketed their salary. He had to leave, of course, but Natty gave him a pension and a public house to run. He had been an excellent gardener.

ally mortal illness — took over the management of Tring Park herself. From these people's point of view another set or two of confusing accounts, addressed to an individual who did not bother to scrutinise them too closely, were very welcome. Furthermore there is a certain satisfaction in aiding and abetting the mild and rather harmless deception of a severe and autocratic father — just as there is some mild amusement in covering up for a flighty husband with a virtuous wife. Moreover Walter's imperious demands, coupled with his good nature and kindliness, and intense shyness, proved an inexplicably attractive combination. "Young Lordy" as he was sometimes called was universally popular, and, unlike his mother, he didn't scare the daylights out of you!

The cost of the so-called Felder affair would probably have been absorbed quite comfortably by the office — split up into smaller units and so forth — only, unfortunately, the Burgomaster of Vienna decided to boast to the press about the prospective sale of his large insect collection to a Rothschild and — hopefully — mentioned that a price of 50,000 guldens (£15,000)[4] had been agreed. Walter was terrified and asked Günther to contact Felder and tell him that if any further publicity ensued he would call the whole thing off. But it was too late. Natty was informed and — outraged — ordered Walter to cancel the deal. It was true that his father could not be expected to appreciate the value of a few thousand butterflies and moths, and it was equally certain that he could scarcely fail to disapprove of a single collection which was said to cost treble the sum needed for building the whole museum itself! And who was going to pay a bill of £15,000 when Walter's allowance, which was the income from Charlotte's settlement, amounted to £1,300 p.a.? Walter tried to comply with his father's edict, but Felder threatened him with breach of contract; he then retorted — quite justifiably as it was subsequently shown by expert witnesses — that Felder was grossly overcharging him. Eventually the matter was settled, for Walter obtained the collection for a greatly reduced and fair price and Emma paid for it. She did this after inviting Dr. Günther round to 148 Piccadilly to discuss the matter.

Shortly afterwards Richardson Carr wrote at Walter's suggestion asking Günther to see him and talk over the best way of organising the Tring Museum. This was imperative, since Walter was now tied to an office desk at N.M. Rothschild & Sons from nine to five, loathing every minute of it, and could not possibly supervise day to day management. Moreover the collections were getting out of hand; there were, for instance, no less than 300,000 beetles stored in boxes, yet to be arranged. Günther recommended the appointment of Ernst Hartert and by 1893 both he and Karl Jordan were installed as curators. This did not prevent Walter in engaging privately in several further spending sprees, and his debts, when discovered, though unjustly attributed solely to his museum, were a recurring source of trouble between himself and his father.

From the beginning the museum was divided into two sections. The galleries, which contained the stuffed examples of the animal kingdom[5] and were opened to the public at first for a total of only 15 hours per week in summer and 11 in winter, and two study sections at each end of the galleries, reserved for students of birds and insects, and for scientific research. The library was also extensive,[6] and the first volume of *Novitates Zoologicae*,[7] the Museum periodical, containing some 600 pages with 15 plates (issued in quarterly instalments, annual subscription one guinea, for which 130 different journals were obtained in exchange) was published in 1894.

All the exhibition cases were arranged personally by Walter. He was very upset when Hartert tried his hand at display. "Now I NEVER told you," he wrote, "to put *Casuarius australis* under the skeletons tail so please TAKE OUT *Casuarius australis* and the so-called *C. beccarri* altogether and put them in the Bird room in No. 1 and put back all the skeletons *just as* they were before and put the *C. sclateri* in the place of the *C. australis*. Then tell Minall to have a man to help him Sunday and I will arrange them myself." Shortly after Hartert's arrival at Tring Walter had got steamed up about the ostriches "*Please* DO NOT put anything in the Ostrich case except the spotted Ostrich and the white rhea until I come myself.... I want to arrange that case entirely myself for I want to prove that one can put in one case 3 times as much as is usually done and *yet* see *all well*." He always had the feeling that only he himself could display these fabulous creatures to their full advantage. He was probably right.

It was a grand year, for the marvellous collection of Humming Birds[8] was added to the museum, and in November 1894 the live zebras arrived at Tring and were, amid great excitement, put into loose boxes in the carriage stables and Walter began the business of breaking them in. This proved no mean task, for the zebras objected strongly to harness and bridles and he had to devise a method of letting down their collars from the ceiling. Loulou Harcourt[9] mentions in his diary that after other guests had departed from Tring he stayed behind to play croquet with Evelina and help Walter drive one of the zebras which was "beginning to go very well". Walter first accustomed them singly to a small trap, but he was eventually able to drive them down Piccadilly as a four-in-hand — three zebras and a small pony was the usual combination — and into the forecourt of Buckingham Palace. Charles told the children that the zebras' camouflage was so good that half way down the street they seemed to vanish, leaving Walter bowling along Piccadilly in a horseless carriage.... Natty was not pleased about the expedition to the Palace which, in his opinion, invited disaster. Walter himself admitted it was rather risky — his heart was in his mouth when Princess Alexandra tried to pat the leading zebra.

Electricity had also been installed at Tring in the early nineties and people came from far and wide to see this magical innovation. The Paris house had sent their own engineer, Mr. Levy, to consult their cousins a few years earlier. Natty had found a ship's engineer, William Morris Thomas, who designed and organised the system for the whole of the estate and was told to spare no expense in doing so. The engineers from Sandringham and Blenheim came over to inspect it, and delighted Thomas by exclaiming together: "Of course we've nothing like this!" Thomas's daughter, Phyllis described it nostalgically as an engineer's dream. At weekends after dinner, every light in the museum was put on — throwing a tremendous strain on both the installation and the electrician in charge — and Walter would give the party a conducted tour. Gladstone, who was shown round in 1893, was so interested that the delighted owner sent him a copy of *The Avifauna of Laysan* which had just appeared.[10]

Many years later N.D. Riley (Keeper of Entomology at the British Museum) recalled that during the last decade of his life, when Walter, annually, entertained the South London Entomological Society and the Entomological Club* at My Museum (he was elected in 1926) the members enjoyed the boyish enthusiasm of the great man even more than the treasures he delighted in showing them. But by then he had become a legend; the members nudged one another and winked as he displayed his quite extraordinary memory for scientific names and the fact that, improbable though it seemed, he knew literally every detail concerning the specimens in his vast collections. Although there was no catalogue, he could point out where every one of the 2¼ million set moths and butterflies was located, and went straight to the correct drawer if someone suddenly asked for a particularly obscure species. They delighted in egging him on and showing his paces for the benefit of the uninitiated. But what we would like to know is whether, in the good old days, Walter's father ever joined these evening tours of the museum with Gladstone and the Prince, Haldane or Milner.... Did he stay behind in the smoking room after dinner drinking his port and discussing the Aliens Commission with Randolph? Or did he once in a while stroll round too,[11] and what did he really think of his eldest son, as he showed the party the exquisite Prince Rupert Bird of Paradise, hanging upside down on a branch — feathers fluffed out like spun glass — skilfully stuffed in the act of full sexual display? Would Walter fiddle for ever while Rome burned? Yes, what *did* Natty think of My Museum?

* This social Club founded and endowed by Verrall, Clerk of the Course at Goodwood races and world expert on flies, is limited to eight members, and up to 1967 when I myself was elected, was restricted to male entomologists only. Once a year the club has to meet for dinner and drink to the memory of its founder.

# CHAPTER 14

# The rump of our only old male
# is rather dicky....

WALTER SHARED certain characteristics with other notable Rothschild collectors. His cousin Henri specialised in *têtes-de-mort*[1] and autographs, his aunt Adèle in Judaica,[2] his uncle/cousin James in books,[3] his uncle/cousin Ferdinand and cousin Béatrix in works of art and *objets de vertu*,[4] his cousin Edmond in precious stones and engravings[5] (40,000 of the latter plus 3870 drawings were donated to the Louvre), his great aunt Charlotte and cousin Arthur collected rings, especially engagement rings,[6] Edmond's wife thimbles and hair ribbons,[7] Walter's cousin Miriam literary[8] and musical manuscripts, his cousin Alphonse postage stamps[9] and his own brother Charles Iris[10] and fleas[11] — one and all possessed zest and persistence and a tendency to become deeply, almost obsessively involved in everything that interested them. But unlike, for instance, his uncles Mayer Amschel or Ferdinand, both of whom also declined to be bankers, Walter had the determination and strength of character, and the streak of egotistical ruthlessness, which enabled him eventually to turn his back not only on the bank but also on the temptations of the conventional family life-style and to withstand both the heavy paternal pressures and his own nagging pricks of conscience. This resolution he owed to the unspoken but powerful support of his mother. Under duress he gave New Court and the House of Commons long, lukewarm token trials which he felt encroached cruelly and to no real purpose on his scientific/museum time — and he retained a sincere though limited interest and concern in Jewish affairs and local Tring activities. But for thirty years he enjoyed — what appeared such a strange choice to the world at large — the rather pedestrian, insulated, abstemious life of a dedicated, professional zoologist. Other members of the family (for instance Baroness James, Hannah Rosebery, Victor Rothschild, etc., etc.[12]), produced or published catalogues of their collections, many of which have since been partially or wholly dispersed or partially fused (as in the case of the contents of Waddesdon Manor) or sunk in vaults of various museums or university libraries. But no other Rothschild, before or since, has combined a collecting hobby with straightforward original research, and turned it from an agreeable, leisured pastime into an arduous life-work.

A passion for detail is another family characteristic, and, owing to Walter's single-minded devotion to My Museum, he could indulge this rather obsessional trait without frittering away his energy and his time. There is no doubt that the amazingly high standard of every facet of his entire enterprise was due not to lavish expenditure — he had a surprisingly small staff — but to a fanatical attention to detail, coupled with his peculiar computer-like mind. In a way his father had a somewhat similar brain, for he had the power to sort and store information and possessed an encyclopaedic knowledge of both financial and commercial developments of the chief countries of the world. This stood him in good stead and was the basis of what a newspaper once described, in somewhat florid language, as "his unfailing comprehension and shrewd and sagacious advice". Baron James and his father (a brother of Baron Lionel) and, to a lesser extent, Natty, were endowed with prodigious memories which, according to Picot,[13] "*ont fait l'étonnnement de tout ceux qui l'ont connu*". This seems to have been an hereditary characteristic in some — but by no means all — of Nathan Mayer's descendants — for James was a member of the English branch* but opted, when 21, to become French.

Minall and Young were both excellent skinners and taxidermists, but Walter considered that certain groups of animals, for example the cassowaries, deserved special individual attention. When his curator objected to the size of the bills, he wrote to Hartert as follows: "I certainly agree with you that £30 is quite outrageous but as to my paying £20 when other people said it was too much, *you know* as well as *I do* that at the present moment Doggett is THE ONLY taxidermist who can do cassowaries as I want them, and it would take any fresh man 3 or 4 years to learn how to do it and meanwhile the chance of many fine specimens would be permanently lost. — He can if asked to do a cassowary like ours make his own terms." Doggett took shocking advantage of the situation until Charles, in 1908, put a stop to it.

Walter was tremendously excited when the skin of the outsize bull sea elephant, well over 18 feet long, arrived packed in a barrel of salt. It had weighed over 4 tons alive "as heavy as the famous African elephant Jumbo" he wrote to Hartert. Could it be stuffed outside the museum and then introduced onto the top floor through the roof? Or was it too large? He rushed down to Tring and measured the doors, passages and skylight himself, and then arranged for the animal to be mounted on a hollow cane frame to avoid overweight. At the same time he had a dozen other matters in his mind, and the letter heralding the arrival of the sea elephant ran as follows:

* Like Charles, who also possessed this type of memory, James developed depressions and nervous instability.

111

# DEAR LORD ROTHSCHILD

*Tring Park*
*July 8th 1900*

*Dear Mr. Hartert,*

   *I did not write to you again because I had little to tell. You will find I unpacked a few boxes in your room. One contained some birds from Alan Owston, another some birds and a few butterflies from Mazagan. The round tin box is the* Gecinus rageni *and then a box of butterflies from Alaska which I told you I had purchased some time ago. Then there are a lot of Batjan & Brunei birds from Watenstradt which you pick and then return to Gerrard Then there is a box from Kühn of Kei birds & a* Proechidua *the latter of course we keep. Then there is a cardboard box & a large wooden box & a box of lepidoptera which please do not touch till I come.*

   *I am not coming down next Saturday so I should like to see you in London the 17th Tuesday. The Sea Elephant will go to Tring Thursday 19th. I have told Minall to be ready.*

*Yours very truly,*
*WALTER ROTHSCHILD*

Walter was interested in all aspects of museum work, but was particularly keen on the most effective display of material and he soon acquired a wide knowledge of cases and cabinets. He himself designed the new type of drawer with a glass bottom as well as a glass top, an excellent innovation, since both sides of a long series of the butterflies could then be examined without removing them from their case. There were no less than 14,000 drawers and 8000 boxes of insects at Tring. He went far afield looking for the best equipment, visiting Dresden especially to investigate glass cases of an interesting and novel construction. He also took a lot of trouble about the type of naphthalene and camphor to be used in the cabinets and sent Hartert various lists with comparative costs. When in 1908, after the Great Row, the new wings of the museum were built for him by his mother and Charles, he fussed about every detail of the cases and demanded from Sage and Co.[14] information concerning the polished edges of the plate glass shelves and the special type of paint required for the wrought iron (he asked for steel but this proved difficult). The metal was to be "bright or cut down dull as preferred". Nothing escaped his attention from the extinguishers and fire bells (which were rung and tested every Tuesday) to the size of their setting pins and brand of writing paper (his hand-made blue linen envelopes were familiar in other museums all over the world) to the type of print and paper used in *Novitates* and the exact proportions of the board — which must be painted white — to be placed beneath the Rhino in the corner.... Walter also took infinite trouble with his exhibits for the monthly meetings of the British Ornithological Club. There was a memorable occasion when he arrived in a cab with the nine foot tall model of the extinct Giant Moa sticking out of the roof,[15] but he also wrote out all the labels of his exhibits himself, and if he was prevented from attending one of these gatherings (a rare occurrence) he sent Hartert detailed descriptions concerning their arrangement, even to the precise size of the labels to be slipped under each specimen.

One of the striking results of Walter's attention to detail was the uncanny lack of dust or rubbish at Tring. For a museum it was almost disconcertingly clean and neat. Walter flew into a violent rage when he loaned some specimens to the Natural History Museum and the spoonbills and storks were returned dull grey instead of pure white. Rather uncharacteristically he gave Dr. Günther a piece of his mind.

In another part of this book I have described the network of collectors used by Walter in building up the museum and some of his own expeditions. But quite a large proportion of the specimens he acquired were the result of exchanges effected with other museums and dealers, and this part of the enterprise was carried out almost entirely by Walter himself. Once again, owing to his freakish memory, he could cast his eye down some massive catalogue and pick out unerringly those species needed to fill a gap at Tring, be they King Crabs or Redstarts that were required. The same applied to boxes of specimens in the museums he visited during his annual holiday. He spent a few days in Florence and made "some profitable exchanges", sending home in two boxes 334 skins of Italian birds, 11 stuffed Italian birds, two antelope skins from Abyssinia, one skeleton of a young Dugong about 2½ feet long and about 30-40 birds from various localities. He thoroughly enjoyed visiting other museums and wrote enthusiastically to Hartert about the marvellous things he had seen. "A group of 9 *Gorilla gorilla beringeri* is magnificent all black, thick black fur like a bear and thick beards." And again in the Milan museum: "The specimen which astonished me most was a White Shark (*Carcharoden rondeleti*) it first showed me what a real shark is." Curators must have enjoyed his visits.

When he wanted to examine giant tortoises to advance the joint monograph he hoped to compile with Günther, (based on the latter's original review paper), he wrote off to borrow specimens from St. Petersburg, Moscow, Sweden, Madrid, Dresden, Berlin, Stuttgart, Milan, Genoa, Turin, Hamburg, Breslau, Guayaquil, Lima, Lisbon, Calcutta, Sydney and various U.S.A. museums. "We will have six times as great a material as you had for your first edition", he wrote triumphantly to Günther. Over a hundred letters written to his friend between 1894-1914 concerned giant tortoises. In addition Walter had an encyclopaedic knowledge of the whereabouts of large collections. For instance while building up the birds' eggs section he negotiated the purchase of the Venturi collection (South American eggs), the Butler collection and the Count Tödern collection. This vast material will take years to register at the British Museum, although the task was begun in 1940.

To someone with a normal memory it is impossible to conceive how Walter kept track of all these parallel transactions which involved at one moment the purchase of thousands of Lepidoptera and birds' eggs, keeping contact with various men collecting in the field, having "first pick"[16] of a whole series of skins or butterfly collections say at Rosenberg or Staudinger, visiting fish markets all over Europe, or nearer home, arranging for two Barbary Lambs to be reared by a goat, the owner of which asked for "three pints of milk plus small charge for depreciation as she missed going to the male through being away from home", and cabling to Richardson Carr to meet six live wallabies arriving at Tring Station at 6pm, and at the same time hammering out with Hartert a complicated article about sub-speciation in pheasants for *Novitates*.

*I must work out the Pheasants of the genus* Phasianus *with you as from what I am bringing back and what I have seen here I begin to believe that here also we have not one but two or three creations of Beelzebub. It will also be necessary among the pheasants to go carefully into the exact geographical limits of the other genera for I believe that several of the* Gennaeus, Calophasis syrmathicus *and others straggle into the Palaearctic area.*

Then we find him writing to Lord Haldane's sister:[17] ".... I have traced the meaning of 'komilas', it is the Sanscrit word for cuckoo and it is used in India both by the natives and English to denote the large black cuckoo which is said to have the sweetest song of any living bird. In connection with the context you sent me Hegel obviously mentions it because of its laying eggs in other birds nests." Nor did he neglect the fellow who wrote and asked if his parrot would suffer from the smell of a freshly painted front parlour.

As for the sale room — Walter found it irresistible, and judging from his library acquisitions he was at the same time an undisciplined, childishly self-indulgent but a very shrewd and perspicacious buyer. Secretly he was negotiating to lease various islands in the Indian Ocean, especially Aldabra, with the object of saving the remains of the Giant Tortoise population and the Ibises, and this he achieved in 1900. In a letter to Ray Lankester in 1906 he describes groups of islands in the Indian Ocean as part of "my private estate".[18] But he always found time to name mushrooms and toadstools for local visitors who made a habit of bringing them to him for identification. Could they eat this one? Yes. Good. He was always genuinely interested.[19] People went away flattered by his enthusiasm and attention.

Walter possessed another rather unusual characteristic for, like one of his nieces, he was always — day or night — aware of the exact time. While most people work by the hour, or half hour, he arranged his schedule by minutes. Not a few people were nonplussed when they received the arrangements for a meeting or interview: "I will meet you at Hanover Square at 3.10 but I must leave at 3.35". Before driving up to London for Trustees' meetings he expected his breakfast — two poached eggs on toast and a pot of China tea — to be ready at 8.10 sharp. He swallowed the poached eggs whole like oysters, drank a cup of tea in a single gulp, and left at exactly 8.15. There is no doubt that the volume of work he achieved was due in part to the speed at which he work-ed — the poached eggs were symbolic. But, in addition, each task was planned to the minute. Even his after lunch nap in old age was limited to 15 minutes. He seemed to be snoring and rumbling in a profound sleep, but would awake with a loud snort as the large hand of the clock reached the quarter. Cecil Roth,[20] characteristically, basing a tale on hearsay, describes how Natty conducted interviews at New Court — with a watch before him on the table — according everyone a set three minutes. He comments that by this method Lord Rothschild got through an enormous amount of work! This suggests that father and son shared another trait in common, but judging from Walter's performance Natty would not have found a watch necessary, nor is it likely that he used one.

The building and equipment of the extra wings in 1908 stabilized the museum and it now assumed its final form with a floor space of 1½ acres. The new show cases had cost almost as much as the original edifice. The staff was increased and consisted of the owner/director, his two curators Hartert and Jordan, Arthur Goodson (a Tring man, son of the local blacksmith) who assisted Hartert, and in the early days also collected butterflies, his brother Fred, and the latter's son who helped Jordan. In addition there was Fred Young (Minall's successor) taxidermist and caretaker (at this time there were still a number of live animals to look after), Gerald Tite, a full-time butterfly setter, James Rance, who was in charge of the public galleries when they were open and of the cleaning and care of the specimens in the exhibition cases, and finally Arthur Poulton, who spent much of his time polishing the plate glass, cleaning the enormous expanse of windows and the care of the lawns and the garden at the back. The nightwatchman was shared with Tring Park. Strange as it may seem, a librarian was not engaged for another ten years!* Hartert,

---

* Phyllis Thomas, the first woman on the staff, insists that Walter was too shy to talk to her for a year after she arrived, and when he eventually plucked up courage to ask for a specific book he did so with exaggerated politeness in a practically inaudible whisper. Colonel Meinertzhagen mistook this shyness for lack of interest: "He had no use for women. Their company bored him" he wrote in his diary.[21]

Jordan and Walter himself looked after the books (they received over 100 periodicals in exchange for *Novitates*, and dispatched about 200 copies per quarter — all the addresses checked by the curators) but kept no card indexes, and all their letters and manuscripts at that time were hand written (although, very exceptionally, Walter typed a letter himself if he required a copy, but all these are lost). Arthur Goodson, it should be noted, became an excellent ornithologist and to-day would have been expected to become joint author of several of the Tring papers.

Walter could now, in the aftermath of the Great Row, (see p. 215) settle down to the life he had always wanted to lead. He had shaken off the dust of the city. His financial worries had for the moment been dissipated although his funds were drastically reduced — since the museum had now to be run as a separate enterprise and the happy-go-lucky relationship with the Tring Park Estate Office was severed. After Walter's debts had been cleared and the Bucks estate sold, Charles had set up a trust, funded by Natty, which brought in £4,000 p.a., but his brother forfeited his annual salary from the Bank. A pension fund for the staff had been arranged, and material outside the scope of the research side of the Museum was disposed of. Thus, rare books on fishes were sold. Only, of course, there could be no more such lavishly illustrated monographs.... Walter had left New Court, and the intense boredom and misfittery of banking lay behind him. He had also decided not to stand for re-election to the House of Commons. While in disgrace — following the Great Row — he had enjoyed a marvellously interesting and successful collecting expedition in North Africa and was planning another one along similar lines for next winter, this time with Jordan. He was now free to work 16 hours a day at Natural History if he so wished, based at Tring, with the odd trips to London, to the House of Commons for the time being (for in 1908 he was down to sit on numerous committees, none of which he attended), to the British Museum Trustees meetings, to the Board of Deputies, the United Synagogue, and with the usual local Tring Council meeting and the Working Mens Club in Aylesbury thrown in. He had escaped from the constant threats and hysterical scenes of Marie Walters and Lizzie Ritchie and as for the blackmailing peeress — Walter felt sure, that, with rumours that had been flying around, and the news of his *debâcle*, she would not see much sense in bothering him for the moment.

So he began his homeward journey via the Alps, collecting assiduously all the way. At Lauteret in the Haute Dauphine he ran into icy cold weather and driving rain, so he sat down to scribble to Hartert who had long since returned home. Walter had written the previous evening, but everyone in the room was talking and he had forgotten about Grant, a fellow ornithologist, and *Novitates*. "I had wanted to figure *Calophasis mikado* myself, but as matters stand now I will let Grant do it, but as the rump of our only male is rather dicky he must send his artist to Tring."[22] There was the familiar imperious ring about this letter but Hartert marvelled. What a traumatic and humiliating crisis Walter had just lived through! Yet look at this — he had not forgotten the condition of any of his stuffed animals during his eight months in the desert.... not even the backside of this Formosan pheasant! The mind boggled — yet it was rather admirable, and Hartert metaphorically raised his hat. But nevertheless it made one uneasy. How could one manage to live one's life with such a split personality? Hartert walked out of his study into the public gallery. Yes, the fellow was right, perfectly right — the Formosan pheasant's rump left much to be desired.

Walter's sheer genius for creating, organising and running a museum was only fully appreciated by some of us after he died, and Tring was taken over by the British Museum of Natural History. (For those who are interested in a comparison between public and private enterprise, Tring could provide a fascinating study, but such an exercise would serve no useful purpose in this context.) The British Museum provided continuity which no one else was prepared or able to do, and this was a matter of vital importance for the Tring collections. Their organisation could hardly be expected to conjure a second Walter or Hartert or Jordan out of their midst. Nor did it. When new and energetic Trustees were appointed to the board they were invariably disillusioned — quite unprepared for its Civil Service attitude and tempo. Lord Crawford, who subsequently became Chairman of the Board, gradually lost his fire under the stultifying atmosphere which prevailed. He wrote in his diary:[23]

> 22 March 1924. Trustees meeting at the Natural History Museum.... But how drab our meetings usually are! I had always expected them to be of entrancing interest, and my desire to be a Trustee of the British Museum was one of the few ambitions I ever possessed: yet now I am a Trustee I confess that disillusion has supervened. It is a dull function. We are told about porters and messengers, about lost keys or missing books, about the leave granted to some clerk who wants to get married, about this that or the other detail: but on great problems of museum progress or development, seldom a word.... I am disappointed — for so much could be done to quicken that dead alive institution.

Walter (elected in 1899 at the age of 31), who was a trustee for longer than anyone else has ever been, which says a great deal for his patience and devotion to the B.M., wrote to Günther soon after his appointment as follows:

*The Natural History Museum is giving me a lot to think about. I fear there are underground currents at work which are a great danger.*

Richard Meinertzhagen, as we will see (p. 120) made a gloomy and highly critical comparison between South Kensington and Tring, yet in the end he, too, decided to leave the British Museum, his wonderful Palaearctic bird skins, crop contents, seeds and plants. O.W. Richards, who donated a superb collection of more than 60,000 specimens of Hymenoptera to the B.M., impeccably curated, was thanked by a Trustee for his great generosity, and replied with a note of flat resignation in his voice: "I am retiring — I didn't know what to do with them".

To-day the two principal functions of a natural history museum are the accurate identification of specimens and the curating and describing of their own collections. Despite the million visitors the Museum can boast every year, the fact remains that stuffed animals or life-like models — even if such models exhibit function as well as form — no longer exert the appeal or excite the interest they did fifty years ago. Films and television programmes, such as "Life on Earth" — are both more attractive and instructive,[24] although the exhibits at South Kensington can still provide a splendid educational outing for school children from the provinces. Undoubtedly there are now better channels for bringing the knowledge of wild animals and the marvels of nature into men's lives. But there are no other organisations so well fitted to provide the continuity demanded by large collections, and to cater for the various sciences which require these collections as tools.

One aspect of Tring remains astonishing to those of us who enjoyed the privilege of working there in the "old days". Tring was run on a shoe string and yet the output per man was at least ten times that of a reasonably affluent museum. The total amount spent per year after 1908 never exceeded £12,000[25] and usually averaged around £7,000.[26] This sum included not only the salaries of the curators (Jordan never earned more than £400 p.a., even after he had been at Tring thirty years, although his house had been built for him by the Tring Estate and Walter paid all the expenses) but also the heating, lighting and general maintenance of the buildings, as well as the purchase of specimens and books. An ex-director of the Natural History Museum was asked to what he attributed the astonishing success and worldwide fame of Tring. His reply was, in one way, remarkably wide of the mark: "Firstly the most important — unlimited wealth, secondly single-mindedness....". The great ornithologist Stresemann too could only account for the Tring success by dubbing Walter a Croesus.

There was really only one "explanation" for Tring, and that was, quite simply, Walter's extraordinary personality. He managed to select (and in several cases train) with almost a sixth sense, a small group of people — none of whom ever left his employ until they retired — whom he could inspire with interest, dedication, enthusiasm, energy, devotion and, above all, the feeling that the museum was theirs — a reflection of his own possessiveness — an organisation they were delighted and proud to be part of and work for. This sense of purposeful identification and enthusiasm was shared by all — from Hartert to Young and Poulton — and was totally devoid of the incentive of pecuniary gain. In 1920, for instance, Arthur Goodson earned £3.10.0 per week, but was given a £5.11.2 bonus for setting 2297 butterflies. A senior scientific officer from the Department of Entomology at the Natural History Museum (F.G.A.M. Smit, curator of the Rothschild collection of fleas) who knew Tring well, and worked there for many years, felt the amount of work which had been produced, not only the 42 volumes of *Novitates*, but the collections themselves, the 1200 papers written about them by the owner and his curators, but also in the form of specialist teaching which had taken place at Tring, was something he could never comprehend. It was somehow so disproportionate to the number of people engaged on this mammoth task. Even more peculiar was this huge, shy, diffident, roaring, solitary man, who had the power to communicate some of his extraordinary boxed up, furious energy and enthusiasm, and profound love of his animals and My Museum to his collaborators, and set them all alight.

One night a few years after Walter's death, the curator of the Rothschild collection of fleas was himself working late at the Museum when he suddenly heard the unmistakable and familiar thumping footsteps echoing along the boards.[27] The so-called flea room, which was also Jordan's room, was separated by some sort of partition from the main entomological hall, but the door was never closed. At first the significance of the sound did not strike Smit, for he was so accustomed to it. Then suddenly he realized with a feeling of awe and astonishment, tinged with fear, that he had heard the footsteps of a dead man. Later he discovered that on one occasion Phyllis Thomas had heard them too. "Unmistakable", was her laconic comment. In the months and years which followed, Smit said he grew quite accustomed to the ponderous tread of Walter's ghost striding through the department of entomology[28] between the hours of 8 and 11 pm, once or twice every few weeks, and the rattle of a distant drawer as it was banged back into some cabinet or other. In fact he found the sounds oddly consoling.

# CHAPTER 15

# The fellow is always right

DESPITE his three "wasted" years at New Court — which eventually expanded into eighteen — Walter at the age of 24, when he opened Tring Museum, was already the only zoologist in England whose name was widely known to the public. Naturalists, especially in those days, were on the whole modest, rather obscure people, usually thought of as "cranky" and who attracted little, if any, attention. Walter somehow stole the limelight. As for his fellow zoologists, he excited, baffled, exasperated and scared them, but few if any of them recognised either during his life time or for some time after his death that they had had in their midst one of the greatest taxonomic *animateurs* of the century

Only Richard Meinertzhagen, with his characteristic perspicacity and intuition, pierced Walter's barrier of painful, crippling shyness. "He is in many ways a remarkable man.... a great scientist...." he wrote in his diary,[1] "[He has] an uncanny memory and a vast knowledge of Natural History, and there is not a single branch in which he does not take a lively interest. His work with Hartert is a model of how to collect and arrange and work in a museum — a model which the British Museum has signally failed to emulate.... He has a great sense of humour and is very human, very sensitive, wonderfully kind.... and I shall never forget those delightful days we used to have shooting up the Tring Reservoir."

M.A.C. Hinton, the Keeper of Mammals at the British Museum, also described Walter's memory and his knowledge of species and subspecies of mammals, birds, reptiles and Lepidoptera as "astounding".[2]

Karl Jordan wrote Walter's obituary notice for the Royal Society of London.[3] He was limited to two pages. Norman Riley,[4] who at the time was Keeper of Entomology at the British Museum, performed the same melancholy task for Karl Jordan, and expanded his contribution to the Biographical Memoirs of Fellows into a 25 page essay. But he failed to point out that it was Walter Rothschild who set alight the Karl Jordan candle and kept the steady flame burning for about half a century. It is no coincidence that both his curators achieved worldwide recognition. Without Walter booming from his study or standing silently at their elbows with his eyes fixed awkwardly on the ground, one would probably never have heard of Jordan and Hartert. As first-rate taxonomists they would undoubtedly have produced a series of excellent contributions to highly respected zoological journals, labouring assiduously in some basement room in a National Museum or teaching in the department of agriculture in Hildesheim. It has been pointed out elsewhere[5] that after Charles's death Jordan failed to produce the monograph based on the Rothschild collection of fleas which the two men had planned — despite the fact that he had the time, the Tring facilities and the material at this disposal. Charles, too, was a great *animateur*, and Jordan, alone, did not feel equal to the task. Similarly without Walter there would surely not have been great zoological revisions, or international entomological congresses; no recognition of the trinomial system of zoological nomenclature (see p. 130-132) — at least not for many years — no fabulous collections with long series of specimens — the tools he gave those men to work with.

Quite a considerable gulf, notwithstanding, separated Walter from his curators and collaborators, with whom he could never discuss anything except animals and his museum. He was an eccentric, creative man with simple childish, yet intense enthusiasms, considerable intuition — more than a touch of Balaam's ass about him, demonstrated by several of his awkward letters to Sokolow and Weizmann during the Balfour Declaration crisis — and with an undeniable spark of genius. Jordan and Hartert, in contrast, were born old souls, gifted but conventional specialists with academic excellence and mature judgement, but lacking in both imagination and vision. Fortunately they were able to generate restraint. It was the combination of their threefold talents which proved so astonishingly productive and felicitous.

The scientific advances, great and small, which emanated from the Tring Museum, the description of the 5,000 new species published in 1200 books and papers (which included the great Revisions[6] and surveys), the enormous collection assembled with obsessional drive and perspicacity, on which the three men based their studies on evolution and taxonomy — the wide and stimulating public relations enterprise which they generated — all resulted from this remarkably closely-knit partnership of three, which never faltered for 37 years. "When one examines the Tring Museum," wrote M.A.C. Hinton[7] — a contemporary British Museum man with the rare vision to appreciate it — "one cannot repress a feeling of deep admiration, not only for the late Lord Rothschild but for his brother the late N.C. Rothschild, Dr. Hartert and Dr. Jordan. It is difficult to believe that, even with riches to assist, four men could have produced such a wonderful temple of science within the space of 50 years. Great indeed is the power of enthusiasm."

Jon Karlsson[8] in his fine book *The Inheritance of Creative Intelligence* has pinpointed the two types of mentality and achievement illustrated on the one hand by Walter Rothschild and on the other by his curators. He particularly stresses two characteristics of creative people which Walter possessed to an unusual degree — namely superior learning ability shown in his early acquisition of an encyclopaedic knowledge of systematic zoology — greatly assisted by his freakishly retentive, photographic memory — coupled with a total lack of both interest and ability in academic matters. It will be recalled that not only Walter, but his father and brother had difficulty in passing examinations at Cambridge. Karlsson also points out a curious but striking resemblance between the results of psychological tests (Rorschach, MMPPI[9] etc.) made with the collaboration of creative people such as writers and architects, and similar results from tests administered to psychopathic individuals in asylums. Unconventionality, novelty, and the ability to tolerate ambiguities are present in both. There is not a shadow of doubt that had such tests been applied to Walter, the same theme would have recurred. Karlsson goes considerably further in his analysis of the "mad genius" and postulates a gene for schizophrenia which in his view should be more appropriately designated a gene for creativity; he regards the frequently associated liability to mental breakdown as one of its unfortunate side effects. While Walter was in a sense extremely sane (Jordan calls him shrewd) he was in many ways so very, very peculiar — and at times irrational — and his toleration of ambiguities was so pronounced that his curators and his peers may be forgiven for abandoning their unequal struggle to understand or assess the man. Most disconcerting of all he was unpredictable. Despite his ordered and regular life, despite his shyness, he gave one a feeling of almost dangerous unpredictability, like the proverbial rogue elephant — as if his wealth and his size and a certain inner lack of convention and judgement bestowed upon him a special sort of unrestrained freedom.

The average individual finds eccentricity difficult to cope with, but quite apart from the idiosyncrasies with which Walter confronted the professional English zoologists, they failed to find an acceptable pattern into which he could be fitted. He was an amateur without any conventional training; he was also wealthy and preserved the maddening independence which money alone can bestow. He drove a four-in-hand of zebras down Piccadilly with considerable *panache*, which attracted enviable publicity; he selected two foreigners — Germans[10] — for his curators, and he was a Jew. Furthermore his collections were so magnificent that on the continent Tring was regarded as a highly successful rival of the British Museum, a dream place to work in. It had also leaked out that the King used to drop in at Tring — not infrequently mistaken for Natty by some chance passer-by, for they were not unalike in appearance — and stroll round the Museum after dinner with considerable enjoyment — something he never did at South Kensington. Finally, Walter's phenomenal memory was disconcerting, and one had the feeling that nothing escaped his goodnatured, twinkling elephant's eyes.* He not only made the zoologists ill at ease, but he aroused their envy. Walter Rothschild had no right, with his lack of basic academic scientific knowledge, and his childish attitudes and inherited wealth, to have — it seemed incredible — somehow achieved worldwide recognition and acclaim, while they struggled obscurely in a cocoon of red tape in South Ken.!

Reading Jordan's article "In memory of Lord Rothschild"[12] one senses that he, too — though for quite other reasons — was, to the very end of their long and fruitful association, somehow at a loss. Clearly Jordan did not wish to write a fulsome, uncritical and conventional hymn of praise, which would have detracted from the real appreciation of the man and his work. Yet he leaves the reader with a sense of mystification and frustration. "His reputation as a zoologist will be lasting," he wrote, ".... The founder of the British branch of the House of Rothschild, Nathan(iel)[13] Mayer, acquired such fame as a financier that the family name still casts a spell over the imagination of man in all quarters of the globe. His great-grandson owed much to that name, but more to his own activity as a zoologist. Passing along a different road, he, too, arrived at world-wide fame."

Although Jordan acknowledges in the first paragraph of the essay that "love of animals was the original motive for the foundation of the Tring Museum", he adds on the following page that Walter's active interest in so many different branches of natural history was "inimical to wise restriction". These are the views of the narrow specialist whose role was to know more and more about less and less — he failed to realize that Walter's great gift was to set his *entourage* alight. All animals and plants were marvellous, exciting, wonderful, unbelievable, fantastic — they were all his — but they were also all yours for the asking. That was his message, and the man in the street understood perfectly. Walter received hundreds of letters from working class people who had become interested in natural history while visiting My Museum.[14] He

---

* Another victim of professional jealousy of a somewhat similar nature was Lord Avebury, also a distinguished amateur zoologist, who neither during his life nor after his death received the recognition he so richly deserved.[11] Even the self-effacing Charles Rothschild, who trod on nobody's toes, assiduously avoided the limelight and specialised conventionally in an obscure group of insects, nevertheless had cause to write to Hugh Birrell, "Fleas are increasing anti-semitism, as the few other students of the Order think I'm too keen a competitor!!"

burned with completely single-minded enthusiasm* and zeal for the animal and vegetable world, and anyone in the vicinity caught fire too. No doubt it can be very daunting and uncomfortable to have a sort of burning haystack flaring away in one's midst....

Both Hartert and Jordan must have suffered: The former wrote a furious, almost savage, letter to Jordan about the careless incorporation of some specimens by his boss — as if he was an undisciplined, obstreperous exasperating child who must be restrained — while the latter attempted to rationalise his uneasiness by means of criticism and mildly patronising reproof. Walter, he notes, was satisfied in the early days with poor curating, especially a lack of precision in the labels attached to specimens, and at first accepted those marked only: "New Guinea" and "South America"! Again, collections had been bought by Walter "as and when they were offered" — as if the whole operation was, alas! entirely haphazard. But Jordan failed to explain that this period of Walter's Museum building was between the ages of 13 and 19, and that even so he had himself collected a large number of the 40,000 Lepidoptera stacked at that period in his bedroom. Then Jordan remarks, with a touch of contempt — for he himself was a fearless traveller, and as tough as old boots — that Walter was "not one of those hardy explorers who can stand any amount of discomfort" and, moreover, as a youth he had contracted pneumonia — a serious disease in the pre-penicillin days — and was "afraid of catching cold". He therefore preferred to take a physician on collecting trips with him. To us, it seemed eminently reasonable to invite a man with some medical knowledge — who was also a keen Lepidopterist — to accompany expeditions into the African desert! Jordan also complains of Walter's artistic, rather than severely scientific manner of describing new species: "The points of difference discovered became so superconscious in his mind that he emphasised them by the employment of superlatives and, in print, often by the use of Clarendon type.... Like an artist he perceived the animals as a whole and not the details which made up the picture...." This is a wrapped up, but quite severe criticism of Walter as a taxonomist. There were times when it was justified. Jordan was a far better morphologist than his employer — that is certain. And he went on to lament the fact that Walter treated butterflies as if they were birds. Furthermore, said Jordan, Walter had never taken kindly to

---

* This was a characteristic he may have inherited from his father. Lucien Wolf[15] recalled an occasion when Natty, in a burst of anger against a certain Gradgrind whom they had been discussing, said "I hate those men who are always discovering that this and that scheme is impracticable. I would rather do something great and fail at it than do nothing at all."

either the microscope or microtome.[16] This was true enough, for he was a passionate lover of whole animals, not bits of them. But nevertheless an entirely false picture is presented by this paragraph — for if Walter was in doubt about the validity of a specific description of some new moth or butterfly, he immediately requested a dissection of the genitalia from Jordan — for who was more skilled and thorough in such matters than his curator? Perhaps Jordan felt that, as usual, he was invited to carry the can, and Walter should have learned to do his own dissections....

There is another quite delightful paragraph of implied criticism which runs as follows: "The library of the Museum contains a number of rare old books on travel and natural history, of little use for systematic zoology, but of interest to a book lover, De Bry's publications particularly taking his fancy." As a description of the Tring Library this must rank as the greatest understatement of all time, for to-day Dr. David Snow (Keeper of Birds at the British Museum) considers the 20,000 books in the ornithological section of the library at Tring the best of its kind in the world, the material selected with such skill and intuition that it almost savours of second sight.

But Walter rarely discussed his purchases with anyone, least of all his curators, whom he sensed would lack sympathy or interest in what they considered the whims of a man with more money than was good for him. Faced with any sort of criticism, Walter shrank into his shell. When he bought the first edition of Audubon's *Birds of America* (in 1896), he left it, wrapped in brown paper on a shelf in his library. Dr. Hartert apparently discovered it there, and we find Walter writing rather anxiously to his curator from London: "Please do *not* touch the volume you have found. It is *not* a duplicate. I put it there on purpose. It is a valuable book...." We do not know what was paid for these volumes (1827-1838, in impeccable condition), but to-day in the sale room these four (bound up together or singly) are worth at least £800,000. It is one of the few occasions when Walter's flair and acumen proved superior to those of his renowned great-grandfather, who had received Audubon at New Court but had refused the offer of *Birds of America* for £200.[17] According to Cecil Roth,[18] Nathan Mayer ventured the opinion that £5 was about the right price for the book. Audubon was justifiably outraged — particularly by N.M's ill-concealed contempt for ornithology. Walter was not, however, a particularly extravagant bibliophile, and, contrary to popular belief, would not pay any price for a book he "coveted" — to use Dr. Jordan's term. For instance, he hesitated a long time before agreeing to pay £14 for Bateman's *Orchidaceae of Mexico and Guatemala*, and carefully preserved a letter from Bernard Quaritch[19] to a would-be purchaser stating that in 1871 the book was worth more. To-day in the sale room it would fetch £10,000.

Walter Rothschild loved books. He had impeccable taste, and an eagle eye for a rarity — he once picked up a first edition of Newton's *Principia* on an open bookstall in Germany. He understood the requirements of a working ornithologist and entomologist and botanist, and encouraged his curators to make the necessary purchases of the periodicals and text books required. Not a few of us would consider the collection of these 30,000 scientific books as a successful life-work in itself. It is significant that in 1933 when Walter was forced to sell his collection of bird skins he refused to include the ornithological library in the deal. He was hard pressed to do so by the American purchaser, who claimed, not without reason, that the library was part of the bird collection. He also refused to sell his collection of birds' eggs — much the largest in the world, which it will take several generations of oologists to "work up". "Did you sell your eggs too?" was the first question Hartert asked him when news of the disaster reached Berlin.

Major Pye-Smith,[20] in his *Memories of Pastor Jourdain*, wrote a most amusing account of Walter in the chair at a meeting of the British Oological Association:

> In due course I was elevated to the post of secretary, which gave me the privilege of sitting beside the great man (Lord Rothschild) himself at the dinners which were held in a large room upstairs at Pagani's Restaurant, that haunt of freemasons, oddfellows' clubs, and quasi-scientific societies, in Great Portland Street. Lord Rothschild's bearded beaming countenance and portly figure, somewhat resembling our gracious monarch Edward the Seventh, was in strange contrast to the keen slender Parson Jourdain, with the great scar across his forehead, said to have been acquired by falling out of an eagle's nest in the Cairngorms.
>
> Lord Rothschild's knowledge in several branches of natural science was encyclopaedic. But I think Jourdain surpassed him in the realm of oology. Each regarded it as a stern duty to correct the slightest deviation from fact made by any member.... The poor honorary secretary used to wilt with agitation when he observed Jourdain rise in his place with the pained offended look on his face which betokened disagreement with some assertion of the noble president. "I think Lord Rothschild is mistaken...." Who but Jourdain would have dared to say it? Who but Rothschild would have thus replied, bringing down the chairman's gavel with a hearty blow that made the glasses on the table ring? "I think Mr. Jourdain has overlooked an article of mine in the "Ibis" the year before last...." But how polite we all were in Edwardian days; and how firmly each man stood his ground with reference and counter-reference; and how excellent a dinner one got at Pagani's for seven shillings and sixpence!

It is strange that Pye-Smith did not refer to the fearsome scar on Walter's head — the result of being shot out of a hansom cab and crashing onto the curbstone of some London street. There was a mystery about this accident ˻. . Walter declined to discuss it in later years. Was it due to a collision or a bolting horse? He would not reply to polite enquiries.

Jordan, somewhere in his memoir, pointed out with a touch of condescension that on the expedition to El Oued Walter, as if he were still a delicate boy, elected to ride a horse while the rest of the party, which included the Danish Consul General from Algiers — the medical-man-cum-Lepidopterist whom Walter liked as a companion on these trips — rode camels. Again, this does not seem an unreasonable choice for a keen horseman, who also suffered from acute travel sickness! One night round the camp fire Nissen asked Walter rather slyly how he got along with Karl Jordan: "We have terrific arguments," replied Walter with engaging candour, "but the fellow is always right." In the way that courtiers exchange gossip about the King, Nissen repeated this tale to Jordan — who was evidently delighted — when he joined the party on their return to Biskra. He himself may have had reservations.

# CHAPTER 16

# Walter's Curators

*Ernst Hartert and the birds*

Sir Edward Salisbury, the Director of the Royal Botanic Gardens, once remarked that evolution was no longer of interest to the public, for the man in the street had grown accustomed to the idea that he was an ape rather than a fallen angel; furthermore the Natural History Museum must now turn its attention to arthropods — for the world of the future would either belong to man or insects, a fact which, willy nilly, claimed people's attention and interest. Inherent in this observation is the knowledge that the burning controversies of science are brief conflagrations — the points at issue are either accepted as part of the facts of life or else recognised as mistaken ideas, and quickly swept aside and forgotten. The chief protagonists are rarely remembered. To the student of to-day it seems impossible that something at once so widely accepted and yet so trivial as trinomials could ever have been the cause of bitter and prolonged controversy. Yet they are wrong.

In 1884, the year in which Walter Rothschild, then aged seven, conceived the idea of My Museum, a group of American ornithologists, headed by Elliot Coues, suggested that a third name should be added to the two names — generic and specific — of a bird species, to indicate constant local or geographical variation. This was a revolutionary suggestion for up to that time the traditional Linnean binomial system of nomenclature had been sacrosanct. In England the idea received short shrift, for the great ornithologists of the day, P.L. Sclater, Bowdler Sharpe (both working at the British Museum) and Alfred Newton were all vehemently opposed to the suggestion — the only supporter was Seebohm. In 1890 the German ornithologists considered the same problem, and a few among them, including Ernst Hartert, then aged 28, were attracted to the American approach. The controversy was not as simple or as trivial as it seems to-day, for inherent in the trinomial system was the concept that geographical races of birds — and for that matter other animals — were evolving species. Thus its acceptance undermined the concept that the species was a fixed entity ("species cannot be taken apart", said the great Kleinschmidt). The geographical variants, far from representing "debased forms produced by climate", were now, by implication, themselves nascent or incipient species. Alone among the German ornithologists Hartert stuck to his guns and refused to accept the watered-down resolution passed by the German Ornithological Society in 1889. He boldly defined sub-species to which the trinomial system should apply as "forms not sufficiently distant from others to entitle them to the rank of a species."

Walter first met Ernst Hartert in the city office of Henry Dresser[1] in 1892, when he and their friend Count Berlepsch were arranging a collecting expedition to South America. Erwin Stresemann (unquestionably the most accomplished, erudite and all-round ornithologist of our day), who devotes a chapter to Hartert in his book on the history of ornithology, allows his imagination full play on the question of Hartert's appointment as Curator at Tring: "Hartert was lucky again," he wrote, "for Walter Rothschild's English curator had been dismissed for gross abuse of confidence". In reality the faithful joiner-cum-taxidermist Minall was still stuffing and mounting and caretaking at Tring, while Walter himself was — as he had always been up to that moment, for he was then only 23 — his own curator. Hartert, accompanied by his bride, had been diverted by Count Berlepsch to the Caribbean Islands owing to the political unrest in Venezuela. Walter now dispatched one of his familiar, imperious telegrams summoning him back peremptorily to take charge at Tring. There is no record of Count Berlepsch's reactions; Stresemann notes that Hartert was "surprised", but over the next 38 years he was to grow accustomed to Walter's lightning decisions. He was also delighted, and so was his wife. Thereafter she did not accompany her husband on expeditions, but became a diligent Tring-based *hausfrau*.

Erwin Stresemann was, like Chaim Weizmann, completely baffled by Walter Rothschild, and, in *Ornithology from Aristotle to the Present Day*[2] dismisses him in one extraordinary sentence: "Two years earlier (i.e. two years before Hartert's appointment) young Walter Rothschild had built himself a small zoological museum at Tring, an hour's train ride from London. Rothschild's wealth quickly made it famous and respected."

Stresemann failed to make any real personal contact with Walter — although he was frequently his guest at Tring but he ardently hero-worshipped Hartert, and was so deeply and emotionally involved with his teacher and mentor (for he spent two years at Tring learning his birds) that he left a testamentary directive to bury his earthly remains in Hartert's grave. This throws some light on his rather naive attempt at attributing what he called "the unparalleled rise of the Tring Museum" and the new ideas relating to classification and evolution produced at Tring, solely to Ernst Hartert. No-one, it must be acknowledged, would have been more irritated by these subjective statements than Hartert himself, for he was a serious man of sound and measured judgement, and was the first to acknowledge Walter's dynamic initiative and Jordan's soaring intellect. No-one knew better than he that wealth does not earn a man "fame and respect" in zoological circles. Where Hartert really shone was in his good-humoured handling of the rogue elephant aspect of Walter's drive and personality, a knack which Jordan somehow failed to develop, for he remained all his life a trifle schoolmasterish and regarded Walter as a spoilt child rather than a mad genius. The two bird men had several important traits in common despite their different circumstances. First of all both were self taught. Unlike Jordan — who was a classical product of the highly-respected rigorous German scientific education, and had taken the coveted *summa cum laude* in his oral examination for a PhD* — neither Walter nor Hartert had managed even a pass degree at the University. Both had spent all their leisure in boyhood and youth enjoying themselves — collecting birds and other animals, and studying wild life in the woods and mountains, and supplementing the knowledge thus gained by voluminous reading at home. Whereas Hartert had to earn a precarious livelihood by selling his specimens, Walter kept his for the Museum. Hartert travelled further afield, and by the time he was 32 he had collected in Nigeria, Sumatra, Perak, Upper Assam and Rajputana. In later years he paid Walter the unspoken compliment of basing his *magnum opus "Die Vögel der paläarktischen Fauna"*[3] (the last part of which appeared in 1922) on the Tring European collection. Quite simply, it provided the best material for their purpose of a new classification based on evolution. Walter's view that a long series of each species was necessary for the study of phylogeny was amply justified, and, in fact, it was only his perspicacious and tireless collecting of European material that made Hartert's task possible. Walter in his "Appreciation"[4] said his director had helped him build up the finest series of Palaearctic birds in any Museum, which no doubt made the book founded on the Tring collections the most complete. "It will serve most thoroughly as a monument to keep his memory green." It is incomprehensible to us that Walter was not a joint author of the Palaearctic Birds, since without him it could never have been written, but he would certainly have been too modest and embarrassed to make the

---

* Jordan always felt faintly condescending towards both Walter and Hartert on account of their lack of conventional education, and when Hartert received an honorary PhD he was distinctly put out, and flatly refused to accord him the title of Doctor, which everyone else hastened to do. In his memorial sketch of Walter, he refers to Hartert pointedly as 'Mr.'. On his side, Hartert was faintly contemptuous of Jordan's social status, for while HE was the son of a German Army officer, Jordan's father was a simple yeoman farmer. They never clashed or disagreed over their work, both recognising each other's sterling qualities, but Hartert remained essentially Walter's man while Jordan became a devoted and lifelong friend of his younger brother Charles. Mrs. Hartert was *openly* contemptuous of Mrs. Jordan.

suggestion himself. Also Walter was not unduly concerned about authorship within the Museum, he was intent upon his curator working up *his* collections scientifically and to the best possible advantage.

In this day and age it seems incredible that Walter remained for 38 years on polite and purely formal terms with his two curators, although working 10-12 hours a day in their company, both in the Museum and on collecting trips. Despite this total lack of intimacy he described Hartert as "a great and valued friend and loyal comrade", and felt sincerely that this was so. The happy exchange of unselfconscious, boyish confidences with the Minall family belonged to another world. Once in a while Mrs. Hartert invited him to tea.... especially when Erwin Stresemann was staying. She procured an extra strong chair for his benefit, and Howletts, the local baker, produced a special meringue which he greatly enjoyed. Similarly, if there was a visitor whom he thought might interest the Harterts, such as the Roosevelts or Selous, they were invited to stay the weekend at Tring Park.

Where joint authorship was concerned, Hartert was not above taking advantage of Walter's extreme diffidence, inarticulate shyness and unspoken fear of rebuff, whereas Jordan was scrupulously fair and even more than fair, in this respect. He never allowed his exasperation with Walter's idiosyncrasies to affect his judgement. Walter himself was generous as well as fair. In fact he was an enthusiastic sharer in all areas of his work, despite his pathological secretiveness, which puzzled people like Günther, who was not himself a great student of character and was confounded by this combination of imperious dictatorship and acute inferiority complex.

However, Walter's violent arguments with Hartert over the text of their various publications startled Phyllis Thomas, the librarian, who confessed that in her early days at Tring she was amazed to hear him banging on the table with his fist and shouting, "Nein! Nein! NEIN!" — for he often broke into German when he became excited. He did not fancy being flatly contradicted any more than the next man, and Hartert could talk in exasperatingly quiet and measured tones, quoting first this and then that authority, while *he* found it difficult to give expression to the ideas boiling in his mind. Walter's ideas, like his actions, came in violent rushes, and owing to his lack of voice control he had either to say nothing at all, or give vent to his trumpeting. He had an engaging fashion of suddenly calming down, lowering his gaze like a guilty schoolboy, and admitting that, after all, his curator might well be right.

But Hartert got on very well with him — as well as anyone could get on with a man of such inexplicable contradictions. He knew Walter admired and appreciated his matchless curating, his sincerity, his unflagging industry and his careful attention to the library, and to Walter's wishes, which he respected, and he was gratified that his boss backed him up to the hilt over his trinomials. What is more, both men understood ornithology as a science, and yet experienced an emotional love of birds. Every day of his life Hartert was grateful to Walter who had given him a unique opportunity to pursue his life's work in security and freedom, who did not require academic certificates, who provided him with unique material for study, and not only valued his work and his theories, but was loyal to him in public discussion and added greatly to his prestige. Furthermore, who could fail to become attached to this unpredictable, boyish genius-cum-bull-in-a-china-shop, who swept them all along on a wave of inexhaustible enthusiasm? Someone who had worked as a student at Tring was asked to recall his most vivid impression of the bird room and replied: Walter Rothschild unpacking a collection which had recently arrived from abroad, and roaring partly in German, partly in English: "Hartert! Hartert! Schauen sie, Schauen sie.... COME AND SEE WHAT WE'VE GOT!"

If Walter baffled his *entourage* he himself failed to understand why his fellow ornithologists did not feel as well disposed and benign towards him as he felt towards them. Both parties were puzzled, and Walter into the bargain was quite hurt. In fact the ornithological *élite* in the United Kingdom were ranged solidly against Tring (Sharpe, only a few months before his death, declared that their trinomial system was "destructive"), although here and there on the continent more enlightened individuals such as A.B. Mayer, von Erlanger and Carl Hellmayr agreed with Walter. It could almost be said that Tring moulded Hellmayr. But in the United Kingdom Rothschild, Hartert and Jordan were regarded as outsiders and when, as time went on, it became evident that their ideas were gradually being accepted everywhere, this increased rather than decreased professional animosity. Hartert received the Godman-Salvin Gold Medal for his services to ornithology but was never made an F.R.S., an honour he richly deserved; Jordan attained the distinction in 1932 at the age of 71, so long after it was due that it became, virtually, an insult, while Rothschild was elected in 1911 for his services to zoology not for scientific achievement.[5] The *Ibis*, the chief ornithological journal of the UK, was edited by Sclater, who would not tolerate the use of trinomials. Walter side-stepped this stumbling block by producing his own Museum journal, *Novitates Zoologicae*, in which all three men pushed the use of trinomial nomenclature linked to the concept of geographical variation. Jordan really contributed more original thought to the whole concept than Hartert, despite the fact that the latter was actually first in the field, and it was he who insisted that the original subspecies must bear a subspecific name like the rest, even though at first sight it seemed strange to designate an animal *Gorilla gorilla gorilla*. Jordan worked with insects while Hartert specialised in birds, but it was Walter who combined both ornithology and entomology. In 1903 Hartert defined a subspecies thus: "We describe as subspecies (of birds) the geographically separated forms of one and the same type, which taken together make up a species. Therefore not just a small number of differences, but differences combined with geographical separation permit us to determine a form as a subspecies". In 1903 we find Walter writing to Hartert from Florence and explaining that where the pheasants are concerned he must adopt the geographical regions he and Jordan had used in their revision of the Sphingidae and *not* use the old terms.

Several years previously (1895) in his Revision of the Papilios of the Eastern Hemisphere, Walter reduced various geographical races, described heretofore as distinct species, to the rank of subspecies (although due to an editorial misunderstanding the first described subspecies was referred to as *forma typica*). This innovation "much clarified the classification" — a typical understatement by Jordan, whose assistance was acknowledged by the author.

Walter was goaded into one brief defence of his use of trinomials in a letter to the editor of the *Ibis*, written in 1904:[6]

*Sirs,*

*Mr. Harvie-Brown, in his letter to you of Sept.˙ 1st, 1904 ('Ibis', 1904, p. 664), criticises my article on nomenclature in connection with Barn-Owls (Bull. B.O.C. xiv, p. 87). I am always willing to accept every criticism so long as I am allowed to defend my own standpoint in return. I therefore venture to ask you to receive this reply to Mr. Harvie-Brown.*

*I would first wish to point out that Zoological Nomenclature does not stand alone in the world in promoting strife on all points of doubt or innovation. From the earliest historical period every innovation in Philosophy, Physics, Mechanics, Astronomy, Locomotion, Medicine, Surgery, in fact in every branch of human intellectual or material progress, has been fought and opposed tooth and nail by those who have been brought up in the schools of thought of the previous epochs. Nevertheless, we have always ended by adopting these innovations. So I feel sure it will be with Zoologists, when once they have grasped the meaning of the trinomial nomenclature as applied to geographical races. Mr. Harvie-Brown has quite misunderstood the nature of trinomials. He says in effect that the trinomial is justified only so long as it has a "distinctive geographical descriptive power". This is an impossibility, for by far the largest number of trinomially treated forms have been previously described binomially.*

*Such is the case with my* Aluco flammea nigrescens, *which has been described as* Strix nigrescens *by Lawrence. Such was also the case with* Dendragapus obscurus richardsoni, *which had been described as* Canace obscurus, *var.* richardsoni. *In both these cases the geographical race had been given a name which no one could be justified in changing into* Aluco flammea domenicensis *or* Dendragapus obscurus montanensis. *The question of calling the forms* Aluco flammea *I., II., III., etc., or a, b, c, etc. is*

133

*only raising a quibble, and, moreover, a very dangerous quibble; for it is much easier for numbers or letters to be accidentally transposed than quite distinct third names. If, as Mr. Harvie-Brown suggests, we were to add the geographical range and the exact locality to each specimen quoted in print, we should be doing exactly what those who use trinomials avoid, namely we should be giving a long, many-worded description instead of a short name.*

*As to the question of different habits of different subspecies, Mr. Harvie-Brown, I fear, has entirely misunderstood the purpose of my instancing the habits of the Robin abroad and in England. But the greatest justification for my calling the English Robin "Erithacus rubecula melophilus," instead of the "dark-breasted English form of* Erithacus rubecula", *is that while our indigenous Robins are all* E.r. melophilus *and easily recognisable in the skin or alive, at Brighton and in other south and east-coast localities continental Robins are blown across, and so we get in England both* Erithacus rubecula rubecula *and* Erithacus rubecula melophilus, *though the former is purely an accidental visitor.*

*In conclusion, I only wish once more to urge strongly that the opponents of trinomials, before they criticise and run down those who employ them, should for once seriously consider what trinomials are. The most ardent opponents of trinomials admit every day names such as "Pelecanus fuscus, var.* californicus", *but exclaim loudly if an unfortunate writer like myself ventures to leave out the abbreviated word "var." for the sake of brevity, and writes the name of the Western Brown Pelican,* Pelecanus fuscus californicus.

> *I am, Sirs, Yours, etc.,*
> *WALTER ROTHSCHILD*

On this occasion Walter probably showed his letter in draft to both his curators — something he often omitted to do, much to their annoyance — for there is an unmistakeable touch of Jordan in the second paragraph.

Ernst Hartert was curator of the Tring Museum for 38 years, and retired at the age of 70. Alone, and with collaborators, he wrote approximately 570 papers and described many new species and subspecies. These included eighteen British resident birds.

He was an intrepid traveller and an all round zoologist — a fact not noted by Stresemann — and he himself actually made a collection of Carabid beetles, a sideline much appreciated in My Museum. His interest in the Tring fauna was something Walter particularly enjoyed — for it was his first great love — and together they wrote several papers on the local birds. According to Richard Meinertzhagen Hartert was the best ornithologist of his day, and a man who had given terrific impetus to the study of palaearctic birds. "He had a great sense of humour and was a charming companion — one never tired of his company", he wrote in his diary.[7]

The death of his only son in the first world war had thrown a melancholy shadow over Hartert's personal life, and his wife could not endure the possibility of a lonely old age in Tring, the inhabitants of which — as we shall see — she had no good reason to love. Furthermore she had no faith at all in English doctors. Following a serious bout of pneumonia Hartert agreed in view of the deterioration in his health and his wife's anxiety, to return to Germany, which secretly he, too, still thought of as home. Mrs. Hartert lived to regret the decision for she could not stomach the Nazi regime and the persecution of the Jews — and said so openly. She had acquired British nationality, but nevertheless the UK Embassy advised her to leave Berlin after Hartert's death, and she retired to a convent in Holland.

Walter did everything in his power to get Hartert to stay — but he could not put his distress into words and thus exerted no pressure upon his curator and made no appeal to him on an emotional basis. His suggestions were of a practical nature. For instance, Hartert could — by all means — spend his winters in Algeria and the South of France and work at Tring in the late spring and summer when the weather was fine. But Mrs. Hartert had made up her mind. It is a very strange fact that the two men never discussed the future of the bird collection. Perhaps they took for granted that it was destined for the Natural History Museum. Certainly it never entered Hartert's wildest dreams that Lord Rothschild would ever be parted from his birds, just as he never imagined that Hartert would abandon them.

According to the Museum Librarian Walter lost his ebullience and aged suddenly when his curator left. Did he talk over this betrayal — for that is what it seemed to be — with his mother? We do not know.

Hartert died in 1933 without revisiting Tring. "This," wrote Walter gloomily, "creates a gap in the ornithological world very hard to fill, and systematic ornithology loses an exponent of an excellence not likely to be equalled for many years."[8] He told Meinertzhagen that he felt the loss keenly — a friend of 42 years standing who was always cheerful — helping in all sorts of different ways and undoubtedly the greatest living ornithologist. He sensed it was the end of an epoch. His epoch.

THE BIRDS

It is an interesting and well known phenomenon that no two collections are ever alike, for even if the objects which have been gathered together are of the same sort, or stem from the same "period", they seem, oddly enough, to bear the stamp of the individual collector. Presumably a combination of taste, knowledge, perseverance and skill, in varying proportions, endows every such entity, not only with a character, but almost with a life of its own. The

Tring Museum
February '26

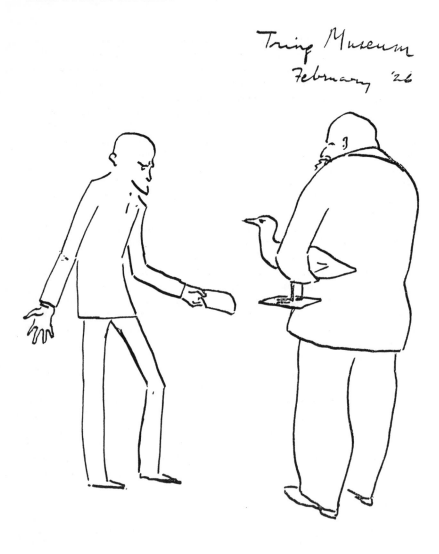

L. Fire.

outstanding feature of the Tring collections was their lively nature. One had the feeling that the owner was a happy man — in fact his capacity for enjoying life was directed towards his animals, for his boyish enthusiasm and love for his gorillas, his quagga and zebras, and every one of the 300,000 birds' skins, 200,000 birds' eggs and 2¼ million butterflies and moths, and each individual snake and sponge, somehow pervaded everything. Walter Rothschild was endowed with a marvellous gift of naive involvement with animals and plants, his amazement at their richness and beauty and endless variety was renewed every morning, and endured all his life. The collections were not only lively, but they carried with them a sense of optimism and were distinguished for both quality and quantity. There was no detail, however seemingly trivial, which escaped his attention. Local talent was employed extensively in the museum. The taxidermists and setters of butterflies were taught at Tring, and while the monographs were in preparation the artists concerned with the illustrations — for instance the Keulemans — lived with Mrs. Minall in her cottage for months, sometimes for years. Each plate was closely scrutinized by Walter, whose eye and memory for the exact shade of colour in any particular part of a bird's plumage was penetratingly accurate. Neither Hartert nor the Keulemans themselves could hold a candle to him in this respect. "No, no, that portion of the breast is brighter in real life — it has faded slightly in the drawer — it is more orange than that...." Consequently the reproductions, however excellent, never quite came up to his expectations, nor for that matter did the cabinets, nor the labels, nor the exhibition cases, nor the lighting, nor the buildings themselves. He was always trying to improve everything. Perhaps one of the secrets of his success as a collector lay in his lofty expectations. There was no limit to his sanguine, grandiose dreams of perfection — of either the specimens or their arrangement or the "revisions" based upon them.

Partly by accident and partly by design the specialist collections at Tring had the great merit of horizontal expansion as well as vertical depth. Walter began collecting when voyages of discovery as well as collecting expeditions were possible. There were many areas of the world which had not yet been explored, and remote islands were of particular interest to the zoologist concerned with evolution. He decided that he wanted material from all five continents. The collection must be global, and in particular made from every accessible oceanic island (see Map first edition). Where butterflies and birds were concerned he virtually achieved this ambitious objective. Furthermore scores of new species were found and described. There were over 2,000 ornithological types at Tring. Walter also realized that one pair of a species was insufficient and that long series were required for the study of geographic variation and evolution. "I have no duplicates in my collection",* was one of his favourite comments when he showed visitors round Tring, for sooner or later someone was sure to comment on the large number of examples of a single subspecies in some of the drawers. Not only must there be a long series, but albinos and all other colour variants and crosses and hybrids should be included. As a young boy he wrote to Dr. Günther:[9] "I have already got 2 fine rats and a WILD ALBINO house-mouse and soon I shall have a whole series ready for the museum.... the price of stuffing mice is 2/- apiece and rats 4/-." Walter had a particular love of albinos — not only for the sheer beauty of, say a snow-white swallow, and the collector's irresistible weakness for rarities — but because long before the rediscovery of Mendelism, he had a hunch that there was a clue to one of the great puzzles of evolution, if one could only read the riddle of colour variation aright, and make use of the material already to hand.

Thus, when he was eighteen, he told Günther that some albinos bred true, which showed that this type of whitening could be inherited. To prove his point he shot, near Windsor, a wild thrush and one of its fledglings which also displayed some white feathers. Günther passed on this piece of information to Lord Lilford, who, judging from the discreet exclamation marks in his letter to Alfred Newton,[10] did not believe a word of the tale: "Günther told me a rumour of the breeding!! of a white thrush near Windsor, and of the shooting of one of the old and one young bird." Walter's observation was corroborated seventy years later by Rollin[11] and Snow.[12]

In 1938 David Lack[13] spent quite a considerable period of time in the Galapagos Islands, and by dint of careful and gifted field work essentially solved the problem of speciation of the Geospizids, "Darwin's Finches". On his return he spent several weeks in New York studying the Rothschild collection of birds. "It's impossible to resolve this question," he explained, "however much information you can glean on the spot — about their ecology, their available food supply, their habits, the size of the different populations — population genetics if you like — breeding the birds in cages and all the rest. You've just got to come back to the Rothschild collection and get down to measuring beaks."

Lack's excellent little book — a very important contribution for theoretical developments, and one which also demonstrated the new found unity between the different branches of ornithology — does not really do justice either to the early collectors, or their concepts and objectives. They are very much taken for granted. Lack wrote: "The drawers of the Rothschild cabinet contain more

---

* See also Chapter 35, p. 316.

representatives of some of the Hawaiian sicklebills than are alive in the Islands today," (his p. 158) — but he does not comment on the men like Henry Palmer or Charles Miller Harris, who risked life and limb and endured untold hardships during their early expeditions (see p. 154, 197-201). Nor does he recall Walter's gloomy prophecies, who declared that the fauna of many of the Oceanic Islands — the giant tortoises in particular — were inevitably doomed to extinction and he wanted to "save them for science". In several cases he succeeded in doing so.

Charles Rothschild privately criticized this attitude, for he considered that his brother's almost childish ardour as a collector conflicted with his love of natural history and earnest desire for conservation. It was yet another of the irreconcilable contradictions in Walter's ambivalent nature. Yet to-day (February 1980) Ron Davies (Chairman of the Department of Entomology at the Smithsonian Institution) agrees with Walter's point of view. "The Lepidoptera collections amassed by Lord Rothschild at Tring are of inestimable value. Not only are the collections among the largest of their kind but they often represent species populations which are now either extinct or are more seriously threatened.... I believe that the need to-day for men of Lord Rothschild's vision is even greater, due to the very rapid destruction of natural habitats throughout the world." Charles Elton, the pioneer ecologist, once remarked "I have tried to express my belief in the debt that ecologists owe to the quiet labours of collectors and taxonomists".

In Ernst Mayr's essay[14] "Materials for a history of American Ornithology" (1975) he describes Frank M. Chapman as the man responsible for the profound impact on ornithology exerted by the American Museum of Natural History. He then adds: "One of Chapman's great triumphs was the acquisition of the world famous Rothschild collection for the Museum." It is a strange fact that Chapman owed this stroke of good fortune to the blackmailing activities of a now elderly but still smiling peeress, coupled with Walter's dogged determination to protect his mother from the threat of scandal. After Emma's death in 1935 her grief-stricken son suddenly realized with melancholy relief that he could no longer be threatened, and in a spell of euphoria, when the significance of this fact eventually sank in, toyed with the idea of starting, once again, to collect a few families of birds. Jordan, to whom Walter mentioned this in passing, not unnaturally decided the loss of his collection had affected his mind.

Stresemann commented in his book that Mayr had been, since 1932, curator of the former Rothschild collection of birds, now housed in New York, and continued: "His *Systematics and the Origin of Species*[15] — a basic work in which he (Mayr) re-examined from the systematist's standpoint the questions concerning the formation of species — will long remain a reliable guide for systematists working in the complicated labyrinth of phenomena through which his predecessors had tried vainly to find their way." Mayr himself — the most competent judge alive to-day — acknowledges frankly that the availability of the Rothschild collection in the Natural History Museum in New York greatly facilitated his work. It contained at that time 280,000 skins (for Walter would not be parted from the Cassowaries and had retained the Ratites at Tring) and was better balanced and richer in new material than any other. So far as New Guinea, the Bismarcks, the Solomon Isles and the Moluccas were concerned, it was superior to the four or five other great collections of the day combined.

David Snow, who is one of the most distinguished ornithologists of our time, remarked that among the first things he learnt, when he embarked on some bird research involving museum specimens, was the value of the Rothschild collection:

*I was interested in the geographical variation of the tits of the western Palaearctic, and I found that far the most extensive series of specimens from North Africa, a key area, then available was in the American Museum of Natural History in New York, being part of the Rothschild collection. Dr. Ernst Mayr, who was then in charge of the New York collections, very kindly made the whole of the North African tit material available to me on loan by shipping it back to England (in two successive parts, in case of accidents in transit). Incidentally, the value of a meticulously labelled collection such as this for quite other purposes than the study of distribution and geographical variation was brought home to me when I found that the Blue Tits could be aged from their wing-plumage. The abundant Rothschild material enabled me to show that the expectation of life of Blue Tits increases from north to south throughout their range, and that it is inversely correlated with their reproductive rate.*

*Later, when I became interested in South American birds, I again had reason to regret that the Rothschild collection was on the other side of the Atlantic, since it contained many of the most important specimens of cotingas and manakins then available. Fortunately I was able to visit New York and study it there; there were far too many specimens for another loan. Again, I was struck by the variety of uses to which a properly documented bird skin can be put, in this case in working out the succession of plumages in different species with age, and the timing of the annual cycle in different parts of the range of a species, as revealed by the post-breeding moult. Almost invariably, one finds that one would like to be able to examine many more specimens than are available; the really valuable collections are those that contain adequate series of specimens from one area. Now that man's increasing destruction of his natural environment has made it impossible, or if possible undesirable, to collect birds on the scale and in the variety that was possible 50 or more years ago, collections such as Lord Rothschild's represent a store of biological data that is literally irreplaceable.*

Hartert played a very significant part in building up the Tring birds. He was the most careful and accurate of the trio and was the first among them to sponsor the polytypic species in ornithology. He and Walter together played a major role in the modernisation of avian taxonomy.

An important angle of the collection, which probably cannot really be fully appreciated to-day, was its timing. Once it had been assembled and curated in such an impeccable fashion there was no other group of animals which could at that particular moment provide the taxonomists, geneticists or biologists with a more valuable tool.

Working conditions at Tring were also felicitous. The rooms were large, light and airy. It was quiet and there was an incredibly well selected library close at hand. The atmosphere fostered creative ideas. Tring was entirely devoid of red tape and petty regulations and a serious student could work round the clock, undisturbed, if he so wished. There was constant attention and willing help from the best bird systematist alive. Munroe remarked that few realized just how much teaching went on at Tring — "for which the entire ornithological community is indebted to Lord Rothschild" added Wesley E. Lanyon — and this applied perhaps even more forcibly to the department of entomology. Walter revelled in loaning his treasures to all the museums who applied for them — Phyllis Thomas insists that not a single skin was lost — and he must in addition have sent hundreds of specimens to individual ornithologists — which inspired the bird man David Bannerman to write that it was impossible to overestimate the influence of the Museum on the systematics of the Order. "Sensational advance was achieved, and Tring became the Mecca of ornithologists all over the world."[16]

Walter welcomed all visitors to the Museum as if they were bringing honour and distinction to Tring by their mere presence — grist to his wonderful mill. And he rattled his drawers open and banged them to again — startling the timid worker out of his wits — and beaming approval as he crashed his way round the cabinets, naively confident that everyone must be admiring the birds and enjoying it all, as much as he was himself. Truly, Walter's profound love of birds was most endearing. His admiration for the aerobatic swifts never left him. From Rome he wrote to Hartert describing the passage of these birds on migration: "....swifts everywhere on houses down streets in the sky and the air and everything else was full of *Apus apus* like a swarm of gnats if I saw one I must have seen 10 million but curiously enough I only saw two *Apus melba*." But at Bordighera he found bird life was nearly extinct — "enough to give one a fit of melancholia for the rest of one's life."

On one occasion only Walter, writing from Venice, became quite lyrical:[17]

*It is delightful here to lie about in a gondola and be rowed about the canals. The sky is so blue and it is so warm and everything is so different to the rest of Europe.*

He soon returned, however, to his usual vein:

*I was very much astonished to find in the public gardens in bright sunshine a lot of bats flying about. They were* **not** *Plecotus auritus nor* Nango *pipistuellus nor* Vespertilia noctula *but what they were I do not know. They seemed to have very long and pointed tails.*

Walter rarely if ever spoke at meals. He ate in complete silence and appeared to be unaware of the presence of his neighbours, which some people found most disconcerting. But on one occasion during lunch a niece mentioned that she had seen the two cranes[18] in the enclosure opposite the museum engaged in what appeared to be some sort of a dance. Walter raised his head like an old war-horse at the sound of a bugle. "They are courting," he bellowed, his eyes beaming with pleasure, and he gave the delighted children round the table a blow for blow account of the wild shrieks, ritual dance and amazing sexual display by the Sarus Crane.

# CHAPTER 17

# Walter's Curators

*Karl Jordan and the butterflies*

WHEREAS WALTER ROTHSCHILD was a crazy lover of plants and animals, and Hartert an orderly, meticulous systematist, Jordan was a naturalist/philosopher. Had Charles Rothschild lived, Karl Jordan would have been persauded to project more of his ideas onto paper, and present them all in book form. He suffered from the same complex as Darwin: he felt — and no doubt rightly so — that he needed more and more supporting data with which to illustrate and clarify his interpretations. Despite his great achievements, he therefore failed to publish and develop some of his best ideas, ranging from novel classifications to new concepts — for instance, those concerning asymmetry[1] in animals. Some of his flashes of insight had been tucked away in *Novitates* in rather obscure systematic papers, and were consequently overlooked completely, or appreciation and recognition came when it was too late. Even his skeleton classification of the Order Siphonaptera (Fleas) was extracted from him under good-natured duress — long after Walter's death — by me. "You need more collecting and less phylogeny," grumbled Jordan, then aged 91 and virtually deaf — this in spite of the fact he had 20,000 tubes of specimens of this relatively tiny Order in the cabinets at Tring. In the end, it was G.H.E. Hopkins and I who had to publish his phylogenetic flea "tree" in a very crude form.[2]

It has been said that where the selection of his staff was concerned Walter had a Midas touch. This is true. It was also extremely fortunate that Jordan's friendship with Count Berlepsch brought him into contact with Ernst Hartert. It was through the good offices of the ornithologists that Walter was prevailed upon to interview Jordan and persuade him to give up his job as master of mathematics, physics and natural history at the Hildesheim School of Agriculture and move to Tring.[3] In fact Walter was lucky as well as perspicacious and gifted in his choice of staff. His influence on both his curators was unexpected, but nonetheless unmistakeable. No two men could have been less alike than Hartert and Jordan — perhaps the only two characteristics they shared were a love of natural history and their rather comical German accents.* Yet the direction they followed in their work was strikingly similar, although intellectually Jordan towered above everyone in the Tring Museum. Both men helped build up fabulous collections, one of birds, the other of insects. Both contributed to the advancement of systematics and to the understanding of microevolution. Both men played major roles in promoting international co-operation through congresses — Hartert was duly elected president of the first post war Ornithological Congress at Copenhagen in 1926, and Jordan, who initiated and founded the first International Congress of Entomology held in 1910, became Honorary Life President after 45 years of unremitting toil. Both men felt strongly that science was international and independent of politics and creed. Both understood that boredom — in the form of meticulous attention to taxonomic literature and correct nomenclature and the production and editing of *Proceedings* — was necessary in the cause of sound contributions to science. It is therefore simple to trace Walter's moulding influence at Tring. N.D. Riley[4] once asked Karl Jordan what gave him the idea of the international congresses. He replied: "It occurred to me in 1904 when I was in the Upper Engadine with the Hon. Walter Rothschild collecting Lepidoptera". Like Charles, Walter was a great *animateur* and a wonderful listener — a good combination for helping to crystallize and then expand other people's bright ideas.

And Jordan's ideas were roughly fifty years ahead of his day, but the fact that he lived to be 97 meant that, very late in life, he received some of the recognition so uncomfortably overdue — for his truly creative period was then long over. As Walter pointed out in his letter to the *Ibis*, the path of the innovator and trail blazer in any sphere is by no means a smooth one.[5] Furthermore the atmosphere in England at this particular period was xenophobic, and Jordan's contemporaries in this country greatly resented that he, a "foreigner", should enjoy a powerfully protected niche in the Tring museum.

In 1955 Zimmerman,[6] a distinguished beetle specialist, wrote: "Jordan is truly a man of extraordinary ability and accomplishment. His powers of observation, reasoning and interpretation have been developed to an astonishing degree of acuteness. His knowledge of general entomology is profound.... He is a genius, the like of which appears only as an extraordinary rare and fortuitous circumstance in time. We have few such minds in Entomology...."

---

* Incidentally this was one of the few areas in which Jordan was anything but objective. When his biographer suggested that his "strong German accent which had never been to the slightest degree softened by time" was an endearing feature, he was quite ruffled. He! A German accent? Never!

Ernst Mayr[7] described Jordan as "one of the great biological thinkers of our time". High praise, but justified. He points out that although Jordan is "not cited as widely as he deserves in textbooks" most of the concepts he pioneered have been generally accepted, especially in the field of the new systematics and evolutionary thought. Furthermore he was the first author who clearly described and defined the biological species.[8]

If you ask any postgraduate student who first invented and defined the term sympatric species[9] — closely related species having the same geographical distribution — he will respond with a blank stare. If you then explain that the description was Karl Jordan's,[10] that it was accompanied by the then new concept of past geographical isolation and was published by him way back in 1916 — no light will dawn, for the "sympatric species" has caught on and has become so plainly one of the facts of biological life, that its origins have been forgotten. It is also true to say that scarcely anyone to-day is aware that the current genetic theory of mimetic polymorphism was first put forward by Jordan[11] at the International Entomological Congress at Brussels in 1910. Such examples could be multiplied *ad infinitum*. Recognition in the scientific field is a funny thing, and it often bears little relation to a man's contribution. It was pure chance that Mendel himself did not sink into permanent oblivion and the inevitable discovery of "genetics" would then have been associated with some other name.

Jordan arrived at Tring with his wife in 1893. It is a pity that there is no record of his first impressions, for he must have been amazed at what he found. His original task was to "arrange, determine and classify" the 300,000 odd beetles (60,000 species) stored in various sheds and outhouses around Tring, for the Museum Cottage and Walter's bedroom could no longer accommodate the gigantic overflow. This incredible feat was accomplished within a year, and 400 new species described during the following twelve months.

The hours of work at Tring were theoretically from 9 am to 4 pm, but as Natty had not at this period abandoned the hopeless task of teaching his son the art of merchant banking, Walter rarely managed to arrive at Tring before 6 o'clock, and consequently Jordan often spent 14 hours or more at the Museum. When, in 1894, he was suddenly asked out of the blue to switch from beetles to Swallowtail butterflies — a group he was not familiar with at that time — he "read everything written about the Papilios — in all languages — available in the U.K. libraries", executed the necessary morphological illustrations (42 in number), and outlined the general section for Walter, so that the Revision of the Eastern Papilios was in print within the year. Incredible — for the Revision was an impressive piece of work and is still extensively used by all Lepidopterists working in this field.

A disconcerting aspect of those early days at Tring, engendered by over-crowding, was the lack of facilities for working with a microscope. Jordan used to retire to the landing window-sill if he required a really good light. He was sometimes joined there by a modest but very enthusiastic schoolboy for whom he instantly conceived a great liking. Thus began the only real and lasting friendship of his life. He always got on well with boys and in his company Charles felt marvellously at ease. Questioned once about the discrepancy in their ages — for he was at that time double the age of his disciple — Jordan brushed this aside with a smile and replied simply: "He was so nice". Walter was delighted too, for he thought Jordan's influence on Charles excellent and was enchanted that his brother should take an interest in natural history. In an obscure sense this was also a score off his father. But where Charles was concerned Walter was always truly generous, for he might well have objected to Jordan starting off on yet another group of insects, and spending so much of what was in fact HIS time on fleas rather than butterflies. He could well have wondered why, after all, he had got rid of the beetles? But Walter never grudged him a moment of Jordan's time, nor of Jordan's devotion and deep affection for his brother, and he himself never forgôt the fleas on his own collecting trips. In 1904 he wrote excitedly to Hartert that Charles, after waiting five years, had at last got the male of the helmet flea, *Stephanocircus*.[12] What is more he felt flattered that Charles admired his curator — "The clever-est man I know" — for Walter identified himself personally with everything at the Museum.

Karl Jordan and Charles had much in common, despite the fact that the younger man was so highly strung and over-anxious while Jordan possessed superb health of body and mind. Both were clear and objective thinkers, possessed of great intellectual honesty. Their broadminded, tolerant views on politics, philosophy, religion and human affairs in general, coincided. They were both reserved, modest, kind and unselfish, devoted to their wives and families, and rather straightlaced. Both were great lovers of nature, and gifted field naturalists. They found each other's company excellent — Jordan, though not really understanding Charles's particular brand of very English humour, acknowledged it "saved him" during his agonising depressive illness. They found keen mutual happiness in collecting their specimens and discuss-ing scientific problems. During the war, when Charles was recuperating in Switzerland, Jordan assisted him with the draft of his beginners guide book[13] to international finance, and one copy is heavily annotated in his handwriting. He seemed to be picking up the principles of merchant banking as quickly as he mastered the genitalia of the Papilios!

When Jordan accompanied Charles to Switzerland in 1916, he left his family and the Museum and his work for two years. Since he was a German by birth, he knew he would not be able to return to England until the end of hostilities, a date that was naturally unpredictable. It was hard, too, on his wife. She suffered greatly from the hostile atmosphere in Tring at that time, and her health was beginning to fail. (Unhappily when Jordan returned to Tring he found an ailing woman who died of renal insufficiency five years later.) Furthermore Charles was a desperately sick man, very difficult to help — the diagnosis was uncertain too — and his symptoms were a source of anxiety. It was an act of truly great friendship, deeply appreciated by the invalid as well as Rozsika. When they eventually returned home in 1919 Jordan thought he had successfully nursed Charles back to reasonably good health, but two years later he suffered a relapse and they once again had to spend six months together in Switzerland. In the autumn of 1923 Rozsika became alarmed and sent for Jordan to come to Ashton. Charles was running a high temperature, he refused to see a doctor, and was again feeling very ill; he complained incessantly of his head. Three days later he suddenly turned the key in his bedroom door and killed himself.

Karl Jordan never wholly recovered from this bitter blow and his relationship both with the Museum and Walter suffered in consequence. To the onlooker it seemed as if he had suddenly grown remote, apart, lost in thought, pondering some insoluble, abstract problem. He often sat for long periods in silence, with his head flung back and his beard tilted at a slight angle, apparently quite unaware of his surroundings, as if a sheet of glass separated him from the outside world. In later years deafness greatly accentuated this impression.

A few days after Charles's death Jordan returned to his desk and the routine of the Tring Museum — piles of correspondence, proofs, new material which continued to pour in, International Congresses, zoological nomenclature, collecting expeditions, species to describe, papers to write, drawings to make, *Novitates* to get to the printers, and all the while tending conscientiously and dutifully to the tongue-tied bereaved and distracted Walter, now, and for the next twenty odd years. But a permanent shadow had been cast over the scene.

Jordan published 460 papers, many jointly with Walter and Charles. He himself described some 2,575 new species, and a further 851 in collaboration with the two brothers. The collection thus became, for the larger Lepidoptera (and the fleas) the most valuable type collection in the world. A little under half of these publications appeared between 1923 and 1954.[14] Nevertheless the truly creative period of his life had ended rather abruptly in 1916.

It is a fact that Jordan never received the recognition he deserved by his fellow scientists in the United Kingdom — and he knew it. During his wearisome last illness he admitted as much. Not only was his F.R.S. awarded when he was over 70, but his "Festschrift" was delayed until he was 94, and the Jordan Medal was founded, not in Europe, but in the U.S.A., and then only after his death! This attitude could be applied to the whole Tring scene. But, notwithstanding, time has vindicated their work for the "true systematics", by which Walter and his two curators set such store. The new concepts which emanated from Tring Museum have been tacitly incorporated in the science of zoology and are now accepted as matters of fact by everyone. Norman Riley could reflect in 1959[15] that it was only on looking back that one realized how much one owed to Jordan: in other words his then new ideas and his particular brand of systematics are now just taken for granted.

Walter concentrated his attention on the butterflies and moths known as the Macrolepidoptera, although he included in his collection the so-called Micros — families which consist of very tiny insects. One of his earliest purchases in Bonn was a cabinet of Micros. His brother Charles made a large collection of these for Walter while he was convalescing in Switzerland.

There are, nevertheless, a great many butterflies and moths among the Macros — about 100,000 species — and Walter collected, set and arranged over 2¼ million specimens. It will take many years for the various groups to be satisfactorily "worked up". As Norman Riley remarked,[16] "The immense series.... are unique and present a field of research so wide that it will take several generations of entomologists to explore them".

Walter's decision in 1895 to shift his main entomological interest from beetles to butterflies was a good one, but he never committed his reasons to paper, nor, as usual, volunteered any explanations for his actions. He did not even discuss the matter with Jordan who, in 1893, had been engaged in general as the invertebrate curator, but in particular as a beetle specialist, of which Walter himself had described a number of new species. He merely told his collectors abroad to hold up the consignments of beetles while he hesitatingly invited Jordan to assist him in a revision[17] of the Swallowtail Butterflies! Was he — subconsciously — afraid of any form of verbal confrontation? During an argument, a close observer had noticed that Walter often seemed confused, embarrassed and tongue tied. Was it due to some childhood hang-up, when he was terrified of the grown-ups' disapproval? Could he not cope with new situations, unless they were clear cut *faits accomplis*, or did he rush in and make decisions from which, subsequently, he could not extricate himself, and went blundering on? It was this curious trait of pathological secrecy and conceal-ment in Walter's character which alienated people in his entourage, for what-ever psychological explanations they proffered — and these ranged from "a touch of insanity" to "a form of childish slyness" — the result was an inevitable loss of confidence in those who felt that, somehow, they had been slighted or misled, if not deliberately deceived. Günther attributed Walter's lack of communication concerning his expeditions — which he greatly resented — to a desire to steal a march on his fellow zoologists. But, although he was a highly competitive collector, Walter's reticence in this instance, sprang from a sub-conscious fear of disapproval and criticism, or even — since Günther was under his father's spell — from a conscious worry that his next venture might be actively impeded. Jordan immediately guessed what was afoot when the stream of incoming beetles was reduced to a mere trickle, and decided that, however aggravating Walter's *modus operandi* might be, the switch to the Lepidoptera was basically sound. At that moment in time the butterflies, like the birds, presented — as a group — greater opportunities for the study of evolution than the beetles. The latter, although easier to collect and more numerous (they are the largest Order of insects) were less well known, less spectacular, thus less well collected, and required more routine cataloguing, with the description of endless new species. Up to this time Walter had hoped to keep both Orders going, but he now realized that without additional staff it was impossible. He therefore sold his beetles, family by family, once Jordan had arranged them, the proceeds being ploughed back into the general Museum funds. He presented the Lampyridae as a gift to the French Coleop-terist Olivier, who was a glow-worm specialist, and the Anthribiidae were retained as a gesture of goodwill towards Jordan, who was deeply attached to the family. During the next 50 years his curator showed his appreciation by writing 145 papers on the group, and described 150 new genera and 1900 new species and sub-species. Meanwhile Walter said goodbye to the beetles by placing some of his favourites, the largest and most spectacular specimens of the Goliath and Harlequin beetles and the iridescent Buprestids, in the exhibi-tion cases in the public galleries at Tring. Thereafter he concentrated on the Lepidoptera.

It is now nearly half a century since Walter died, and his collection of butterflies remains "one of the greatest collections made anywhere in the world". This is E.F.B. Common's assessment — one of the leading Australian entomologists of to-day. The sole judges of the excellence of such scientific assemblages are those who have to use them. Common writes:

> *First it* [the Rothschild collection] *made available to the scientific world an immense quantity of well preserved specimens from many remote areas which were then difficult of access, and now often greatly changed or impossible, for political or other reasons, to sample adequately. Our knowledge of the occurrence and distribution of many of the larger Lepidoptera, for example, is based on the Rothschild material. The collection is therefore rich in type material of great taxonomic significance. Second the representa-*

*tion of so many species is sufficient to permit reliable conclusions about variation. Consequently the collection is of great value as a source of hypotheses on zoogeography and genetics, and a reference point for such studies. Third, the collection provided the basis for the valuable work of Rothschild himself and of Karl Jordan. It is highly significant, I believe, that the major contributions of these two scientists are still the standard reference works.*

*These comments, it could be said, support a case for the historical value of the collection but do not necessarily mean that the collection has a great scientific value in the modern world. On the contrary, the present-day value of the collection lies in its providing an irreplaceable source of voucher material for the taxonomic and other conclusions that have been based on it. It is a well accepted principle that voucher examples of the specimens or species upon which biological work is based, whether taxonomic, biochemical, physiological or ecological, should be deposited in a reputable collection, so that future workers will be able to determine objectively the reliability of any identifications. The Rothschild collection not only provides these voucher specimens, but a vast amount of additional material upon which much valuable work can still be based in the future.*

Keith Brown, an entomologist working in Brazil, combines a knowledge of chemistry, ecology and systematics, and therefore represents the modern approach to the Lepidoptera. He fully endorses Common's opinion:

*At least for my own work, it is probably the single most valuable collection in existence — especially as it is housed together with what is surely its only competitor, the large composite collection of the British Museum (Natural History). The long series which compose the Rothschild collection, carefully organized and labelled and representing areas in the Neotropics very poorly collected in general, are exactly what is needed to analyze variation and gene interaction in different subspecies. This in turn is the base of much biosystematic work, which "takes off" from the museum material and leads to experiments in the field and in controlled conditions on live populations. Any person who wishes to make sense out of the ecology of Neotropical butterflies today will have to visit the Rothschild collection at the beginning of his systematic studies. For that reason, the enormous treasure which this collection represents to modern entomology and ecology in the Neotropics needs to be more widely known.*

Keith Brown thus supports the opening paragraph of Walter's Memorandum to the Trustees of the British Museum (see p. 316) when he offered them his collection: "One point strikes me as particularly important: the systematist is constantly hampered by the fact that the material at hand is not sufficiently large...."

At Tring the British Lepidoptera were kept in a separate room — up a small ladder-like staircase from the main entomological hall. Walter had taken an immense amount of trouble in obtaining large series of variations — he had himself collected intensively in the field and bred a lot of British material, and also spared neither time nor money in chasing up all the rare aberrations which came onto the market. In the tradition of the Tring Museum, E.A. Cockayne, a geneticist and paediatrician, who was also an amateur entomologist, was invited to work on the collection, and when he retired from the Middlesex and Great Ormond Street Hospitals he moved to Tring and concentrated for the rest of his working life on the genetics of the British Lepidoptera. Meanwhile Bernard Kettlewell, a field entomologist — also a medical man by profession — was drawn into the magic circle and added his own excellent collection of British material, together with long series of moths obtained in carefully controlled breeding experiments, to the Rothschild Lepidoptera. Eventually, after the Tring Museum had passed into the possession of the Trustees of the Natural History Museum, Cockayne formed a small trust and endowed the three amalgamated Collections, now known as the National Collection of British Lepidoptera (Rothschild/Cockayne/Kettlewell). This would have delighted Walter, for there was nothing he liked better than the knowledge that other entomologists made use of his butterflies and appreciated the marvels he had brought together in the drawers of his cabinets.

Jordan considered Walter's greatest contribution to the scientific scene was his unflagging, persistent, eager generosity, with loans, gifts and information — maintained throughout his life. Nothing was too much trouble.

Cyril Clarke, one of the few fellows of both the Royal Society and Royal College of Physicians, who has used butterflies as tools in his solution of the prevention of haemolytic disease of the newborn and his elucidation of the mechanism of the supergene in the inheritance of polymorphisms, wrote the following paragraph on Walter's collection of British butterflies and moths.

*Marvels indeed they are, and a feast to the eye and intellect of anyone interested in polymorphisms, local races, rare aberrations, teratological specimens, gynandromorphs, intersexes and mosaics (one of Cockayne's specialities) and every aspect of mendelian genetics.*

*The melanic form of P. machaon, of which Cockayne first worked out the autosomal recessive inheritance, is a magnificent sight, as are the Tiger Moths and the Thorns. Wherever possible Cockayne's beautiful handwritten labels have been retained and in many drawers the genetics of the forms are summarised and the references given.*

*H.B.D. Kettlewell is industrial melanism, and Biston betularia his animal, and there is telling evidence about the evolution of dominance in carbonaria Jordan, the earliest specimens in the last century being much paler than their present-day descendants. There are also amazing aberrations and mosaics in this species, including a fascinating one with white and dark stripes, and a truly astonishing buff recessive bred in the 19th century. That the dominance of the melanic form has not everywhere become complete, is shown be crosses with B. cognatatia from the north American continent. On the other hand, it has been done so in Kettlewell's herefordi ( ♀ carbonaria × ♂ strataria) hybrids.*

These remarks serve to emphasise the point made by Keith Brown. Scientists, whether they are systematists, morphologists, geneticists, ecologists, ecological chemists or students of evolution, find the collection exciting and stimulating. For, curiously enough, it is only against the panorama of modern research that the full value of Walter's butterflies becomes apparent. Its importance lies in the *unfolding and presentation before your eyes of a whole order* — in all its variety and complexity, culled from continent to continent, from one far flung oceanic island to another, from desert and forest and prairie and mountain range. There is also an indefinable factor about these collections, a *Walterian* factor — call it what you will — a whiff of zest and wonder, which must somehow have been pinned in among the butterflies. Suddenly the outlook broadens, the horizon expands — a penny drops, new ideas materialize, the mind "takes off".

The joint monograph with Jordan on the Hawkmoths[18] was undoubtedly Walter's *magnum opus*. This is a classic. In the book market to-day his *Extinct Birds* of which he was sole author, and the *Avifauna of Laysan* have become collectors' pieces, and at auction fetch prices of three or four figures. But in the scientific world he will always be remembered for the *Revision of the Sphingidae*. This massive work, in two volumes, contains nearly 1000 pages of text and 67 plates — seven in colour. Each plate includes between 13-58 figures of whole moths or their morphological details. The drawings were all executed by Jordan. Contemporary reviews[19] ran to fifty pages. "It is the wealth of detail here accumulated in regard to the structure and anatomy of all the outer portions of the Sphingid imago that marks this Revision as something definitely beyond anything of the sort that we have yet had on any family of the Lepidoptera. The care and minuteness as well as the accuracy of the observations and the large number of species to which they relate enforce one's admiration.... Not only [is it] an advance in the study of the Sphinges but it is a step forward in the systematic treatment of the Lepidoptera and must inevitably command the admiration of all who can appreciate it."

The Introduction and General Subject — to which Jordan contributed a major share — are delightful. They are replete with quaint reflections: "Science is knowledge of Nature", "Science is a republic where everybody may do as he likes", — and a plethora of the Clarendon type so beloved of Walter but frowned on by his curators: "The SIGNIFICANCE of corporeal characters is established by biology.... Anatomy and morphology gives the QUANTITY, biology determines the QUALITY...." But, in the best Tring tradition, this section of the Revision gives us clear, unequivocable definition of species and subspecies — the geographically separated entities. And the coverage of the literature is awesome: for instance approximately 400 references are provided for the Death's Head Hawk Moth alone — about 28,500 altogether. Various new genera and species are described.

We are told that Tolstoy's wife copied out *War and Peace* seven times. Walter had neither wife nor a secretary, so he and Karl Jordan between them made a longhand fair copy of the Revision for the printers.[20] For a slow writer this must have been a feat of considerable endurance. In a footnote Jordan comments that in the "catalogue" which they added as a kind of supplement, bound in at the beginning of the volume, they corrected "several slips of the pen".

The year the Revision appeared Walter published approximately twenty other notes and papers, including the last part of the joint Monograph (with Karl Jordan) on *Charaxes*, and 50 pages describing Lepidoptera collected in North-East Africa.

He was nothing if not industrious.

# CHAPTER 18

# Walter's Collectors

THE WORLDWIDE WEB of collectors with whom Walter kept in touch is further evidence of his tremendous drive and burning interest in the whole animal and plant kingdom. He lived in a period when communication was not as simple as it is to-day. There were no telephones at Tring when he first opened the Museum to the public. Letters took a long time to reach their destination abroad — two months or more, for instance, to Australia before the advent of air mail. In fact he often lost contact with individual collectors for months, if not years,* once they had reached the Pacific Islands or the interior of Africa.

In a way Walter possessed some of the qualities which characterised the gifted financiers of the family, but he had reached the stage of using, rather than making his fortune. As we have seen, his father had evolved along somewhat similar lines, for although a gifted and highly skilled man of traditional business and finance he was more deeply interested in social problems, civil liberties, agriculture, and especially in the fate of the Jewish people, than in big deals. It is not possible to compare the various activities in which the Rothschilds were engaged, ranging as they did at this period from colonising Palestine,[1] to building the Théâtre Pigalle,[2] but looked at objectively in 1980, it would appear that Walter had been among the most successful members of the family. His major interest — the exploration of the animal kingdom — seems to take up a lot of time on the television screen to-day, and his favourites — the Galapagos tortoises and Birds of Paradise — are much in evidence. It is probable that long after time has completely obliterated N.M. Rothschild & Sons, the Suez Canal deal and the various glamorous myths attached to the acquisition of fortune and fame, 100,000 people or more will still be visiting his Museum at Tring every year, as indeed they are to-day.** But

---

* The occupational hazards of collecting in those pre-penicillin days, and before the development of a host of other useful medicaments and inoculations; were great. Alfred Everett, a Tring collector, died from fever; three men died of yellow fever on the Galapagos expedition, Doherty died of dysentery, Ockenden of typhoid fever, Webb of an unspecified illness on his way home, Stuart Baker had his arm bitten off by a leopard. The list of deaths and disasters is a long one.

** In 1981 126,000 people visited Tring Museum.

it was in his relationship with his various collectors that Walter's Rothschil-
dian talents can be seen to full advantage. As far as we know he kept no notes
or card index, and he made no detailed plan in advance. He played everything
intuitively by ear, and his grandiose schemes were pushed ahead like a gigantic
game of rollicking chess. He possessed a magic touch and single-minded
tenacity of purpose. None of his letters to collectors survives, for in the absence
of a secretary no copies were made before 1917. But during his most creative
period, say from 1890-1908, the business letters[3] to Hartert, dealing with such
mundane matters as invoices, receipts, the sale of duplicates, etc., show that he
employed, *at this time*, over 400 collectors.

But this is a gross underestimate since the large majority of entomologists
who collected insects only, wrote to Jordan[4] or to Walter directly (not Hartert), and all but a handful of these letters were destroyed. Thus, for instance,
those from some of the most successful field collectors of insects for Tring,
such as Faroult, le Moult, Stein, Eichorn, Corporaal, Weiske, Riggenbach,
Ruckbiel, Jettmar, etc., are missing. A count in three randomly chosen
volumes of *Novitates Zoologicae* (1890, 1901 and 1902), reveals the names of
twenty-six collectors who were not included among the 400, since they had not
written to Hartert.

"The world with a severe attack of measles", was how one cartographer
described the distribution map (facing p. 152) showing the immense area and
the innumerable localities visited by the men while working on behalf of Tring
Museum, but the records of the insect collectors are lacking, except those who
included both birds and butterflies in the material they sent home.

Walter could keep track of all his collectors simultaneously, knowing where
they were located at any one moment, at what stage their enterprise should
have reached, what species or sub-species of bird or butterfly or giant tortoise
they should be looking for, and in which particular valley or river bed. He
knew the equipment they were using, and had given them precise instructions
how to preserve and pack their material. Those of us who are endowed with
more "normal" memories find this difficult to believe.

The collectors could be divided into four main categories, although each and
all of them possessed qualities and characteristics of their own. First of all
there were men like Palmer[5] and Harris,[6] who were put in charge of specific
expeditions large or small. Sometimes they were temporarily employed at a flat
rate and/or a salary, or their expenses were defrayed and the collections they
sent home were paid for specimen by specimen. This was a risky business for
the collectors, but a risk they were prepared to take, for otherwise they would
have had few patrons willing to pay for the cost of distant expeditions. It was
no easy matter to find such men, for they not only had to possess the qualities

of the explorer, but also had to know what to collect and how to prepare skins. These expeditions were planned, arranged and financed by Walter, and he proved fantastically successful in this domain — often against all odds, the dictates of common sense, and everyone else's advice. Thus Alfred Newton thoroughly disapproved of Palmer although his judgement was affected by jealousy, for, as we know, he considered that he had staked out a claim for collecting the Hawaiian Islands himself. Time proved Walter irritatingly right about both the man and the expedition. Palmer collected 1832 birds — relatively few for a three-years' slog — of which fifteen were new to science, but it was the completeness of the collection which made it so valuable. He had been sent alone to Laysan and the neighbouring islands in the South Pacific to collect the birds, and was told to find assistance locally. He was there three years — whereas Harris (see Chapter 22) had a ship and a crew of five men to assist him.

It was the usual practice to find skinners and trappers locally. Charles Rothschild, for instance, collected round Kandy for three weeks "scooping up 3000 butterflies". "There," he wrote to Birrell, "We [he and Gayner] picked up an Eurasian Youth who catches bugs. He was a thoroughly depressed being, but we took him along and finally sent him to Loo Choo where he got a fair lot of things...."[7]

Secondly, the highly-skilled, professional, whole-time collectors, such as Meek, Doherty, Riggenbach and Boden Kloss made their own travel arrangements, but put their services at the disposal of museums or private collectors like Count Berlepsch and King Ferdinand of Bulgaria. They were prepared to go to any part of the world, into wild, uncharted seas or virgin forests, in search of specimens for their various patrons — picking up further valuable material, such as minerals and native artifacts on the side. Before Hartert came to Tring he earned a precarious livelihood in just this fashion, thankful that the cost of the journey was defrayed, thus enabling him to see and describe new fauna and flora which would otherwise have proved utterly beyond his reach. Over the years Walter developed a peculiar sort of friendship with several of these men, and a few love/hate relationships — rather like a general who develops special relationships with some of his most able officers. When Doherty died, he wrote[8] unhappily to Hartert:

> *I fear we are doomed to failure this year. First Doherty is dead then Meek fails, then Wilson fails then Martinique birds are all exterminated and now worse luck the enclosed appeared in today's papers; I fear Riggenbach's expedition & Newton's are done for.*

The third group consisted of genuine explorers, such as A.F.R. Wollaston, who were usually delighted to collect *en route* for museums and friends, and combine such activities with travel. These men were adventurous, tough, intrepid and resourceful, generally with only miniscule private means, who voyaged solo, or with a couple of companions or a party of native bearers, or with a museum expedition, into relatively unknown territory. Some were geographers, others ethnologists or naturalists (who quite often worked out and published papers on some of their own material), or medical men, many were missionaries and clergymen (like the famed butterfly hunter the Rev. Miles Moss), or serving in the Army or Navy, empire builders or prospectors, or merely *bona fide* scientists like Percy Rendall and Heim de Balzac, who exchanged material with Tring. Such fellow scientists are not included among the collectors, for although it is true that these zoologists collected for each other, this amounted to no more than a by-product of their own researches.

Finally there was a fourth category — a motley crew of professional dealers, with businesses of varying quality and size, selling natural history objects and equipment — like Watkins and Doncaster, and W.F.H. Rosenberg of London, and Derolle and le Moult of Paris (a particularly expensive dealer who presented his bills on lovely decorated paper), and Staudinger and Bang-Haas of Dresden, who had built up a network of innumerable contacts and suppliers in various foreign countries. Rosenberg and le Moult assembled fine collections of their own in the process. Many of these dealers considered Walter one of their best and steadiest customers, always ready to do business. On receiving interesting material they would send him the whole collection from which to select anything he required.

Occasionally Walter stumbled upon an exceptional individual collector like the Algerian, Victor Faroult, who, after 1908, was employed continuously collecting and breeding Lepidoptera for him in Algeria. Or, for example, he might sign a contract[9] — as he did with George Forrest who was to collect for two years in West and North-Western Yunnan, for an annual sum of £375, half to be paid to his wife at home. Albert Meek,[10] when he set out for "Woodlark, St. Aignans, Sudest, Rossel, all islands of the Solomons except Guadalcanar, the Shortlands and Rubiana, Holnicott Bay, or other place north of Cape Nelson in New Guinea, and, from there inland, Matthias and Admiralty Islands, one larger island of the Santa Cruz and New Hebrides Islands, Western Australia, and other places we may arrange afterwards" agreed to send his collections of birds, butterflies and beetles to Tring for first selection. Walter agreed to pay for:

Birds *(except Birds of Paradise for which the prices are to be agreed upon with Mr. Gerrard). Six specimens of every species at 6/6 each, with a bonus of £4 for every new species.*
Birdsnests & Eggs, *one clutch with nest @10/- or of large birds 3 eggs without nest for 10/-. Nests & Eggs of Birds of Paradise to be valued by Mr. Gerrard.*
Of Lepidoptera [*Butterflies and moths*]. *Eight specimens of every species @ 2/- each (Ornithoptera and Morpho tenaris to be valued by Mr. Janson) In cases where there are less than eight to take less i.e.: — One of two of same sex. Two of three, Two of pair. Three of five, four of Six and Seven, five of eight & nine, Six of ten, Seven of Eleven, Eight of twelve.*
Of Coleoptera [*Beetles*] *Eight Specimens at 9d each without restrictions.*

In fact it was Walter who started Ernst Mayr[11] on his scientific collecting by suggesting he replaced Eichorn, who had had a stroke, and tried his luck in the poorly worked mountains of Papua New Guinea. In April 1928, Mayr began collecting in the Arfak, Wandammen and Cycloop Mountains — and eventually amassed 2700 specimens, including two new species and thirty new subspecies of birds. Mayr said that he "....liked Lord Rothschild tremendously for his simplicity and gentleness" and added some comment on his shyness and great modesty. "He had a completely extraordinary memory* which on many occasions surprised me by its precision and comprehensiveness."

One day Mayr and Hartert were discussing a rare New Guinea bird which was not in the Tring collection, and Walter remarked: "That bird is illustrated on plate 87 of Gould's Birds of New Guinea." It was.

He also remembered details about plants as well as animals, and we find him writing to Hartert (in 1903): "I now understand why I could not guess to what tree you referred. The flowers are not white but green and orange and yellow and you will find I think a figure of the flower and leaf on one of the plates of Abbot & Smiths 'Insects of Georgia'." Walter appeared to have stored in his head a mass of unsorted data which he could apparently retrieve when it was required. This was a useful gift. On one occasion he wrote a hurried note to Hartert from New Court: "Please at once take one of the supposed new Petrels from the Galapagos Isles and relax the feet and legs and open the toes and make a note if the inside webs between the toes are yellow. I suddenly remembered either Beck or Harris in their diaries saying something about yellow webs. If this is so it is Wilson's petrel *Oceanites oceanicus* or *Oceanites gracilis*." On another occasion he received some Dodo bones from Dr. Sclater and suddenly remembered that Sclater had gone to the Mascarene Islands with

---

* The Archbishop of Canterbury remarked on Walter's "amazing memory" in a speech in the House of Lords (5 July 1938).

the Transit of Venus Expedition and their station was Rodriguez and it was they who had brought back the Solitaire skeletons. Perhaps the skeletons came from Rodriguez after all. That would solve a conundrum.... He asked Hartert to send the bones up to London *immediately* and he would compare them with those at South Ken.

His museum was so well known that many people in the United Kingdom would spontaneously send material to Tring, and no collector was too modest or too inexperienced to be brushed aside or ignored. Walter was not only interested in everything connected with natural history; he did not begrudge the time nor the effort spent in encouraging and helping these stray correspondents. Some of them were hoping for large payments like the man who asked for £200 down (equivalent to, say, £2000 to-day)[12] for a "golden tinted blackbird", and the lady who claimed she had seen, in the St. James's Gazette, that Mr. Rothschild had given £1,000 for an Ecuadoran butterfly.... Others asked for a modest remuneration, but many gifts figured among the specimens offered. A high proportion of these were freaks or abnormalities like a six legged bullock, a kitten in a glass jar with eight legs, two tails, four ears and one head, a stuffed double calf's head (20/-) and a three-legged colt. Then the colour varieties figured frequently among the animals tendered — a white kiwi, a white woodcock, a black guillemot "caught by means of a snickle" and so forth. Often there were objects which the owner hoped would be enshrined at Tring.... Someone offered the skull of a New Guinea native, an iron spear embedded in it, and — improbably — enveloped by an elephant's tusk, "objects from the abortion of a cow," a luminous beetle, a synistral oyster and red velvet worms, "A dab and a lemon sole". Then there was a continuous stream of live animals for which a kind home was sought in anguish — tiger cubs, live cassowaries, white donkeys, Nigerian monkeys, even a pet spider, and laughing owls "in the pink of condition". Walter acquired quite a number of specimens from these chance correspondents (he paid the owner of the black guillemot £20) but in addition he commissioned a string of individuals at Billingsgate and various fish markets to keep a lookout for especially large and unusual species brought in with the daily catch. Telegrams would arrive at Tring which read, for instance, "Barracuda fish caught South West coast of Ireland about 9 feet long 5 hundredweight remarkable scales fins breastplate shark Tell your own price." A traditional tale describes how a collector turned up at Rothschild's bank leading two live bears "For Mr. Walter".

Finally, Walter himself, his brother Charles and his two curators were themselves experienced and skilled field collectors.[13] All except Jordan, who did not carry a gun, were good shots. On the whole, however, their own expeditions took the place of holidays — to be "fitted in" as and when they did not interfere with the more serious task of developing and running the museum, or in the case of Charles, the Bank, the Royal Mint refinery and his conservation enterprises. At the time of Walter's fall from grace and the Great Row in 1908, when he left N.M. Rothschild & Sons, he prudently absented himself from England, and he then enjoyed eight months continuous, highly successful collecting in North Africa and the Alps. Similarly, while Jordan was caring for Charles in Switzerland between 1916-18, both men collected Lepidoptera and fleas continuously for over a year. Jordan during this period reported regularly to Walter, but only three of his letters have survived in the one box which escaped the bonfire. It is worth reproducing one in full, for it demonstrates the detail that Walter expected from his men in the field:

*Fusio, 25/7 1917*

*Dear Lord Rothschild,*

*I have to thank you for your kind letter of July 14th, which reached me here at Fusio. We have been staying at this place since the middle of last week and intend to leave for Bignasco to-morrow. As the weather has been fine all the time, we have been able to explore the neighbourhood. The fauna in the valley is not rich, but as there are many species which do not occur at Bignasco, it was quite worth while coming to Fusio. On the 22nd I went up to the Campolungo Pass, which is well above the tree-line. It is a most beautiful alpine spot, whence one looks down into the Airolo valley on one side and the Fusio valley on the other. The fauna and flora are truly alpine. Marmots are there in abundance, and of plants I noticed especially* Primula longiflora *(similar to* farinosa*),* Gentiana brachyphylla, Solanella pusilla, Saxifraga oppositifolia, *etc. When we visited Fusio for a day about a fortnight ago we met a Swiss collector from Chiasso, Signor Fontana, who told us that* Erebia florafasciata *was almost over; he kindly offered to take me up to the locality where it is found; but warned me that I could not expect to get more than a dozen specimens. As we wanted to go back to Bignasco the next day I could not avail myself of Fontana's offer. Fontana told us that he had accompanied Rene Oberthür as well as Mr. Jones (treasurer of the Ent. Soc.) on a hunt for* flavofasciata. *Well, when I arrived at the place high enough,* flavofasciata *met my eye at once. The species was quite common, although I was too late, most of the specimens being worn. I took 50 altogether, among them some good ones. I had sometimes 4 in the net. As I wanted to get some specimens of all the butterflies that fly on those slopes I could not devote all my time to* flavofasciata. Erebia lappona *was quite abundant,* Melitaea cynthia *not common,* Arg. pales *very common,* arsilache *absent.* Pieris bellidice *(which Fontana had not found) in small numbers, it is so difficult to catch in the wind on bad ground that I got only 7 or 8. No* Parnassius, *no* Cokias palaeno. Zygena exulans *in swarms. It was a most enjoyable excursion. There is only a cowtrack in many places, often deeply washed out, and sometimes altogether disappearing where it strikes grassy slopes. It is not a dangerous walk but it requires good* *legs. At the camp we got lately two ♂ ♂* Leucania taraxci, *the ordinary yellow form. Of rodents we got here only two species of Arvicolidae; a long-tailed small shrew is common.* Mus silvaticus, *which is common at Bignasco, does not seem to occur at Fusio.*

*I am afraid I shall have some trouble with Melon if he goes on sending the same things over and over again in quantities. Our arrangement with him is as follows: (I am writing about it also to Dr. Hartert).*

*The price of the collections is to be fixed by us according to the contents of the collections.*

*Each consignment is to be valued by us and the money to be sent to Melon.*

*If sufficient specimens of a species have been received, the additional specimens are to be put aside for Melon, & he is to be told not to send any more of them.*

*It is left to us how many specimens we wish to take; the price for a small series being per specimen higher than in the case of a large series.*

*As Melon appears to send only small things he should be told to look out for the larger kinds. We do not take Micros from him. As you see from his collections, a fixed price all round was not advisable, and he was quite agreeable to leave the matter in my hands.*

*Your brother is much better, but he must not yet think of going home. I shall probably come home before him.*

*With all good wishes,*
*Very sincerely yours*
*K. JORDAN*

On his own collecting trips Walter regularly reported back to his curators.[14] Returning from Algeria he stopped at Grenoble, Lautaret (2079m), La Grave, Digne, etc., in July (1908) and summed up as follows: "By the time we get back if all is well we shall have collected in 6 months over 15,000 Lepidoptera, about 4000 other insects, 80-85 mammals, and 376 birds — an achievement that cannot easily be equalled in the Palaearctic region, especially as we had no professional assistance." He wrote to Hartert from each stop *en route.*

*Digne*
*4 July 1908*

*Dear Mr. Hartert*

*One line before leaving this morning. Please tell the Goodsons and Smith when setting the specimens just sent home to be extra careful with the Zygaenas, those from Algeria because they are pressed very flat, and those from Digne because we to prevent this pressure have put cotton wool in the papers and if they are not extra careful in moving this cotton wool the antennae & legs will be smashed. In fact Zygaenas ought to be set the day they are caught but alas unless one has a setter with one how is it possible*

161

*to hunt for bugs 6 hours a day and kill and prepare 307 specimens and set 93 Zygaenas oneself which would have been one day's work here. Also no insects other than Lepidoptera are to be pinned or set till we return.*

*Yours Sincerely,*
*WALTER ROTHSCHILD*

*Grand Hotel*
*Thibaud, Propre*
*Grenoble*

*We have just had hatch*
*3 fine ♂♂ of Orgya*
*splendida from Blida cocoons.*

*July 5th 1908*

*Dear Mr. Hartert*

*Will you please tell Goodson & Smith to set first the paper sack marked ""Digne 1.7.1908"" & two small envelopes marked Special Zygaenas Hammam R'hira & after that the rest of the Hammam R'hira lots & the sacks marked respectively ""Dourbes Digne"" & ""St Michel de Cousson"" & ""Digne to St Michel de Cousson"" as these contain the things I wish to see as soon after my return as possible.*
*We are sending you today by "Sample post" 2 larvae of a* Saturnia *which please have fed on Lime tree leaves (Lindenbaum Blätter). We stayed over Sunday at Grenoble as the tramway to La Grave is too full on Sundays. If we have electric light at La Grave as they say there is, we ought to do well but if as in most of these "Blessed" French holes the electric lamps are* incandescent *and not arc lamps we shall not have much to do THERE. At Lauterets on the contrary as long as we have no snow, we can work our two 56 candle power (new acquisition) acetylene lamps whether there is arc light or not & the result is out & away better than electric light as on the large 1 metre square white muslin cage in which the lamps are placed the moths sit down and are boxed without a net. (Dr. Nissen's patent) But when we get home we propose to invent a better one combining Dr. Nissen's & Captain Hall's, as the former's is not easily carried about.*

*We got 2769 at Digne.*

*Yours Sincerely,*
*WALTER ROTHSCHILD*

*Grand Hotel de la Meije*
*La Grave*
*le 22 Juillet 1908*

*Dear Mr. Hartert*

*Thanks for sending on Sclater's letter. I fear except Godman and Harmer he has got the wrong people to sign my application.\* Last night I wish you had been here. It had been a brilliant day but at 5 fog came down and filled up the whole valley. We set up the lamp at 8.15 & up to 8.45 we had caught our usual 80 to 100 but at 8.45 it came with a rush & Dr. J. Harmon and I could not cope with it at all in spite of having 7 Cyanide bottles, over 200 glass boxes and a large chloroform tin. The gauze cage round the lamp which is 1 metre long by ¾ metre broad and ¾ metre high was covered on all sides with a seething struggling mass of moths such as I suppose was never seen before in Europe. We caught up to 10.45, when we were too exhausted to go on 1167 moths and when we came to prepare them this morning, we were able to pick out to keep 1082 of about 70 species. I do not suppose outside the tropics this has ever been done before. We have postponed our journey to Lauteret for two days as the mountains are all in clouds & it is raining & therefore up there 1800 feet higher icy cold.*

*Yours very sincerely*

*WALTER ROTHSCHILD*

After collecting at La Grave for thirteen days (three of them blank) he had captured 5191 Lepidoptera among them 78 Burnet Moths (Zygaenids). "I think a record anywhere as Wallace's collection during his Malay Archipelago trip of several years was not 6000."

Walter liked record bags. He never quite grew up.

Together he and Hartert described their "Ornithological explorations in Algeria"[14] in an article in *Novitates* (some 90 odd pages), dealing with the 1926 bird skins collected. The Introduction appears to have been written up by Hartert and lacks Walter's inimitable style. One so much regrets that the mass of letters written to his mother and Charles from Algeria were destroyed. A few extracts illustrate a number of typical episodes which characterise the classical collecting trip. The weather is always capricious: — "The next night we pitched our tents at a place called Nza ben Rzig.... and a very heavy sandstorm made

---

\* To support his election to the Royal Society. His fears were justified, just as they had been ten years earlier when the subject of his election was first raised. At that time Haldane made strenuous efforts on his behalf, but too late to retrieve the situation.

the night terrible. Every minute it seemed certain that the tents would be blown down, and though they stood the storm excellently they suffered, especially Hartert's own small tent. The roar of the wind, together with the banging of the canvas, the snorting and roaring of the camels (there were 17), which crowded round the tents, and the sand which penetrated everything, made sleep impossible...."

The beauty of the desert captivated Walter like everything else. "It is impossible to describe the simple beauty of the rich yellow sandhills, the clear blue sky overhead, the great quiet, here and there interrupted by the wonderful melancholy notes of the "Muka" (*Certhilauda*)". Then there was the inevitable, unforgettable fiery sunset behind the purple mountains, and the "fata morgana".

> *From Kel el Dor, where there is a heliographic telegraphic station of the Government, one descends into a vast plain of hard soil saturated with salt and saltpetre, very tedious and with very little bird-life, dry enough at that time, but covered with water after heavy rains and dangerous for camels, which easily slip on the greasy surface. This district lies, like the whole Chott Melrhir, below the level of the sea. In dry times the salt crystalises on the surface, and glitters in places like snow. No vegetation is visible, with the exception of isolated, thick, roundish tufts of halophilous plants of a greyish green colour, all of two species,* Halocnemon strobilaceum *and* Limoniastrum guyonianum.*
>
> *This is the country of the "mirage" or "fata morgana". There appears constantly in the distance what seems to be a vast lake, dotted with islands and towering clusters of trees, and on pushing onwards nothing but the same eternal grey-green clumps of bushes or some kind of stipa meets the eye.*
>
> *In spite of this wonderful spectacle one is quite pleased when at last palms appear that remain trees and do not vanish and turn to little stunted bushes, but, taking shape, become a fine oasis.*

Every expedition suffers some calamities. This time it was the medical man of the expedition, Dr. Charles Nissen — whom Walter had originally met by chance in Algiers and who became a life-long friend — who fell ill and delayed the travellers. Then, on the return trip they hired a ramshackle motor vehicle which not only broke down, but shook off the boxes of material from the luggage rack.

> *On the way from Laghouat to Boghari, before we came to the little caravanserae of Guelt es Stel, we had a great misfortune. Probably some one cut the rope with which two cases, containing all our birdskins and insects brought together during the journey to the M'zab country, were tied on to the back of our motor, or else it got cut through by*

*the sharp edge of the box; anyhow, when we arrived at Guelt es Stel it was seen with indescribable horror that both cases had fallen off! In less than one minute Hartert was on the seat by the side of the chauffeur and rushing back over the same road with all possible speed. After a long drive he saw a big white mass in the halfa-grass — the box containing the birds broken open, and the skins strewn over the desert sand. Fortunately the nomads, who evidently had broken it open, could do nothing with the birds, and not a single one was missing; but some cigar-boxes containing sixteen clutches of eggs were gone — probably only because of the wooden boxes, which the Arabs covet very much. Among the eggs were properly identified clutches of* Galerida theklae carolinae *and* Cristata macrorhyncha *and a splendid series of eggs of* Ammomanes deserti algeriensis *and* Emberiza striolata sahari.

*The other box, with all the insects and many other things, after searching in vain all the tents of a nomad camp, in spite of great unwillingness of the owners, was found hidden away some distance from the road among some tamarisk bushes. As it was we got off luckily enough with comparatively little loss.*

Walter was fascinated by the M'zab, a peculiar tribe of small Arabs.

*....with square shoulders, wide chests, and strongly developed calves, as compared with the — on the whole — slim, thin-legged true Arabs. They belong to a different religious sect, and build different mosques, are much more industrious, quieter, fond of trade and agriculture — also, judging from the orderly look of their streets, gardens, and cemeteries, cleaner. Nevertheless lice, especially* Pediculus vestimentum, *are exceedingly numerous, and bed-bugs are not unknown, though apparently not overplentiful; human fleas, however, which are absent from the whole of the Sahara (and tropical Africa except where introduced), do not occur. The cemeteries are truly beautiful as compared with the entirely bare Arab ones. They are covered with green bushes and (at this time) with innumerable lilac and little white flowers, and each grave is ornamented with pieces of broken pottery, broken instruments, or other broken things evidently to signify the broken life of the deceased. There are also special little houses near the cemeteries for ablutions, without which prayers may not be said; and for the grand prayer, said on certain days in the year, large platforms like big elevated barn-floors are built.*

*The most extraordinary feature, and a remarkable proof of the industry of the people are the deep wells, of which over 1100 are said to exist in the M'zab valley alone. They are being worked almost the whole day, by mules, horses, cattle, and camels, bringing up bucket after bucket of the fluid element without which the palms and gardens cannot exist. Far superior to all Arab oases we have seen are the M'zabite ones, and the beautiful appearance is chiefly due to the vines which are trained like garlands from palm to palm and along the walls.*

There were other problems not mentioned in *Novitates*. The Arab assistants they engaged locally proved amazingly careless and smashed the nets and glass boxes. Steinbach, the man whom they had taken with them as skinner, suddenly refused to tackle any more birds, and Hartert was faced with the impossible task of skinning after collecting — a race against time, as putrefaction set in very rapidly. Eventually it transpired that Steinbach had advanced consumption, and was not, as Walter had been led to believe, a man in need of a warm climate after a bad bout of pneumonia.

And then, of course, despite the atrocious weather, the fevers and discomfort and the disappointments, there were the great moments — for instance the capture of a huge, brilliantly coloured specimen of the spine-tailed lizard (*Uromastix*) and a rare butterfly *Euchloë pechi*. Walter and Nissen caught the latter species while Hartert was searching for the last box of specimens. On several occasions they had climbed the mountains near El Kantara in search of it — all in vain.

And now here it was — under their very noses — a wonderful discovery.

The cost of these various collecting expeditions and enterprises was relatively enormous. One method of defraying part of the expenses was to sell duplicates to other interested parties. Thus the Tring Museum and Natural History Museum — whichever financed the expedition in question — would have first choice of the material available and the rest would be sold either to other institutions, to individuals, or to dealers. Sometimes, for instance, when Charles helped finance a British Museum expedition such as the Ruwenzori Expedition of 1906 the fleas collected would, as a matter of course, be offered to him. Considering that 404 mammals (23 new species) and 2470 birds (24 new species) were captured, the five fleas he received — of which four were new species! — indicated the difficulty of procuring these insects — even when the collectors were professional zoologists.

How could anyone find the time and energy to deal with this army of collectors, becoming personally involved — as Walter often did — with a tremendous load of detail, examining and sorting the material which poured in, and yet relentlessly churning out paper after paper, publishing *Novitates Zoologicae* regularly every quarter, and directing the Museum and the Library? This, commented Ernst Mayr, was truly remarkable. Walter's phenomenal memory was his greatest asset, but he also had other fortunate characteristics. When it came to selecting his staff or his work force he had a magical touch. They all proved superlative at their particular jobs and he had the power both to delegate and enthuse. He was, above all, a great *animateur* and for some rather mysterious reason his employees worked three or four times as industriously as ordinary mortals; they felt that what they were doing was important and interesting and enjoyable. Walter never left anything for tomorrow — except his accounts.* This not infrequently led to tremendous confusion and a feeling that everything was getting out of control.

Some of his telegrams to his curators were a hotchpotch of varied information and instructions. Thus from the pier at Bournemouth he wired to Hartert:

> *Please tell Sharpe Keulemans can make sketches of the 3 rails and send Dresser the Madagascar rollers I took all those new finches out and have got them here Huge beaks and heads like nestor parrots Write to Palmer to collect carefully on Oahu and get all he can and on Nühau also I saw Castang* [sic] *and settled about pheasants*

He frequently became involved in minutae. Some idea of the detail — rather puzzling for the ordinary reader — can be gauged from the story of the White Rhinoceros, which included yet another acrimonious exchange of letters with Alfred Newton.

In 1893 Walter brought off a great coup by procuring the first entire specimen of the White or Square-mouthed Rhinoceros (*Rhinoceros simus*) and he included these glad tidings in a letter to Alfred Newton. He never dreamed that behind his back Harmer (the Director of the University Museum, later of the Natural History Museum) had written at Newton's instigation to Walter's own collector Mr. Coryndon and asked for a White Rhino for Cambridge! Actually Coryndon had already left on the expedition and was out of reach of mail by the time the letter arrived at Salisbury. Newton, on hearing that, contrary to instructions from Tring, two bulls had been shot by the hunters, wrote and suggested that one might be given to Cambridge. "I have been trying for more than 25 years to get one," and added ingenuously, "I think we have been for some time in communication with Mr. Coryndon, but Harmer who is now superintendent of the Museum carried out the correspondence and I am not acquainted with the details". Walter wrote and explained that the British Museum had accepted one entire animal, but he himself would be satisfied with only a skin if Newton would like to buy the skeleton for £120. The price was reduced because "unfortunately it appears to have 3 or 4 caudal vertebrae missing and the large boss on the thigh bone which fits into the socket of the ilium was smashed by a bullet.... However the bones can be replaced in

---

* Walter's juggling with his finances deserves a book to itself for he combined wild and insouciant spending in some directions, with extreme care and attention in others. He spent almost nothing on himself, and was frugal in all ways, but his livestock was pampered and indulged. While he was in Algeria, Charles was trying to deal with the contents of the linen baskets. There is one telegram from Walter — a *cri de coeur* — which has survived: "Tell my brother not to sell snake".

plaster." Newton passed the hat round, collected £162.1.6 and accepted Walter's offer "providing that it had no other defects than those you mention". Newton added he would greatly prefer having the maceration done by his people. "I need hardly point out to you the importance of looking after the bones of the feet, and especially the terminal phalanges of the toes which are too apt to be left in the *sabots* unless care be taken."

In reply Newton received a depressing letter from Walter, written from the Bank:[15]

> *New Court*
> *Dec 13th 1893*
>
> Dear Professor Newton,
>
> *I warmly regret being unable to send you the skeleton unmacerated for a reason which affects me as much as you. F.C. Selous & Coryndon are the two scouts of the company and as you know went to the front as soon as war was declared. The consequence was a man named Maclaurin brought the 2 Rh. Simus home & packed all the bones indescriminately in one case & the two skulls in a 2nd as the British Museum insist on mounting their own & I could only pick out the correct number of bones to each skeleton & not fit them till macerated they have put them in soak & I, till they are fitted, have nothing more to say.*
>
> *I used the term* boss *and not* condyle *(Newton had made a sarcastic remark about this term) as I had when I wrote not seen this injured bone & so merely quoted Coryndon's expression. The toe bones I can assure you will be carefully dealt with, for I have seen they are all there & have given my taxidermists orders to carefully take them out.*
>
> *Yours sincerely*
> *WALTER ROTHSCHILD*

Newton then demanded to know where the specimens were located, for as he said "If only that is to assure ourselves in the allotment of bones the Brit. Mus. may but take the opportunity of giving themselves the benefit of any doubt should a particular bone be missing from their skeleton." He certainly did not trust the B.M.! He suggested that Harmer should visit Tring forthwith having apparently forgotten that maceration was taking place elsewhere.

Walter explained that since one beast was larger than the other there was no need to worry about the "owner" of any particular bones, but it would be 6-8 weeks before they could be on view. The suspicious Newton was not satisfied and continued to fret and pester, demanding to know where maceration was being carried out and saying even if it was only *the outside of the macerating tanks*, he wanted Harmer to see them.

Walter patiently replied, in the most conciliatory manner, again on New Court notepaper:

*Dec 18th 1893*

*Dear Professor Newton,*

*As you specially ask me where the* White Rhinoceros *skeletons are I must tell you they are being macerated by Alfred Brazenon 39 Lewes Road Brighton but when the bones are clean & dry you or Harmer shall see both skeletons before the British Museum take theirs. Of course if there should be any defects not known to me you will then see everything yourself & form your own judgement.*

*Yours sincerely*
*WALTER ROTHSCHILD*

As we have seen, Walter did not bar natural history from his working day at the Bank, and quietly carried on a considerable flow of Museum business from New Court.

This letter apparently satisfied Newton, since the correspondence seems to have ceased, but Walter had to wait a year for his cheque for £120. Nevertheless, in June 1895 the two Museums were still wrangling over the White Rhino's tailbones. Walter decided to call in Günther to play Solomon, and to supervise the mounting of the two bulls side by side at Gerrards, the taxidermists. He left Hartert and Harmer to deal with the correspondence. He had had just about enough of both Newton and the tailbones of *Rhinoceros simus*.

The brief sketches of the five of Walter's collectors, Harris, Wollaston, Doherty, N.C. Rothschild and Cunningham which follow give some idea of the varied and unusual individuals who engaged in this highly specialized type of activity. Although so different in background, life-style and character, all five shared a profound, even emotional interest in the marvels of the animal kingdom — including those men who, like Harris, depended for their livelihood on the specimens they caught and slaughtered. All five of them contributed — in a greater or lesser degree — to the unique character of My Museum.

# CHAPTER 19

# Walter's Collectors

## A.F.R. Wollaston and N.C. Rothschild

WOLLASTON — whose friendship with Charles dated from their undergraduate days in Cambridge — was an explorer and naturalist who studied medicine rather as a stepping stone to expeditions and not for its own sake, although he told Charles he was "really keen". He was a man who loved wild country and nature and he also craved harsh truths, and insisted on facing hardship and deprivation on his journeys of exploration, as if he drew consolation and a sense of personal achievement from hunger, fevers, fatigue or acute discomfort and danger. Wollaston never learned that his fellow men were not so keen on facing up to hard facts or home truths, with the result that he had the reputation of being delightful, but quarrelsome. Charles affectionately nicknamed him "Bear"[1] on account of his prickly exterior. Wollaston was a good correspondent and wrote an enormous number of intimate letters to Charles, but the two men agreed that they could only be frank on paper if they could be sure that their letters would be destroyed after their deaths. Rozsika carried out this melancholy task most unwillingly in 1923 and Wollaston himself burnt Charles's letters at the same time.

Their first serious collecting expedition outside Europe was a joint one to the Sudan in 1901. The animals, all except the fleas which Charles kept for himself, were destined for Tring. One of the principal objectives was to obtain wild donkeys for Walter's museum. Wollaston wrote home to Francis Gayner (who had accompanied Charles on his previous collecting trip to North Africa):

> *My journey out was a great success.... at Port Said it was simply magnificent; what with flamingoes and pelicans and kingfishers — not to mention the sun and the desert — I nearly went off my head with delight.... I am sure we have got a very good collection of birds of the district even if none of them are new, and after all that doesn't matter in the least. Of mammals we have a very good lot on the whole, considering the difficulties of trapping. Altogether this has been a very great success and it has certainly been far and away the very best time I ever had in my life — I never expect to have another half so jolly.*

Charles was equally enthusiastic in his letters to Birrell:

*Wollaston and I had a grand time up at Shendi some 100 m north of Khartoum. We camped out here on the Nile bank and collected the districts round about.... will only say I never had such a lovely time in my life. We were all extremely fit in fact in camp; none of us were ill at all. We bagged 600 mammals and birds, 500 fleas and a fair lot of Coleoptera and Lepidoptera.*

*The faalin Arabs are weird folk and candidly I do not like them. They SEW their unmarried women folk up to ensure their keeping virtuous. They also preserve the penis of the crocodile in honey and eat it as an aphrodisiac.*

Furthermore the depression which had been threatening Charles was now dissipated. Both men were delighted with the wealth and variety of wild life. They kept six gazelles, two adults and four fawns, a young hyaena, two baby foxes the size of kittens — very soft and playful — and several hedgehogs and a large monitor, in the camp.

There is no doubt that this was the most notable as well as the most enjoyable of Charles's expeditions, for at Shendi he discovered the plague-carrying flea — at that time a new species, which he later named *Xenopsylla cheopis*.[2] He intuitively knew this was a great find. He wrote to Birrell: "One of them I fancy is a good thing.... and five of them belong to a group which in my opinion will become famous as to this group probably belong the plague carriers in India". The good collector not only has a magic touch. but also a sort of second sight, which, suggested E.B. Ford, combined to make "a work of art rather than a science". It is very unusual for one and the same person to be lucky enough to collect, as well as describe, an insect of prime medical importance. In the case of Charles, his experience was backed up by his good memory (not quite, but almost, as good as Walter's, and more precise) which provided massive background information. Wollaston, writing home, remarked that when he found a nest of the Desert Bullfinch under a stone Charles at once told him that this bird's nest had never been found before. "Bear" was the explorer, crosser of deserts and rivers, trail blazer through virgin forests, climber of unscaled peaks, almost masochistically attracted by hardship and danger, with a profound interest in nature, especially birds, but he was no scientist. Charles loved animals and plants and wild country, but, although a devotee of the simple life, he was far from tough and suffered more than usual from the effects of rough seas and fevers. He lacked the qualities of the explorer which Wollaston possessed to such a high degree. But he was a scientist, a master of detail, meticulous as well as keen.

Charles, in a letter to Birrell describing the projected trip, said: "If we do not quarrel...." It was well known to all his friends that Wollaston awoke like the proverbial bear with a sore head, and it was said that the good natured Charles was the only one of his close friends with whom he had never quarrelled before breakfast. Charles had once gone so far as to tell Wollaston that he fancied he could, if he so chose, be rather unpleasant.... As one might have predicted, the irascible Bear lost his temper early one morning and walked out "not to return" as he announced. The temperature was 110°F in the shade, so Charles decided to write up his notes, on the assumption that conditions did not favour a very long tramp. While he was waiting in his tent the flap was pushed aside, and to his delight a pale sandy coloured, almost white, fox poked his head inside and gave him a long quizzical stare. A little later Wollaston returned with an impish grin.

Neither of the men was an experienced hunter, for this was Wollaston's first trip to the desert and Charles's second, and they had considerable difficulty in bagging Walter's wild asses. Wollaston kept a diary:

*Wady Halfa: We rose early and rode out about 16 miles into the desert to-day to shoot gazelles, but saw none. Tracked a big lizard for about 2 miles through the sand and eventually ran it to earth in a hole at the foot of a small tree; immense Arabic excitement when we dug him out.... Slept out under the stars.*

*Left Shireik on thirty-three camels and rode to the Wady abu Sellam where we expected to find wild donkeys but there was no sign of life; it was a most God-forsaken spot.... On to Nikheila, and while waiting for our baggage camel to turn up we had to take shelter from a strong wind by a palm tree. Made a glorious bonfire of the fallen palm leaves, the warmth of which was very comforting. Our camp was on the edge of the Atbara battlefield of April 8, 1898. Beyond the Dervish camp is the river, with a beautiful fringe of palm trees and palm scrub and yellow flowering acacias full of birds.... I rode about four hours into the desert with two natives and slept out at a spot where I hoped to shoot wild asses in the morning, but no sign of them. Later came across three but couldn't get a decent shot. Went back to camp by moonlight.*

*February 3. Hunted for wild asses but could not get near them. At last Charles gave his rifle to a native who was with us and he stalked one very skilfully, killing it at about 100 yards: it was a very old male of great size. Henley and I went out in the evening to see the beast. It was a fine moonlight night and it amused me to think we were walking 4 miles just to see a dead donkey by moonlight. The shikari who had shot the animal was in tremendous high spirits at the prospect of good backsheesh, and broke into a wild chanting of Sudanese songs....*

*February 7th. Crossed the river on camels and went due west for a few hours and then saw a small troop of donkeys. In the middle of stalking them they were frightened off by some gazelles, but we saw them again later on some ground as bare as the palm of your hand. By that time it was sunset so we sought out a bush and dined sumptuously by the light of the stars off sardines, bread, and dates. Had a discourse about stars with the two natives, and greatly astonished them by being able to point out the way to Nikheila, Atbara, Kassala and other places.*

*A native shikari has just shot two wild donkeys, so now our mission to Atbara is finished, as we only had leave from the Sirdar to kill two or three donkeys. Our skinner went out in the afternoon with an acetylene lamp, skinned all night, and brought back the skins in the morning.*

Charles and Wollaston discovered that in the great heat of the desert birds must be skinned immediately they were shot. This was a limiting factor they had not anticipated. As we have seen, they finished up with a collection of 600 mammals and birds, 500 fleas and a fair number of beetles and butterflies. Among the Lepidoptera taken was a new genus and several new species which Charles named in honour of Gayner, Wollaston and their assistant, Waters. But it was the discovery of the plague flea which immortalised the expedition.

Apart from the plague carrier, Charles Rothschild described approximately 500 new species and subspecies of fleas, and wrote alone (or jointly with Jordan) 150 papers on the group.[3] Three genera and seven species were named in his honour. When Fabian Hirst demonstrated the relationship between plague and the geographical distribution of rat fleas, he wrote, "The discovery is but further testimony to the essential unity of science and its bearings on the welfare of the human race, for it is the natural outcome of the purely zoological researches of Rothschild and Jordan on the systematics of the Siphonaptera."[4] It is worth noting that, although the plague flea is a parasite *par excellence* of rats, N.C.R., in the Sudan, obtained it first on a Spiny Mouse *(Acomys)*. Although Charles first became interested in fleas at the age of 12 when he acquired a slide of the Helmet Flea from W. Farren, the taxidermist, we find him four years earlier thanking his mother for the mouse trap she sent him to Bentley Priory and announcing his first capture.

Wollaston, in due course, served as the doctor to several major explorations — the Mountains of the Moon, the Congo, the Himalayas and New Guinea — where a peak was named Mount Wollaston in his honour. Although his greatest love was for the birds, he was an excellent botanist and brought back numbers of new species of plants and wonderful collections of seeds. It was taken for granted that naturalists at the turn of the century knew some botany, in fact a man who did not know his plants was held in contempt. For instance Walter[5] not infrequently sent Hartert lists of interesting plants seen, as well as the birds and insects. Charles was also an excellent botanist, specialising in orchids and the genus Iris. He brought home corms and seeds of the blue water lily from Lake Victoria, which excited Walter as much as the Nubian ass. A special greenhouse was provided for them at Ashton, where for some obscure reason their flowers grew to a considerably larger size than in their native homeland. In order to assist Dykes in the production of his beautiful paintings for the illustration of his monograph ("The Genus Iris"),[6] Charles collected and grew at Ashton and Tring almost all the then known species, so that the drawings could be made from life. Claridge Druce[7] helped him complete his English series by obtaining *Iris spuria* from Lincolnshire and *Iris foetidissima* var *citrina* from Dorset. Collecting together at Ashton, the two men found a new wild rose which Druce named *Rosa rothschildii* in Charles's honour.[8] His last collecting expedition for plants was in 1922 when he and Rozsika spent four weeks at Bettyhill in the north of Scotland (an occasion, Druce noted, when it was she who found a rare variety of *Rhinanthus minor*, the Yellow Rattle).

A. von Degen, a Hungarian botanist, another of Charles's collecting companions, described him as "one of the most amiable and honourable characters I have ever met.... with an extraordinarily thorough knowledge of Natural History. The sight of a rare plant or insect gave him almost child-like pleasure." All his life Charles was greatly excited by the beauty of moths and butterflies. His ambition as a child was to own an Oleander Hawk Moth — "such a beautiful creature", he wrote, and, convalescing in Switzerland he collected "gold spangled moths for each of the children. They are such lovely things. I believe they will be pleased with them". In von Degen's opinion the intense attraction which the mysteries of nature and the unknown exerted over Charles, together with his patience, thoroughness and determination, combined to make him into a crack collector. He was amazed how tenaciously the fellow searched for the larvae and food-plant of a rare Hungarian butterfly — the life-cycle of which was then unknown — year after year, night after night, at one particular spot in the woods, scouring the grass stems with a handlamp until at last he found the caterpillar in question. A strange way perhaps to spend your brief holidays — rather hard on Rozsika — but something which she understood and in which she shared his happiness. Von Degen remarked that Charles by his own energy and the perspicacious direction of a skilled assistant[9] whom he had trained, amassed Hungarian material[10] which surpassed in abundance and excellence all other collections from that country. His zeal knew no bounds, said von Degen, and he once pulled the communication cord on the Dabas railway train because he had seen a rare butterfly through the window. He ran down the line and caught it. It is worth noting here that Charles promoted the establishment of a Nature Reserve at Puszta Peszer where the prohibition of collecting was strictly enforced. Von Degen named an interesting hybrid *Iris rothschildii* in his honour.[11]

During the war, Wollaston served in the Navy, but on one occasion he visited his friend who was convalescing in Switzerland with the faithful Jordan in attendance. At that time he had apparently little understanding of Charles's nervous condition, and told Rozsika bluntly that her good man was malingering, but later he changed his mind and tentatively diagnosed Charles's illness as *Dementia praecox* rather than encephalitis. Rozsika privately marvelled at this curious streak of compulsive truthfulness in Bear's nature — for he believed that an honest opinion was always a sincere act of friendship.

When Charles eventually recovered sufficiently to come home, he gave Wollaston a "legacy" of £25,000. At first Wollaston flatly refused to accept the gift, but changed his mind when Charles explained that he had left a similar sum to him in his Will but felt it was foolish for Wollaston to have to wait, probably for many years, for his death. Knowing that Charles was the victim of suicidal depressions, the Bear felt this was a good omen. Unfortunately, as we know, it proved not to be the case.

Wollaston married relatively late in life — at the age of 47. He wrote to Rozsika saying he owed the possibility of marriage to Charles's generosity, and he could never express his gratitude adequately — for he was unimaginably happy.

On June 2nd 1930 Rozsika received a letter from Wollaston which caused her acute apprehension and distress. It concerned a prospective visit to Ashton with his wife — for he was then living at King's College, Cambridge, only 29 miles distant. The letter ended with this sentence: "Yesterday while I was strolling across Newmarket heath I was once again joined by Charles, who walked across the downs with me". There was no explanation — it was just a bald statement.

"Wollaston has gone off his head," said Rozsika in considerable agitation, "You know — he has always been an honest — at times almost a militant — atheist. He never wavered in his total and harsh disbelief in anything which savoured of religion or spiritualism or life after death. But look at this." She passed the letter to me and I read and reread the sentence. It was impossible to misunderstand it. "Charles" was written with such complete clarity. But Rozsika was never able to ask Bear for an explanation — whether — as it seemed from the words "once again" — the episode had occurred on previous occasions, or whether Charles's ghost walked silently, or whether he spoke to him. On the very afternoon on which she received the letter, Wollaston was assassinated by a mad undergraduate in his study at King's.

Wollaston's son recalled that on one occasion his father (leading a New Guinea expedition) described how a mysterious man walked ahead of him, steadily and helpfully during a particularly arduous and dangerous return trip through the jungle. Wollaston never caught up with this man who always kept some distance ahead, but he retained a very clear-cut mental picture of him. On his return to England, when he was buying a coat in a clothes shop in London, he glimpsed in a long mirror the familiar, unmistakable, figure of his unknown friend. Very startled, he turned round quickly, and realized it was his own reflection seen from behind. It would seem, therefore, that the strange occurrence on Newmarket heath was not the first time he had experienced a psychic phenomenon. The frightening part of the story was the matter-of-fact description — as if Wollaston was not at all surprised by the presence of Charles. He seemed to take it for granted.

Certainly collectors are unusual people.

# CHAPTER 20

# Walter's Collectors

## William Doherty

DOHERTY, whom Walter admired greatly, met Hartert on the road to Perak in the Malay Peninsula in 1888, and they remained friends ever afterwards. The relevant entry in the diary which he kept for his sister reads as follows:

> *Unsuccessful collecting in Borneo, expedition to the Riam Kanan. Loss of my papers on Perak butterflies, etc. Leave Borneo in bad health and low spirits. Return to Singapore and Perak, where I meet Hartert. Collect together in various parts of Perak, and then go by Calcutta to Assam. Collect in Mishmi and Khamti country above Sadiya. The Sadiya annual fair. Coal-mines of Margherita, visit to the naked Nagas. Great collection of moths made with our invented baits. Concerts of howling monkeys. Needham and his government of the wild tribes. Lancashire miners at Margherita. Canadian oil-well prospectors. Farewell to Hartert. I leave for Calcutta in December. Write a paper on Assam butterflies for the Asiatic Society.*

Dohery was an indefatigable collector — for instance on his fourth trip to Perak he took 30,000 beetles — and these were the days before ultraviolet light traps. The material he sent to Tring was immensely valuable as a contribution to systematic zoology. Like Walter, he was too delicate as a child to go to school, and it was during this period he developed a profound interest in living animals. Surely this is no coincidence. He, too, was spared the formal education which seems to knock the wonder and love of nature out of the pupil. Alfred Newton was educated at home, and Hartert, too, escaped early from conventional lessons. Howard Hinton, one of the finest entomologists of our time, could not read before he was 14, and it was not until he went to University that he learned that the snakes he harboured so lovingly in his bedroom were deadly.

Doherty, who was an American by birth, started out as a traveller, interested in new countries and the languages, history, religions, manners and customs of the nations of the East — from Asia Minor to the Sunda Islands. He was widely read himself, and possessed an impressive knowledge of English, French and German literature. He only took up collecting seriously at the age of 25, when by chance he met a German Lepidopterist in Persia who instructed him in the art. Even then he did not sell specimens for about three years, but circumstances eventually compelled him to switch from a hobby engendering pure pleasure and interest to commercial collecting. In 1893 he visited Tring, and Walter persuaded him to add birds to his butterfly catches — something he had sworn he would never do, owing to the frustrations and irritations connected with guns and customs regulations. Furthermore capturing and papering and packing butterflies was child's play compared with shooting, skinning, labelling and transporting birds. However Walter's enthusiasm was infectious, and he fired Doherty's imagination. Indeed, he eventually made some wonderful ornithological discoveries on the various remote islands he explored, although he wrote that he "became horribly discouraged about birds" especially over the time lost — sometimes two or three days — arguing and pleading with the Customs about his guns and ammunition. From the Settima Escarpment in East Africa Doherty sent Tring over 3,000 skins (all prepared by an Indian skinner, for he never acquired that particular skill himself) and an immense number of Lepidoptera. In fact Tring obtained its most important East African and Eastern Archipelago material from him, and certainly the first representative collection from a single area in Africa. Doherty also supplied insects and land shells to many museums and private individuals in America, Europe and India, but all the birds were reserved for Walter. Furthermore he himself wrote several papers on the butterflies he collected, especially the Blues which were his great favourites.

One entry from the diary gives a vivid picture of his restless journeying.

*Return to Calcutta. Write papers on Sumba, Sumbawa, and Engano, on new* Lycaenidae *and on the anatomy of the* Danaidae. *Leave for Java with four men. Eruption of Sméru (my Lepchas lose their reputation for veracity, because they tell of flying fish, sea on fire, smoking mountains). Trip to Pulo Laut, bad weather there. Then to Banjermasin, Borneo. Drag canoes up the Martapura river and shoot the rapids down it. Serious illness of our whole party. Miss the Surabaya steamer, which leaves ten days before its time. Moths at Sungai Tabok. Chedi has cholera. Departure for Surabaya, Java. Tungkyitbo in hospital there. His death at sea. Reach Macassar, Celebes, and all recover. Great catch of butterflies in Maru and Bugis country. Robbed at Petunuang Asné. Trip to Cayi. Scenery at Bantunurang. Leave for Sumbawa. Ride on horseback over the Donggo mountains. New butterflies and moths at high elevations. Short visit to Sumba, etc. Reach island of Alor; curious fair at Moru. Attacked and burnt out at Kalabahi by the Leindola savages. Escape and go to island of Pura, between Alor and Pantar, a volcano 4000 feet high. Drinking the most horrible water there, stay on mountain top. Voyage to Pnadai and down the wind to Adonara — almost surrounded by fighting. Reach Flores. The Catholic monks and nuns at Larentuka. Pambu has D.T. Cross over to Timor. Many new insects from the mountains. Leave Kupang for Banda and Amboyna, where I spent Christmas. Arrival at Buru.*

Once again one marvels at these explorers' avid pursuit of hardship and danger, and their total disregard of discomfort and disease. Collecting birds for Walter on Jobi Island (Papua) — an area of about 1000 square miles — Doherty found the people were unusually handsome and not overtly unpleasant compared with many Papuans. But Ansus, where he was stationed, was notorious for its continual massacre of foreigners. Here he had the great misfortune to lose his right hand man Pambu ( a Lepcha) who was murdered by the savages. Two of his other assistants were shot at, but escaped. All the Ansus men protested their innocence of the crime and, to prove it, volunteered to undergo the boiling water ordeal! Doherty himself, however, was not attacked, and collected alone and unarmed in the jungle. The climate was deadly — it rained all the time Doherty remained on Ansus. On the journey to Takar the weather, if anything, was worse. The trip took three weeks by sea and their vessel rolled and pitched more viciously than any he had ever sailed in. He enjoyed it, but his men were seasick every day for the entire voyage, and held the ship in perfect horror. After collecting hard for three weeks on Takar, they lost a whole box of bird skins in the surf and another got so wet the contents were seriously damaged. "But for the wonderful skill of the Masi Masi as canoe men," remarked Doherty, "We would have remained for ever on Takar, as prisoners of the surf...." They had only lost one third of their birds, which under the circumstances he thought was not too bad.... On the return journey his men were continuously seasick for another three weeks. He described the trip as "trying".

Doherty found the Escarpment at Settima in East Africa, which he reached in October 1900, a grand place. It proved to be his last, but most memorable collecting expedition. The forest trees were enormous — junipers 10 feet thick and 120 feet high. But here again half of his men were permanently down with "jiggers" (the sand flea, *Tunga penetrans*) and their feet were awful to look at. Doherty himself was infested, but not so seriously. The cold seemed intense.... Twice, in the first few days, they had encounters with lions in broad daylight, a rogue elephant haunted their best collecting ground, a herd of rhinoceros were omnipresent, and twice charged the men, and a leopard stole into their boma every night, in fact "the usual adventures". They were on peaceful terms with the natives, but Doherty was afraid of the Masai. But to move to Mau which was considerably safer but devoid of edible vegetables, meant risking scurvy, and in the gloomy, pathless woods at Mau one's chance of recovery was small. And the weather was AWFUL.

Most of Walter's collectors of the Doherty type — if they survived the occupational health hazards of their trade — seemed to lose much of their *joie de vivre* as time went by. Bouts of fever and other tropical diseases undermined

179

the constitution, and they lost their sense of well being. Furthermore they were rarely able to save any money for an honourable retirement, and the life they had led did not fit them for so-called civilisation. Although Doherty was only 44 when he died of dysentery, he had become intensely gloomy, nervous and fatalistic. Hartert tried to persuade him not to embark on his last voyage since he had not properly recovered his health, but he said that all his plans were made and he must go.

Walter was appalled by his death. He felt the loss deeply and he was especially distressed to think that a man who had braved the Moluccas and New Guinea for twenty years could die within eight months of arrival in East Africa. He described Doherty as "unquestionably the best collector for the last fifty years". Posterity has endorsed this assessment.

# CHAPTER 21

# Walter's Collectors

*The King of Bulgaria is coming on Friday....*

DOCTOR CYRIL CUNNINGHAM — an "ace" collector, greatly appreciated by Walter — presented any Palaearctic and Nearctic bird captures which came his way to Tring, and received specimens of tropical material in exchange. He loved wild and difficult country, especially steamy jungle terrain, and thus frequently came across rarities and made many interesting and original observations. With an eye like a hawk and skilled and accurate shooting, he could pick out and kill a bird on the wing, glimpsed for an instant only, among a tangle of bushes, or skulking in a lofty tree. He prepared admirable skins with true artistry, every feather in its right place. His meticulous and patient scientific approach produced notes and labelling which were clear, accurate, detailed and legible. But Cunningham also possessed boundless drive and imagination. Thus in addition to the birds themselves he collected eggs and nests, faecal pellets and blood samples, but he never wrote up either his collections or his observations. In this he resembled Lord Hillingdon, a fine entomologist who never put pen to paper. The curating of every specimen — weight and wing measurements also provided — was impeccable, and his preserved material proved both aesthetically satisfying and scientifically valuable. It radiated a special charm of its own. Tragically his collections were lost during World War II.

Cunningham and Walter were both true bird lovers, and identified themselves with their collections in a peculiarly personal manner. Walter, as we have seen, always spoke of My Museum, with a special emphasis on the 'My'. This is characteristic of many collectors who are known for their possessiveness and strange complexes, but in Cyril Cunningham's case they reached staggering proportions. Fred Young was at this period the caretaker and taxidermist, a man of considerable talent and integrity, with a tremendous sense of his own responsibilities. He walked with a deliberately measured tread, and his long, neatly waxed moustaches were a never-ending source of giggling speculation for Walter's irreverent nephews. One morning he received an unexpected call from Cyril Cunningham, who explained that he was lunching at Tring Park with Lord Rothschild and his mother, and he would come to the museum afterwards to compare some of the tropical and South

181

Sea types.[1] He named a few series including some Birds of Paradise which he would like to examine, and checked with Young that Dr. Hartert was, unfortunately, away. Young mentioned that there was no-one else in the Department of Ornithology at the Museum, but he would of course pull up the blinds in the bird room and get out and open the drawers which were required.

Cunningham arrived punctually at 2.30 and Young showed him the relevant specimens, and left him at work. It was a rule at the Museum never to serve tea or refreshments of any sort, either to guests or "regulars", but at five o'clock Young went along to see if the visitor required further assistance. He found the bird room deserted. The lids had been replaced and the drawers had been put back in the cabinets. Young, who was the most conscientious and reliable man in the world, was distinctly put out. What cheek! He had specifically told the man not to put the birds away. That was *his* job. He pulled out one of the drawers to see if the skins had been replaced satisfactorily and noticed immediately that the type specimen, with its bright red label was missing. Probably put into the wrong drawer.... He began to look for it. He opened drawer after drawer. It was just not there. Suddenly, in a flash, he realized that *ALL* the tropical types had gone. For a moment he was stunned. Then he went through every drawer again to make sure. Cunningham must have *borrowed* them without permission. This was inexcusable with Hartert away, and placed him, Young, in an intolerably awkward position. He strode out of the Museum and into Park Road with a faint hope of seeing the car. He looked up and down the street, but it was completely deserted. As he turned to go back into the Museum something lying in the gutter caught his attention. He picked it up. It was the type label off one of the missing birds. Scarcely believing the evidence of his own eyes, he realized that there, in the gutter, lay all the type labels — they had been ripped off the birds he had so carefully laid out that morning. Young swept them up, and for once in his life forgot his measured tread and rushed to the house telephone. It was a very long time — or so it seemed to him — before anyone answered. "Speak to his Lordship?" asked a puzzled voice at the other end of the wire, "But Mr. Young — the family is in London."

"In London!" echoed Young, aghast, "But didn't Doctor Cunningham have lunch with his Lordship?"

"How could he?" replied the voice, "They are all in London until Friday. There was no lunch served here to-day."

When Walter finally heard the bad news he was completely at a loss, and sank forthwith into gloom and despondency. It was the sort of improbable, crazy situation one couldn't cope with.... He decided his types had gone forever — they would never be returned. Cunningham, of course, would deny all knowledge of the matter, and might even accuse the faithful Young. Reason would be on his side, for no sane man could have perpetrated such a crazy act. The fact that the labels had been found in the gutter would seem to *prove* it couldn't have been Cunningham. When events got too difficult for Walter, he felt that there was only one thing left to do — ask his sister-in-law for advice. Trembling with agitation he drove round to Palace Green. When he was shown in by the maid, he found Rozsika, as usual, writing at her desk, with a cigarette clinging to the very edge of her lower lip — an inch of ash defying the laws of gravity — with one eye half closed against a wispy rising spiral of smoke. She listened attentively to Walter's story, and noted his trembling hands. "I think I'm done," he finished unhappily.

"Not necessarily," said Rozsika very sympathetically, but decisively, "You must write to Cyril Cunningham as follows: 'Dear Cunningham, As you are well aware, I am always glad to lend you any specimens of mine that you may wish to study. Unfortunately Hartert was away on Tuesday when you visited Tring, otherwise he would have told you that the King of Bulgaria is arriving at the weekend and has especially asked to see all the types. So regretfully I must ask you to return them immediately. Could I please have them before Friday?'"

The King of Bulgaria was a keen ornithologist, known affectionately to the bird men as Foxy Ferdinand. He rather liked Walter — they enjoyed the same sort of jokes. Bringing him into the picture was an astute psychological move on Rozsika's part for the good doctor Cunningham was a tremendous snob.

"But," objected Walter, almost in tears at the thought of Foxy's reaction if, indeed, the King had been due at the weekend. "What about the labels?"

"But no-one need know the labels were found," said Rozsika briskly, "You are lucky to have a man like Young — if he hadn't looked in those drawers immediately, the situation would be very different. But as it is, I think the letter is worth a try."

Walter sat down obediently at the desk which had once been Charles's, and wrote the letter. Like a child he checked the wording with his sister-in-law.

The posts were considerably faster in those days than they are now. On Friday morning a large anonymous cardboard box bearing a London postmark arrived at the Tring Museum. It contained no covering letter, nor a word of explanation. But, wrapped neatly in new tissue paper were the label-less types. All of them. What a wonderful woman Rozsika was!

A few years later a row blew up at the British Museum because a junior member of the staff at South Kensington had been caught appropriating museum specimens and secretly adding them to his own collection. No member of the staff was allowed to make a collection of the same group of animals upon which he was working, since it was thought the temptation and opportunity for theft was then too great. The curious psychological quirks of collectors had to be catered for. The man was thus guilty on two counts. Soon after this episode Cunningham met Walter at an evening party, and to the latter's acute embarrassment, immediately asked him what he, as a Trustee of the Museum, proposed to do about it.... Cunningham was a marvellous talker. He had a beautiful speaking voice, and a way of choosing the evocative *mot juste*. Without waiting for an answer, he launched into a devastating attack upon the luckless offender, and passed harsh judgement on the despicable nature of his crime. Stealing types.... The wretched little pip-squeak! Walter felt a prickly sensation at the base of his scalp and said to himself that if he had had a single hair on his perfectly bald head, it would now be standing bolt upright. His embarrassment had changed to a sort of clammy fear as Cunningham talked on gracefully and easily with his usual charm and candour, demanding the culprit's instant dismissal from the Museum. For a second Walter almost felt confused himself — had the Tring affair really happened, or was it a dream? — It was uncanny — not to say disconcerting — that the whole drama of the tropical bird types, the fictitious lunch at Tring Park with Emma and himself, which, of course, had never taken place, the missing labels, Walter's own letter which had been neither acknowledged nor queried — all, all had been completely erased from the Doctor's mind. Cunningham simply knew nothing about it.

The staff could never understand why, after a brief interval Doctor C. was allowed back into the students' department at Tring, following what Phyllis Thomas described rather quaintly as "a breach of confidence". But Walter understood, for privately he considered that different standards applied to collectors: were they not all just a trifle eccentric? And furthermore, he sensed that the types were now quite safe in the presence of Cyril Cunningham.

CHAPTER 22
# Hurrah! We are off....

A SERIES OF BOOKS could be written about Walter's collectors, and the brief synopsis in the previous chapter scarcely does the theme justice. The reason the Webster/Harris voyage to the Galapagos Islands has been selected for a more detailed description is because a portion of their notes and diaries has survived the British Museum's tidying up operations at Tring. The great bulk of the papers relating to other expeditions was destroyed between 1938 and 1972. Thus all Palmer's diaries were lost. But in any case the Galapagos expedition was an excellent example of the type of adventurous enterprise inaugurated and pursued by Walter, and it concerned among other things the giant tortoises, of which he was so inordinately fond.

We do not know how Walter first made contact with Frank Blake Webster — a professional collector — since all the copies of his letters to this likeable individual are lost. As we have seen, Walter's interest in Island fauna flared up while he was an undergraduate at Cambridge, and he immediately determined to send an expedition to this area of the Pacific. It was part of his grandiose scheme for the investigation of Island fauna of the world. Neither of his curators took any part in planning or organising these larger expeditions.

Walter, when he decided to ask Webster to equip and man a collecting trip to the Galapagos Islands, must have impressed upon him from the start that his funds were limited, for the question of costs seems to have obsessed everyone concerned. Webster, who apparently owned a one-man company which specialised in collecting Museum specimens and objects of natural historical interest, wrote on unheaded notepaper, in rather illegible longhand, from Hyde Park, Mass. He was obviously the most honest and conscientious of men, mad keen to prove Walter's confidence in him was not misplaced, desperately anxious to make a success of the trip, and, notwithstanding, to keep within his far too sanguine budget. He shared with Walter energy and drive, a gift for unorthodox organisation, and a sort of confident, crazy optimism.

The first letter we have from Webster is dated February 26th 1897. It reads: "I have since last writing read all your letters carefully and I can only repeat that I understand your wants, that all your proposals are satisfactory and your wishes shall be carried out. For our first taxidermist I have selected C.M.

185

Harris...." He then goes on to give a brief account of Harris's past collecting experience — a man who had gone on his first expedition at the age of 17 — and adds: "He makes as nice a skin as can be wished for (capacity say 50 skins per day size of thrush); is also expert in such work as mounting tarantulas and Horned Toads. He is ABLE BODIED and will practically take his chances on the results — which I considered my best plan. I feel highly satisfied." This brief description of Harris's capabilities must, in the light of this man's subsequent achievements, rank as the understatement of all time. It is extraordinary that three such ebullient personalities were somehow swept together within the compass of this tiny expedition.

The party sailed from New York at 10 o'clock on March 29th, aboard the Valencia, bound for Panama. It had taken Webster 25 days to prepare the expedition. The letter of instruction from Webster to Captain Robinson is a remarkable document.

As a "Last and Final Instruction" to the Captain, Webster wrote "should you or your successor receive any orders from Hon. Walter Rothschild.... you will consider it from us, and to be of more importance than any order handed to you from here."

He noted especially that "In the case of Albemarle, each section will be considered a locality", for Walter had sent him a sketch, in which he had divided the island into seven sections, anticipating that some sub-speciation had occurred in the northern and southern parts. In his letter to Walter "On the statement of the use and disposal of funds" Webster remarked that he had "told him [Harris] what you wrote about the Red Flamingoes and that I thought it was a warning not to run after birds because of the colour and let some obscure thing of more value slip through." This phrase makes one sigh for "all the letters you have sent", as Walter's discursive and lively style and his schoolboyish *tournure de phrase* remained most endearing.

Webster wrote that the "matter of alcohol [for preserving specimens, not for consumption by the crew] was the hardest thing I had to contend with. The price here is about $2.35 per gallon and I did not think it safe to take less than one barrel." He was fairly confident that the party would be able to find a vessel in Panama but in this he proved sadly over optimistic and lacking in judgement.

It should be emphasised that simultaneously with the Galapagos expedition Webster was attempting to satisfy various other requests for Walter. First of all the latter was financing a small expedition to San Blas and the Socorro or Benito Group Isles which was undertaken by Professor J. Hennrick from the Department of Biology in Denison University who wrote very long, detailed, illustrated letters covering the various sorts of skinning and preserving techniques and who, not unnaturally, felt that he should be paid extra for a sojourn on an uninhabited island. He eventually received $554 for 272 skins. Then Walter had asked for both an outsize sea otter and an outsize bull walrus which involved Webster and himself in a voluminous correspondence with various hunters and collectors. Thus Mr. Sheard, a dealer in raw hides and skins, wrote from Tacoma in 1897 "Such a sea otter as you describe is one in a thousand.... I have three schooners doing nothing but hunt sea otters and we have never seen such a one." Webster decided that Sheard was "making too much of the walrus" and passed the order on to someone else. In due course Walter, who was marvellously persistent in such matters, received at Tring two animals of exactly the dimensions upon which he had set his heart. Highly delighted, he went to London to see them prepared by Rowland Ward. Webster also sent Captain Nicholls out to Hudson Bay to collect gyrfalcons, moles and mice. The Captain was in addition provided with 24 vials for fleas and lice. Among the letters dealing with the Galapagos expedition, there are casual references to an albino porcupine just come to hand, a whistling swan, condors, a collection of outsized beetles and black-capped tern from other areas.

While the Webster expeditions and the search for individual species were in progress, Walter had his usual web of collectors working throughout this period, supplying him with material. Meek was after birds in the Louisiade Archipelago and at Cape York in Queensland, Dr. Doherty in the Sula Islands and New Guinea, Captain Giffard on the Gold Coast, Mr. Heinrich Kühn in the Timorlaut Islands, Mr. Everett in South Flores and Timor, two Japanese collectors in Guam and Saipan (Marianne Islands), Mr. John Waterstrade on Lirung (Talant Group) — to mention only a few of the bird collectors.

Meanwhile both Robinson and Harris were struggling to find a ship in Panama, while the rest of the party hunted birds and mammals and prepared skins. Robinson complained bitterly of "the rapacity of this Godforsaken country" and added that the ship-owners "seemed to think we were Mr. Rothschild himself".

Harris privately considered the Captain inept at the job. He wrote a longer letter with detailed accounts of the ships they had inspected and lists of the species of birds seen. In the afternoons he chartered a whale boat and shot a few of the thousands of sea birds crowding the Bay of Mexico, but found them in poor plumage. The great heat "took hold of us at first". He ended his letter: "Keep up good cheer. We will do everything to see her through." Harris's letters show without a shadow of doubt that he was not greatly — if at all — interested in the financial aspect of the trip, except to satisfy Webster. All he really cared about was the expedition and the collecting of rare and interesting animals and earning just sufficient to carry him on to the next voyage of discovery. He was anxious that Webster should not lose heart, and above all things, he must cajole their patron into sending the additional financial help they required on account of the delay in finding a suitable vessel. At this point, after cabling Webster for his approval, Robinson and Harris decided to travel back to California and give up the attempt at finding a ship in Panama.

Meanwhile another problem arose: Otis Bullock, the 21 year old assistant taxidermist and third in command, proved to be an alcoholic. In his letters to Walter, Webster declined to give details of this lad's misdemeanours while "drunk and disorderly," but after several warnings he had eventually been dismissed by Robinson and Harris and advised to make his way home. On the voyage from Colon he fell ill and, on arrival in New York, died of yellow fever. This created a considerable stir in the press and several totally unwarranted accusations were made against the expedition. Bullock's parents must have known about their son's weakness because no word of reproach was directed towards Webster, nor did they enquire about the details of the tragic affair. Instead they asked, pathetically for a keepsake — one of the birds their son had shot if one could be spared. In his diary Bullock had apparently made no mention of returning home before December, and his father felt very anxious about the rest of the party and asked for news. Webster picked out three birds shot by Otis Bullock and dispatched them forthwith. In a letter to Walter he said: "I trust you approve. They were not rare. Bullock had disgraced himself which was very painful to me, but death seems to have closed the matter."

But a far greater disaster was to befall the expedition. Harris wrote a letter to Webster describing briefly what had happened, but entrusted the delivery to Bullock. For obvious reasons this communication was greatly delayed and Webster heard nothing until Harris arrived in San Francisco about a month later.

*The poor Old Captain [Robinson] is laid away. It has broken us all up. We feel terribly for his wife and children & all of you who knew him so well. He passed away about 1 o'clock, at 3.30 he was in the Vault. In Yellow Fever cases they have to be burried at once. We did all that we could for him poor man. Mr. Bullock will bring you this. He has done wrong but dont blame him. He has no control of himself.*

*This has saddened us all very much. The Yellow Fever is beginning to be bad here, it was brought from Guayaquil, all ports on each side are infested badly. Got your Cable saying Go Best North. Nelson, Cornell and I will start for some Port North as soon as possible. Possibly California, unless we get different instructions. Can fit out best in California can you get money anywhere? So far it has been bad luck, but keep stiff upper lip & I will carry this through If I dont Kick. Meantime Bullock will tell you all.*

*I am too much worked up to write more.*         Yours truly

                                                          *C.M. HARRIS*

*Captains body can be brought to the States after 2 Years. I am afraid that the funds will be reduced to £174 by the time that I get the funeral Expenses and the Doctors Bills settled.*

Webster sent a copy of this letter to Walter and added a footnote "By stiff upper lip *Expression means* "Do not get discouraged."

Subsequently Harris sent a more detailed description of the Captain's last days.

*Mr. Robinson was taken sick April 7th in the afternoon, but was not very sick (apparently) for 48 hours and then began to sicken and weaken. He was attended first by Dr. Serpa and then by Dr. Cooker — Dr. C. being called the best physician in Panama everything which could be done for him was done. Mr. Cornell stayed with him most all the time and waited on him and the last night both of us being worn out we had a nurse. Mr. Robinson seemed to like to have me with him and if I was away would often send for me. He died about one o'clock in the presence of Mr. and Mrs. French of the Hotel, Dr. Cooker, the Party and several others. Mr. Cornell held his hand and I closed his eyes for him. I dont think he suffered greatly during his sickness. He gave me his watch for his eldest boy (aged 8) to have. He was buried at 3 the same afternoon. He had a first class hearse, 4 nation pall bearers. The Party with Dr. Bradley, W.S. Custom Inspector followed the hearse to the Native Cemetery where the body was put in Vault 201. Dr. Bradley reading the Church of E. services; us, the Party, and pall bearers stood about with uncovered heads. The body was sealed up and can be removed after 2 years.*

Even in this dire predicament Harris was worrying about the doctors' bills — although, secretly, he suspected that he himself had already contracted the disease since he was now suffering from malaise and fever. He mentioned that "with the Consul's help he got Dr. Serpa to accept $35 instead of $70 which he charged me at first. This makes $51.70 Columbian money that I have saved on Bills presented for Captains sickness and funeral simply by kicking and fighting...." He also complained bitterly that one had to pay men double for handling Yellow Fever corpses and that they took occasion to rob right and left. "I suppose you will blame me," he went on, "for taking cabin passage [for Cornell and himself to San Francisco] but the steerage on this coast is a living Hell."

Webster commented apologetically in his letter to Walter that he thought "they had done the best, as in case of sickness they would have better attention." Harris went on to urge Webster not to let Walter's enthusiasm wane: "DONT LET THIS FALL THROUGH if you can help it. $1000 more I think will carry the expedition from San Francisco and Return in Good shape.... I will carry this through and make it pay in the end if the money is finished."

That night (April 16) the three remaining members of the party sailed for

189

San Francisco. Both Harris and Cornell developed sickness on board, and the latter died at sea on May 2 of Yellow Fever. Harris recovered. To the latter's fury Nelson decided to go home, but he stubbornly refused to give him enough money for his fare. Nelson had been "pretty blue" after the Captain's death but Harris thought he had pulled himself together and got over his fear of Yellow Jack. He advised Webster to persuade the lad's father to write and tell him to *brace up* and be a *man*. Nelson, however, had secretly made up his mind, although he successfully pulled the wool over Harris's eyes until, at his urgent request, his parents sent him enough money to buy himself a ticket. Harris wrote to Webster:

> *I now cant prevent Nelson from going I am* very sorry *as I liked him well and he would have given good service.... he is perfectly well and able to continue and fulfil his contract and nothing but cowardice is deterring him.... I am disappointed and disgusted and I have talked very plainly and forcibly to him and I hope in after life it may do him good. He* begged *me to let him go home with Bullock but I* flatly refused *and the next day he came to me with apologies for showing his fears. If I was his father I would not send him money to get home but send to parties who would do his bidding a kettle of tar or a bag of feathers with instructions to apply them liberally; such* rank cowardice *is beyond my conception.*

This was true. Harris, apparently, had no understanding whatsoever of the fears to which ordinary human flesh is heir. Webster was also very upset because among other things he felt this sad story reflected on his judgement in team selection, but he wrote his usual rather soothing, if not strictly accurate, account to Walter. "Harris did not attempt to force him. There is no room for cowards on this trip. The only excuse for him is that he weakened after the Panama experience." Thereafter he referred to him as "the deserter Nelson". Not unnaturally Webster, who wrote his account in June, was very anxious to assure his most valuable patron that the series of disasters which had befallen the expedition were not due to neglect on his part. It must also be remembered that at that time the transmission of yellow fever was not understood and Webster firmly believed that Harris and Nelson had survived on account of their temperate habits and because "they took care of themselves". He added that "Robinson and Cornell were both Whaleman of Experience and Good Fishermen but sailors ashore seem to lose their reckoning. Had I put Harris in Charge I think we would have been on the Islands now." We do not know what Walter thought about the matter, nor what he said in two cables to Webster, but 100 years later one can heartily endorse these reflections.

Harris landed in San Francisco on May 10th, and immediately informed Webster that he was willing to try once again to reach the Galapagos and collect as originally planned, providing he was put in charge of the expedition.

On paper the idea seemed completely mad. Harris was an excellent collector and taxidermist, with considerable experience of travel, but there was no reason to suppose that he had the necessary knowledge to select or obtain a vessel, choose a crew, equip the ship and command the expedition. Moreover he had never been anywhere near the Galapagos Islands. Walter had a sixth sense where the selection of personnel was concerned, whether he was engaging a budding scientist like Jordan, or a valet, and with his usual flair he decided to take a chance with Harris. His confidence was based on the copies of Harris's letters to Webster and the latter's personal assessment together with a very slight personal knowledge of the man. He cabled his agreement, and the $1000, for which Harris had asked, followed on May 23rd. Harris was delighted that "Mr. Rothschild will carry things through" and thereafter he signed all his letters "C.M. Harris (GALAPAGOS or BUST)". However he made the mistake of trying to persuade Walter to buy, rather than charter, a suitable vessel. He fell in love with the "Lily Lo", a 65 ton schooner built in 1887, fast, seaworthy and fitted to carry a large supply of water. In fact just what they needed — a "dandy", which into the bargain had weathered violent storms in the Japan seas, and might have been built for their 20,000 mile sailing trip. The owners were asking $2500 and she only required a new set of sails, scraping, a little corking, the copper painting and her boats put in order. On their return she would sell for at least $2000. The Lily Lo was "a peach".

Walter was not impressed.

The extraordinary ebullience of Harris was well illustrated by this incident, for although he was clearly nonplussed by what he considered Walter's ungrateful or unreasonable lack of enthusiasm for the matchless Lily Lo and the economic advantages of his scheme, he was in no way discouraged and himself suggested a Ships Broker as the best solution to their problem, since it had proved impossible to find a suitable vessel to charter. Both he and Webster realized that time was running out and only 5-6 months of favourable weather for collecting on the Islands remained before the rains set in. There followed a series of long, tedious letters concerning the technical details of Bonds, the cost of wharfage clearance, pilot, broker and so forth and the difficulty of obtaining the right type of taxidermist. Walter's rather inexplicable delay in providing the necessary bonds (for example guaranteeing the value of the vessel if lost through fault of the charter party, or providing for the salaries paid to the crew's wives during their absence) may have been due to his own misgivings*

---

* As we know, one of Walter's unusual characteristics was his ability to forget or shelve anything which involved a difficult or unpleasant decision. This streak in his nature verged on the abnormal and it was something his father could neither understand nor tolerate.

about the whole enterprise — he was by no means reassured by Harris's assertion that "this business of bonds is only a matter of form providing nothing happens to the vessel". But more probably he feared that if his plans leaked out at home the whole expedition would be dubbed dangerous or crazy, and immediately cancelled out of hand by his irate and autocratic father. Harris wrote to Webster on May 27th:

> *All we need now to make her go is for Mr. Rothschild to consent to Bonds and to secure 2 men. If I can get away from Frisco by June 15 this will permit of doing a good deal of work by Dec. and I presume we could arrange to stay there 2 months or so later. On receipt of this letter telegraph me if you have $1000 by code word ONE, if you think you can get bonds say TWO; if you are going to send one man THREE and if 2 men FOUR — to receive a cipher reading one, two, three, four will put me in most jubilant spirits — my heart and soul is in this trip. I want to pull it off with success. I have felt from the beginning it would eventually be a success. Believe me my Dear Mr. Webster when I say I will do everything in my power to carry things through.*

This was Harris.

Webster now seems to have risen to the occasion and shown considerable skill and determination in handling the situation in general and Walter in particular. By June 19th the two masted-Schooner Lila & Mattie had been chartered (through Bunker and Co., Freight and Custom Brokers) and equipped with officers and crew who were, in the words of Mr. Bunker, "Young and active men imbued with the spirit of enterprise". The vessel was insured for $10,000 despite the fact that most companies would not insure boats sailing in unexplored regions. By a strange freak of fate this Policy (issued by the Alliance Assurance Company, of which Natty was chairman) escaped destruction. It had been placed in an envelope on which Walter had written the one word "keep", and was in the box which had toppled off the truck on its way to the ill-fated bonfire, lit eighty years later by the tidy-minded civil servant. Webster presented a bill for $844.80 to cover Storage, Drayage, Clearance fees, outward Pilotage, Provision and Grocery Bills, passing effects and dispatch of Vessel. Meanwhile Harris provided a list of provisions and checked them aboard. Since there is no mention of the guns, harpoons, fishing tackle, camera (the latter especially purchased following Walter's specification), films or the general collecting gear originally listed by Webster, and sent to Panama, they presumably travelled back to San Francisco still packed in the original crates. It was a considerable burden to Harris to have to correspond with Webster and Walter by post. Letters took 10-15 days, and cables were costly. Furthermore he considered he was a poor letter writer and only put pen to paper with great reluctance.

In the interest of economy only four pillows and four pillow slips, two hats and two bathing trunks were to be shared by a crew of five.

Webster now obtained the services of two taxidermists and an additional collector R.H. Beck to support Harris. He sent Walter a skin of an albino Song Sparrow as evidence that Galen D. Hull, aged 30, could make a fine skin. In fact he could make such a skin in fourteen minutes, which indicated good workmanship, He was also a geologist. The second taxidermist, Frederick Peabody Drowne, was only 18, and sample skins of a Red-Eyed Vireo and a Cedar Waxwing were sent along to demonstrate his ability. He had also made a speciality of preserving marine animals. Beck was engaged at $25 per month without a commission and with no rank. Webster, anxious to save expense, arranged that these three men should arrive in San Francisco (after a 5½ day train journey) as near as possible to the sailing date of the Lila & Mattie, and they could stop on board rather than wait in the City, running up unnecessary bills. "I believe this time you have a good party," he wrote to Walter. As usual, we do not know the Patron's thoughts on the subject. It is clear, however, that he had again sent detailed collecting instructions with sketch maps for Harris, and orders to photograph each island. But these letters are lost. Of the three cables of which copies were made, one told Harris to collect South Albemarle "despite Bauer";[1] another sent on June 11th in reply to one asking for a missing guarantee and suggesting that delay was costly, read: "Have cabled Calbank to guarantee Charter Order Harris to start at once".

In a long letter summarising the enterprise from February to June, Webster says: "I have impressed it on Harris from the beginning to the end to make a clean and complete job of it when he gets to the Islands. That you insist that time be taken so that there will be no use for another party ever going there. I know of nothing that I have omitted to instruct them. I do not see any chance of failing to understand what you want done in every particular."

Once again the question of rising costs worried Harris to death and he wrote page upon page to Webster explaining his difficulties in great detail, but, as usual, he ended on an optimistic note: "The trip will cost nearly twice as much as first contemplated but at the prices you say the stuff is worth I will bring back $20,000 worth for you unless the vessel sinks and then we all have something to lose."

Walter was, for zoological reasons, pressing Webster to make certain of collecting in the Cocos Island. He felt that the secret of the origins of the Galapagos fauna would be revealed by the birds from Cocos. Harris did not want to go anywhere nearer to Panama than he could help because of the stories of the Yellow Fever epidemic which he heard in the port. Every ship reported the loss of some passengers, and the captain of the "Para" had died.

193

His mother, who was in touch with the consul in Panama, was also worried. Finally Harris decided to postpone the visit to the Cocos Island until they were homeward bound. "I may never come back," he commented, "but if I don't it won't be a great loss — In case I don't I would like my mother to have the trunk of things I left with you. But I don't fear any casualties. We may go down in a blow but that is too far off to think about."

Apart from the costs he was also worried about a rival expedition. He reported someone working the Tres Marias and collecting birds from Santa Barbara Islands for the Smithsonian, and added, "We are not getting to the Galapagos too soon." Webster was also afraid of rivals: "There is just now an expedition on the Coast of California running about in a way I feared that if their attention was attracted to our destination might take it into their heads to go there." Furthermore Harris was having great trouble with the press who were trying to discover, by hook or by crook, the ultimate destination of the mysterious expedition. He marvelled at their persistence. Someone (the name is illegible) posing as a friend, had called while he was out, got into his room and "using their eyes to great advantage, got up an amusing article". Harris assured Webster that he would set matters straight before sailing. He expressed great appreciation of the zoological books and notes Walter had sent him, although he could not be sure of copying them correctly. He was also well satisfied with the two taxidermists at this point, but he was beginning to feel exhausted. He also had a recurrence of fever, a fact he decided to keep quiet. "This thing," he wrote to Webster in a sprawling note in pencil, "has been quite a strain on me and I am so excited I have hardly slept for 2-3 nights." Small wonder.

His last epistle before departure (twelve sheets of foolscap) was written to Webster at 11 pm on June 19th aboard the Lila & Mattie. He confided he was "completely played out" having worked round the clock for ten days. The fact the ship was ready to sail only five weeks after Harris had landed from the ill fated Panama diversion was a major *tour de force*. His letter as usual was mostly concerned with costs and their ramifications, although he somewhat grudgingly agreed — for he still hankered after the Lily Lo — that the new vessel was "a good one, 105 tons, 93 feet long, built in 1888, staunch and well equipped". He was pleased with the bench and platform he had had constructed under the afterhatch. It would provide quite a fine workshop. He sent Webster a pencil sketch of the deck and announced they would begin to collect as soon as they had left and got over sea sickness. Harris also gave a brief description of the Captain and his mate who were Germans — the latter tough and rugged and both first class-navigators. They had a good steward and two able seamen. The ship's articles called for six men, but Harris decided their party of four could take the place of the additional seaman in an emergency. He therefore "persuaded Mr. Hull to sign as second mate merely to blind the shipping commissioner" and thus save a little money. This rather dubious subterfuge was concealed from Walter.

Harris was so seriously worried about his own ability to spend the funds wisely that he became "half sick and horribly tired".

*I hope in spite of the large expenses you will believe that I have done fairly well — someone else might have done better, some might have done worse — at any rate I have been honest and have done my best, — that is all I am capable of. In doing business here I have done nothing hastily and I have consulted people of all kinds in regards to different matters.... many things have had to be bought which were unlooked for and not calculated on — the vessel had a keel boat and I have had an Alaskan seal skiff built for $35. All the provisions had to be especially packed in air tight packages to keep them, which added materially to the cost.*

He went on to list scores of "extras", from the casks and tanks which carried their 2000 gallons of water, to the ballast and monthly allowances for the crew's wives. He was also worried, and with good reason, about the time left for collecting the Islands.

He ended with a melancholy *cri de coeur*: "I am sorry to think that I have had the responsibility of handling the funds for as far as the collecting was concerned I could have made a success and I dont want this to spoil it." Then his natural optimism suddenly flared: "....but keep up your courage until we get back and see what we bring you!"

The last 24 hours in harbour were full of exasperation. Harris expected to sail that afternoon but the surly German captain merely "failed to show up" evidently unwilling to go out with a southerly breeze. One feels that had Walter himself been on board the captain might have sent a messenger to advise Harris of this decision.* It also poured with rain all day — a most unusual event at that season — and the photograph of the party, which Walter had especially requested, could not be taken although it is itemised on a receipted bill for $2. Harris battled to straighten out the accounts. In four pages of foolscap scrawl, he described his unsuccessful efforts to make duplicate copies of all the bills. "Everything is mixed up to-day, but it will do when we get back. The money has gone away and I have tried to use it wisely." The flat-bottomed, high-nosed skiff for landing in surf was delivered at the eleventh hour. At 2.20 pm the Captain arrived on board but was doubtful about going out.... The first entry in Harris's diary records: "We left S.F., A southerly wind took us out to

* Webster in fact billed Walter for seven cables and five telephone messages at $16.

the Farralones. At about 4 pm the wind died out and at dark we were drifting about North of the Islands. Four big California whales were seen sporting about the vessel."

Webster was immensely relieved when he received a telegram on June 22nd which read:

*11d paid 21/6/1897 San Francisco, Calif. To F.B. Webster, Hyde Park. Hurrah are off. Have written. See you nine ten months. Goodbye. the PARTY.*

We will never know if anyone cabled Harris that he was a bloody marvel. Somehow one fears this is one of the details which may well have been overlooked.

*Yours truly,*
*C. M. Harris*

*{Galapagos or Bust}*

# CHAPTER 23

# The Giants

WALTER was obsessively fond of giant tortoises, for these marvellous beasts displayed several features which particularly appealed to him. First and foremost they were of enormous size and had apparently developed these proportions in response to an island environment. They were survivors from a past age of reptiles and thus infinitely romantic and — had one the wit to read their message correctly — held the secrets of evolution, past and future, beneath their massive carapaces. Walter was not, in the early days when Günther first introduced him to these animals, deeply interested in their systematics, classification and relationships — he just cherished and admired them — although he ultimately became so, and wrote seventeen scientific papers and notes on the group. He always hoped, up to the time of Günther's death, that the latter could be persuaded to join him in a massive revision of the Land Tortoises (see p. 113 and p. 204). Walter described several new sub-species, expended a lot of time and energy, and, as we have seen, showed amazing tenacity and zeal, chasing up types and specimens in Museums and Zoos all over the world. Moreover the giants were clearly threatened by extinction, and this engendered specific emotional reactions....

It was an excellent idea of his to rent the Island of Aldabra between 1900 and 1908 and thus to attempt to preserve at least one breeding population. Whether his plan of introducing some specimens from the Galapagos Islands ever materialized we do not know. It seems rather unlikely, although he told Günther he would do so. On Duncan, where extinction seemed inevitable, he tried to collect the remnants for his Museum. In a letter to the Keeper in 1897 he wrote:

> My chief reason for telling Dr. Harris to bring away every tortoise they saw, big or little, alive or dead, was that the Orchilla moss hunters had already reduced them by more than half since Dr. Bauer was there in 1892, and they would have eaten them all in 2 or 3 years more; and I wanted to save them for science. Bauer found over 100 Testudo ephippium on Duncan Island in 1892 and Harris carried off all those that were left which was 29, so some 80 tortoises had been eaten in 5 years.

Walter, as we have seen preserved all his life a naive sense of wonder and amazement about animals and plants — a sort of childish delight and personal involvement, constantly renewed, and these emotions were aroused with special intensity by the giants. He really loved them. His ambition was to collect the maximum number of species *alive* at Tring, and he was one of the first zoologists to compare the differences in behaviour between closely related tortoises, although curiously enough he did not incorporate these observations in his published papers. Probably at that period such comparisons would have been considered "unscientific" and out of place in a systematic work of this type. When, in 1898, he decided to stand for Parliament after the death of his uncle Ferdinand, he wrote mournfully to Günther that this would curtail his time for the tortoises.

In Webster and Harris he discovered two men after his own heart, for they found his enthusiasm infectious and understandable, and were easily carried away by his passionate desire to own some living Galapagos giants. To-day, when nature is everywhere tamed, threatened, and in need of protection, it is difficult to realize that only a hundred years ago there were still many unexplored areas in the world and that expeditions in sailing vessels to remote islands presented great hazards. Man was still pitting himself against the elements. The quest for the giant tortoises presented such a challenge to Webster and Harris, quite apart from their mutual interest in a satisfactory scientific and financial outcome to their expedition. Both Hartert and Rothschild thought their efforts were crowned with success, for the collections they eventually sent to Tring solved the problem of the origins of the Galapagos Island fauna — except for the Giant Land Tortoises which, according to these authors, still constituted a puzzle.

The Party, according to Harris's diary, arrived at Duncan at noon on September 4th, but could find no anchorage. Harris took a short trip to the island but failed to locate a camping site, so the Lila & Mattie lay to. Next day, together with Hull and Drowne, he explored Duncan and, in an immense crater, about half a mile across, full of vegetation, discovered his first tortoise. In the course of the afternoon the three men found seven more, turned them on their backs, weighted them with rocks and returned to the vessel, very tired, after dark. They planned to collect the tortoises two days later. "It will be almost killing work, but must be done". On Tuesday all hands duly landed on the east side of the Island, leaving their lunch in the boat, anticipating a return with three tortoises by noon. "On getting to the crater," wrote Harris, "we found one big tortoise dead; one of the big rocks that we weighted it with had shifted and fallen on its neck and shut its wind off. Rats had gnawed out one of its eyes and had also gnawed a piece out of one hind foot of the living smaller specimen. Several of the others had got loose, but all were found. By noon we had just got the tortoises secured and were two miles from lunch and our water was short. Two men each took a tortoise lashed to a pole and started for the coast. It was the hardest work I ever did for my part and I guess that the rest thought the same. At 4 we got to shore above a high bluff. We tied them here for the night and started for the boat two miles across the Island. This was very tough work. No dinner. No water. The sailor Charles was completely exhausted after reaching the boat at dusk." Beck, it seems, was made of sterner stuff, for he then "secured a rat".

Drowne's diary emphasised the difficulties of tortoise transport. "It was very hard getting up the side of the crater, walking being so rough and thorns so plentiful. But this was nothing compared with going down on the other side, which was very steep and *terrible* walking. The sailor had on a pair of wooden clogs which soon began to chafe his feet. After a long time spent in stumbling over larva blocks, tearing through thorn bushes and other such pleasantries we reached a point as near the shore as we could and tied the creatures up securely and left them.... that distance seemed terribly long.... The trip was very hard on the tortoises too, and they acted as if "played out". Two of them being set close together got their poles somewhat tangled up, and, by the way, opened their mouths at each other as if they were going to have a fight."

Next day the tortoises were put in separate sacks and lowered one at a time over the top of a bluff 75 feet high to the skiff waiting below.

The party remained anchored off Duncan until September 25th, collecting every day on the island on which they had established a base camp. The mate cooked supper "doves, fruit, bread, butter and coffee. Short-eared owls hooted about us continually during the night"; the boys complained they kept them awake with their screeching round the camp. Everyone enjoyed the eruption of the large "extinct" volcano on James — there had been no previous activity reported, at any rate since 1855. It was a great sight to see the mountain pour out huge streams of molten larva.

Harris's life was beset with problems. First of all he had a return of his Panama fever, and decided to remain on the vessel for a day or two. He wrote to Webster that he felt it had "rather broken him up". He noted (October 15th) that a 20 mile walk and 15 miles on horseback, with 6 hours collecting, was a hard day's work in the tropics! The Captain was over cautious and when ordered to land on several occasions refused to do so. A lot of time was lost "drifting away" and then spending all day "beating up — very disgusting". The rough walking had reduced everyone's clothes to rags and their boots were completely worn out, and Harris decided they must make new ones out of leather. For the moment they wrapped their feet in canvas. All the auxiliary gun barrels were worn out and one gun broken. Water was a great problem and the only place where they could eventually take a 1000 gallons on board was

Chatham (on October 17th) where it had to be hauled five miles from the Cobos farm. (Walter, on noting this, immediately wrote to Senor Cobos to start collecting for him!) Harris also experienced the inevitable problems so familiar to all collectors. The climate was such that the lizards would not dry and had to be preserved in alcohol; the large number of fish defeated him — they refused to be baited; the Geospizids were unexpectedly difficult to skin. But his chief problem was time. On one occasion (on Abingdon) he instructed the boys to collect about 20 birds and be back at the boat at 11. Drowne failed to show up. A search was instituted and at about 3 pm he was found by Harris on the opposite side of the Island "well exhausted and scratched. He had got lost, and lost his head," he recorded irritably, "This escapade knocked us out a whole day." They were, in fact, always in a hurry: they would be obliged to leave the Southern Islands before the end of December, as the Calms began in January and there would then be no sailing for three months. They were even now becalmed for 2-3 days at a stretch, and the extent of this problem can best be appreciated if one considers that the true distance between, say, Tagus Cove and Tower Island was 110 miles, yet due to drifting they sailed 1400 miles. Harris now considered that, in any case, they would have needed at least two more men to carry out the programme adequately.

The Party left Duncan on September 25th with 29 live tortoises on board. Drowne thought, maybe, only 2-3 giants had been left alive on the Island, but this later proved to be incorrect. Harris now had a new worry — obtaining enough cacti to feed the tortoises. It was only after reaching San Francisco that he found bananas were an acceptable substitute. Harris's affection for the tortoises was to be strained to breaking point before he got them safely off to Tring in May — not, of course, without casualties. The largest individual died on October 20th.

On Indefatigable they met Mr. Thomas Levick, an Englishman from Charles Island who was on a trip around the Islands in a small boat with two men, a South American and an old Portugese. To the question "Speak English?", Mr. Levick replied "I used to". It was the first sign of another human being the Party had seen for four months. He told them among other things that tortoises were common on Albemarle, but reckoned they were extinct on all the other islands — killed off by fierce feral dogs, cats, rats, and so forth. In due course Drowne and Beck collected one specimen weighing about 15 lbs and a lot of tortoise eggs on Albemarle, but confirmed that many had been dug up and eaten by dogs. Blowing the eggs was hard work. They made an omelette out of the yolks which was voted superior to the usual kind.

On Charles Island Mr. Hall had a pen containing several tortoises obtained from Albemarle. Walter surmised that there had been a considerable exchange of tortoises among the Islands due to human agency, which confused their relationships. About November 12th Webster wrote him a progress report, explaining that the expedition had already collected about all the birds that had ever been taken in the Galapagos up to date, as well as some new species. Short of a shipwreck "we can count on a WINNING TRIP".

Drowne's published diary was illustrated with delightful drawings by Fro-hawk, all made, according to Walter, from the photographs which, unfortu-nately, are lost. In one entry Drowne remarked that Beck took a picture of Clarion Isle from N.E. to S.W. We also know the Party took 144 films in their baggage, but no photographs were ever published. What accident befell them remains a complete mystery. Only the pictures made by Beck in the subsequent 1901/2 expedition have survived.

Harris's worst fears about the calms materialised, and he was forced by the captain to abandon the last part of their programme, including the vital visit to Cocos Island. They left Tower Island on December 28th after taking a lot of cactus on board for the tortoises, and reached San Francisco on February 8th.

Shortly after arrival Harris wrote to Webster complaining that his health had been ruined and, as usual, his money hadn't arrived. He had collected successfully on Culpepper, Wenman, Abingdon, Bindloe, Indefatigable, Dun-can, Jervis, Charles, Barrington, Chatham, Hood, Albemarle (Walter was demanding an explanation for the omission of North Albemarle!), Narbo-rough and Tower. He had packed[1] and dispatched 60 crates (insured for $7000) containing 3075 beautifully prepared bird skins, 400 birds' eggs, 13 seals, 150 iguanas, 65 tortoises, 40 tortoise eggs, 8 turtles, several hundred lizards, 6 centipedes, several hundred sea urchins, starfish, holothurians, shells, etc., etc. He now wanted to come east and here he was stuck in San Francisco with the tortoises! He was, understandably, browned off.

On this point, however, Walter was adamant. Harris was NOT to come east. He flatly refused to allow this. At all costs Harris was to stay in San Francisco and tend to the wants of the tortoises until they had completely recovered from their arduous sea voyage. Harris stayed. He rented a heated greenhouse to avoid any chance of chilling the animals, and experimented with different diets (including cooked vegetables) until he found they were not averse to bananas and liked squash. By the end of the month his iron constitution and built-in optimism had triumphed and he announced that he was feeling better all the time, and in a few weeks would be ready for another expedition if Webster could place him "with a regular salary". He had been working with the tortoises and had provided them with several layers of sand, and they were now eating and drinking better than at any time.

It is sad that we have neither written nor verbal accounts describing their arrival at Tring or the Zoo, but several photographs survive which show Walter in a felicitous mood with the Giants. He certainly enjoyed their company.

Harris made one modest request — which one fervently hopes was granted. "I wish you would be sure and save *two large black Iguana* skins for me *out of those in salt*". The flightless Cormorant — the most extraordinary ornithological discovery of the expedition — was named *Phalacrocorax harrisi* in his honour.

The story of the Giant Tortoises is an excellent example of Walter's fanatical interest in his various animals, and almost phobic persistence in their pursuit. Before the tortoises reached Tring he was arranging for Beck* to return to the Galapagos Islands. This time to search both North and South Albemarle — for he expected (quite rightly) that he would find another sub-species of giant there — and on this occasion there was to be no mistake about the Cocos Isles either.

In 1897 he wrote to Günther: "I believe I am on the verge of an interesting discovery.... The largest tortoise in Sydney was presented to Mr. Alexander MacDonald, its present owner, by "Rotumah" one of the chiefs from the outlying Pacific islands; now in 1813 Porten turned out a lot of Galapagos tortoises on one of the Marquesas Islands and I expect the Sydney pair comes from there." This guess proved to be correct, although the animal came to Sydney via King George of Tonga, not direct from Madison Island (i.e. Rotumah Isle). Walter was now determined to acquire this beast and set the wheels in motion. On February 5th, 1898 he received a letter from Robert Ogilvie from Sydney advising him that he had, as requested, secured the tortoise "Rotumah" — reported to be 150 years old, a most erotic and savage individual not withstanding, and the largest in the world. It was travelling to England on board the Oceana, sailing on February 12th, and due at the Albert Docks on March 22nd, when it must be met by Walter's representative. This tortoise had been living in the grounds of a lunatic asylum in Sydney, and by a strange coincidence Dr. Manning, who was the Government Inspector of

---

* In 1899 Walter wrote saying he was afraid he could not join the expedition himself. He was such a wretched traveller that he was "as sick as a dog" travelling by train to Braemar. His instructions for this trip were to get good sets of birds from Gardner Island and Hood as Bauer's were bad, to try for tortoises from two sections in North Albemarle inland from Tagus Cove, and to urge Beck and Anthony to get rails from all islands *other than Duncan*. Walter never gave any reasons for his requirements, and Beck, despite instructions to the contrary, re-collected on Duncan which had aready been over-collected by Harris. Several years after Beck's expedition Harris and Walter met in London, but we have no record of what either thought of the other. Harris recorded that the interview ended by Walter saying "Well, that's settled." They had been discussing another collecting expedition, which however never materialized.

Lunatic Asylums for New South Wales, was also a passenger on the Oceana and had promised to keep an eye on him during the trip. The animal's diet sheet was enclosed. £125 was paid for "Rotumah" (equivalent to about £2500 to-day) and it was possible another £5 would have to be paid to a party who claimed to have an interest in the animal. Dr. Manning, unfortunately, must have been less gifted at handling Chelonians than his human charges, for the tortoise was housed on the top deck and, in Walter's words, "arrived quite stiff and motionless with cold and another 12 hours would have killed it". It was safely ensconced at the Zoo by 11 pm and next morning with Dr. Günther and his son and a veterinary surgeon, Walter visited the animal which was visibly improving. He wrote an account to Dr. Hartert from New Court: "It fed freely out of our hands and stuck out its long neck as far as it would go. It is certainly a species quite unknown and undescribed". Unfortunately Rotumah's mate had been left behind in Sydney. No doubt she was sorely missed, but all that could be sent to Tring by way of comfort was a picture of the centenarian "caught in the act". Two years later Walter wrote to Edgar Waite in Sydney to say that Rotumah had died of sexual over-excitation.

At this period Walter was himself travelling about Europe examining all Museum specimens of Giant Tortoises, and was haggling over an exchange with a museum in Milan. He wrote several letters to Hartert on the subject, one of which suggests that he was rather a sharp negotiator when it came to the Giants:

> So far my researches in Italy have found out that there are three specimens of giant tortoises in Italian Museums: 1 in Turin, a young Testudo daudinii which will shortly arrive at Tring, 1 in Florence from Chatham Isle, Galapagos, probably an undescribed form which also will be sent to Tring, and lastly 1 at Milan labelled Testudo elephantopus, but which is most likely T. guntherii or T. microphyes; this they will not lend but I think I will succeed in "swindling" them out of it by giving them a much larger T. elephantopus = vicinia in exchange. I think if Dr. Günther can come over to Paris to look at the type of T. daudinii and I can get all the tortoises from other museums lent to me, we can produce a really good revision.

Then Walter had a flaming row with Professor Boulanger, the Keeper at the Natural History Museum. He wrote again to Hartert from the Bank: "When you go to the Museum please tell Thorpe and Sharp that I consider Boulanger treated me so shabbily that I do not think I shall let anything go to the Museum (of course in the end I will, but I want to give them a good fright) and tell Boulanger from me that I consider it a mean and *ungentlemanly* trick to describe a tortoise of mine without my permission." A week later, following some explanatory note of apology, he sent "a most polite letter" to Boulanger asking him to bury the hatchet and to work amicably with him in the future. However he wrote to Hartert: "You have not evidently understood the question of my tortoise *at all*. I was only angry because Boulanger described it without asking me first as I thought he knew it was mine.... In 1883 there were about 209 tortoises catalogued and after 1896 there has been *not a single* new species described for 35 years and as I am specially interested in tortoises I certainly want to describe any new species I get myself, and not let others do so." Boulanger agreed to bury the hatchet and as a gesture of good will Walter called his new species *Testudo boulangeri*. The recipient of this signal honour wrote and thanked him and added: "What you say of your having felt that I thought you too much of an amateur does not apply". However in 1902 we find Walter throwing a final dart: "Without in the least wishing to criticize this admirable work [Catalogue of Chelonians etc. in the British Museum] I feel obliged to correct what I consider a most serious error.... If he [Boulanger] had, as I have, examined the type of Quoy and Gaimard's *Testudo nigra* he would never have done this...."

By this time Walter had added to the Tring Museum eighteen specimens of Giant Galapagos Tortoises collected by the Hopkins Stanford expedition, eleven collected by the Johnson-Green expedition, and fourteen collected by Beck. The ex-Queen of the Sandwich Isles presented him with a particularly fine example over 100 years old. Eventually he amassed 144 giant tortoises at Tring, which he considered quite as important scientifically as his birds and Lepidoptera.

In 1894 he had invited Alfred Günther to join him in writing a monographic revision of the Giants, for his collection at that time was already the most comprehensive in the world. Because of the row with Boulanger and its implications, and also because — according to young Günther's recollections[2] — there was a difference of opinion between his father and Walter on the need for dissecting the specimens, the invitation, although originally accepted, was eventually declined. This was a sad error of judgement on Günther's part, for there is no doubt that, had he played Jordan's role, applying the brake when necessary on Walter's excitable and impetuous flight during such enterprises, one of the great classical Tring monographs would have resulted. But Günther was not a man of Jordan's calibre and he failed to grasp the golden opportunity which K.J. seized with such alacrity and insight. Also he was apprehensive of the rogue elephant quality in his would-be co-author, and afraid of the possibility of annoying Natty by running up large bills for illustrations. Günther stood greatly in awe of Lord Rothschild, whom he had met on various visits to Tring. Somewhat hurt and deflated, Walter abandoned the plan, although later (in 1915) he published a scholarly review of the Giants, which included 6000 careful measurements of individual specimens.

Another fifty years elapsed before interest in the giant tortoises and Aldabra revived in the zoological world. Owing to the threat of military activity in the region, the Royal Society organised an expedition to the Islands in 1966-67, and over 100 scientists from seven countries have put in 50 man years of research on their fauna and flora. D.R. Stoddart, in his report on these scientific studies,[3] mentions that Rothschild's 1915 review is still the most important work on the Giant Land Tortoises, and adds that "for a number of years they were conserved [on Aldabra] by his private philanthropy". Suddenly Dillon Ripley and Julian Huxley were saying what Walter had said 70 years ago: "Certainly the most scientifically interesting atoll in the world.... an ideal location for the scientific study of evolutionary processes...." The present Keeper of Zoology at the British Museum considers that these animals have proved to be "extremely valuable for studies of adaptation to insular environments".

At the time Günther turned down the offer of co-authorship, Walter moved a number of his living tortoises to the Zoo, simultaneously subscribing generously to the building of a new house at Regent's Park to ensure their comfort. Some of his Galapagos giants outlived him, and at the outbreak of World War II were still ponderously cropping the sooty grass in Regent's Park, which they no doubt compared unfavourably with the cacti on Albemarle.

Walter was always incapable of discussing or even mentioning matters which were near and dear to him. A perspicacious observer could see him on occasion turning something over in his mind, struggling but failing to put his thoughts into words, until, somehow, the moment would slip by.... In 1908 — due mainly to the machinations of the blackmailing couple whom he allowed to shadow his life — he was forced to abandon the lease of Aldabra. But he never so much as hinted at this tragedy, and he also sorrowed in silence for the 100 year old Giant presented to him by the ex-Queen of the Sandwich Isles, which died at Tring the same year. It is possible that the inevitable impending doom of the tortoises in their Island homes filled him with gloom and despondency too deep for words, for he never once in our presence discussed the fate of these animals. Or it may simply have meant that the protracted love affair between himself and the Giants had, over the years, lost some of its ardour. In any case it came as a complete surprise when in 1937 I received a letter from the London Zoo announcing that I had recently become the owner of several Galapagos giant tortoises.

# CHAPTER 24

# The Rothschilds and animals

THE GREAT SUCCESS story of My Museum was based on a combination of two Rothschild family characteristics, a collecting mania and a great love and admiration for animals. And into the bargain Walter found them all intensely *interesting....* As we have seen, he went up to Cambridge with a small flock of kiwis — for he was entranced by these quiet, strange little birds and liked to have the troop with him. At the end of his first term he added an enormous St. Bernard dog to the party — the symbolic expression of his streak of megalomania—which remained his constant companion for twelve years. When Bonny died of old age Walter was so grieved that he decided not to have another dog — a decision he adhered to for 38 years until he moved to Home Farm after his mother's death. He then procured himself a gigantic snow-white Pyreneean Mountain Hound, Monné, whose unborn puppies were the subject of a law suit after his death.[1] "Lord Rothschild's bitch in Court," ran a headline on that occasion — somehow, where Walter was concerned, whether it was his hat in the House of Commons, his wolf at the pub, or violence with the staghounds — muffled laughter seemed to pursue him, even beyond the grave. His heirs, however, successfully resisted the demand of £8000 for the pups which Monné never conceived.

Dr. Jordan recalled that Walter's passionate love of dogs proved somewhat embarrassing on collecting expeditions in the wilds of North Africa, for he insisted on establishing friendly relations with the savage feral curs scavenging round their camp. Curiously enough the dogs never bared their teeth at Walter, and he was never bitten — the attraction was mutual. Charles also loved dogs, and when he became engaged to be married he wrote[2] to Rozsika that Biddy, his champion smooth collie, was his most precious possession. From his desert camp at Shendi, when he was collecting for Walter's museum, he wrote letters to his mother[3] full of the wonders of animal and plant life. "I have three young gazelles tethered outside my tent and I keep a goat as foster mother for them — they are such sweet beasts and lie in your lap like Snip."

Some years previously Snip had saved Charles's life. For, in the middle of the night, the little dog, whining in frantic excitement, had pulled the bed clothes off Natty. On following the animal out into the dimly lit passage

— puzzled and only half awake — he suddenly saw the shadowy figure of Charles balanced precariously on the top of the wide balustrade surrounding the well at 148 Piccadilly, just about to plunge to his death in the marble hall thirty feet below. Stealthily Natty put his arms round the somnambulist and drew him down to safety. Charles experienced only a brief period of determined sleepwalking during his teens, but none of the usual dodges he himself, and the nightwatchman employed to wake him had worked on this particular occasion. Somehow he had navigated past the bath of cold water outside his door, and the sheet of paper glued across the exit crackled in vain as he pushed through it. How much all this worried his parents* we do not know — but after the Snip episode the nightwatchman, was, for a time, restricted to a vigil on the top landing.

Natty and Emma were both great dog lovers, and a delightful miniature of Snip painted on porcelain stood on Natty's bedroom table. During his last illness he asked, one morning if two of his granddaughters could come downstairs and talk to him. "I think," said Rozsika, very seriously, to the two little girls, aged five and six, "Grandpapa may ask you to choose something off his table to keep — but on no account are you to ask for the little picture of Snip. Is that quite clear?" Yes, that was crystal clear. The elder child, who was allowed to choose first, selected a pale blue Wedgwood matchbox garlanded with a white frieze of flowers, but the younger, impelled by some irresistible impulse to disobey those in authority, chose the portrait of Snip. "What a splendid choice," said Natty, smiling wistfully, and the elder child — shocked to find how the wicked seemed to flourish — noticed that the old man's face was the same colour as his beard and his pillow. In fact one could hardly make out his face at all....

Snip had been buried in the dogs' cemetery on the edge of the Park, and an appropriate epitaph was carved on his tombstone; "Snip — a faithful friend". One obituary notice[4] only, among hundreds written for Natty, remembered Snip and added a sentimental reference to "the tenderness and gentleness of spirit from which sprang the affection that he bestowed upon a small and devoted four-footed friend that for years trotted by his side and accompanied him on all his drives".

An affinity with animals and a profound interest in animals are not characteristics usually associated with Jewish people, but the Rothschilds possessed

---

* Nervous parents are full of contradictions. Both Emma and Rozsika, for instance, were extremely worried by the possible exposure of their children to infectious diseases, but the former taught Walter to ride at the age of three while the latter encouraged her daughter, aged four, to milk a cow every afternoon, although one careless movement on the part of the docile animal could have smashed her child's face to pulp.

both qualities to a marked degree. Three slightly different expressions of this family preoccupation are well illustrated by Walter, his second cousin Harry Rosebery, and his first cousin Henri. Harry's grandfather, Mayer Amschel (Nathan Mayer's fourth son) as we have seen, showed no inclination for banking and business although he was a Member of Parliament for Hythe (1859-1868): his main interests centered round his country estate at Mentmore, the Rothschild staghounds (which hunted a carted deer) and breeding animals. Although racehorses were his major concern, he was also much involved with animal husbandry and he had bred, among other animals, a famous Jersey bull which he named Mentmore.[5] He was also a most successful owner, and 1871 was known as the Baron's Year,[6] for he won the Derby, the Oaks, the 1000 Guineas and the St. Leger, while another Rothschild horse was first in the Cesarewitch. The East End of London loved it — the Rothschilds always gave you a winner. "Uncle Muffy's" brothers must have been delighted with such excellent publicity. Harry, who seemed even more Rothschild than Walter — with his protruding lower lip, his tremendous zest for life, his endearing bursts of horrendous indiscretion and his personal identification with *my* animals and their outstanding achievements — had similar but wider interests than his grandfather Mayer Amschel's. He not only won the Derby (twice), the Oaks, the 2000 Guineas, and the St. Léger, but he also swept the board with his farm stock. Supreme champion Aberdeen Angus, supreme champion black-faced ram, supreme champion Hunter Class, and he even won the Waterloo Cup with a greyhound called Danielli. When it came to breeding winners he had the Midas touch, and seemed to know intuitively the sort of conformation the judges would fancy two years hence — and then produced it. But Harry's Rothschild characteristics were most in evidence when he grew old — for he was then almost a family caricature, always fussing over his Siamese cat, despite the fact that it rarely left his side, and his blue and yellow macaw — which seemed to give him more immediate pleasure and satisfaction, and on which he lavished more patient care and attention than on any member of his family.* Extolling the cat's virtues his eye twinkled almost as brightly as Walter's.

The Baron Henri was more articulate than his cousin Walter, and he wrote a marvellous account of his boyhood[7] and his tyrannical and deeply neurotic mother, who in several ways presented a caricature of Emma. Thérèse's rule

---

* After Baron Guy de Rothschild (of Paris) gave an endearing television interview about his relationship with his dogs and horses, someone exclaimed "If he had cherished his wife like that there would never have been a divorce!"

inflicted virtual isolation on Henri and his sister Jeanne, who, like Walter and Evelina, lacked all companions of their own age, and also introduced into their household a privileged German governess, with even more sinister qualities than those ascribed by Charles to Ellie Glünder. Henri tells us how, when he was 17, his mother refused to allow him to walk in the park after dark without the governess, in case he fell into the lake!* This is reminiscent of Emma who cancelled the fifteen-year-old Walter's drive to Richmond for fear of sunstroke. But Baroness James used the lake as an excuse — what terrified her was the opportunity of sexual experience which she thought might be offered in the park — whereas Emma's over-protective attitude was probably based on medical advice received in the dim and distant past.

It is fascinating that, like Walter, Henri's first love was for animals — enormously encouraged by Thérèse, who was anxious that her son should study natural science and readily acquiesced to the numbers of stuffed monkeys, rats, birds, otters, hedgehogs and the like with which he completely filled his schoolroom. By the age of ten he had a small museum which rivalled Minall's shed. As ill luck would have it, his tutor was allergic to the taxidermist's preservatives and the Baroness James — who always listened to the pedagogues — was prevailed upon to banish the collection (which became a permanent exhibit at a communal school at Gouvieux) and the child was persuaded to collect autographs instead, five thousand of which — ranging from the Middle Ages to the 20th century — were later donated to the Bibliothèque Nationale. Henri was, in fact, a Rothschild animal lover *manqué*. Anxious mothers have curiously specialised and illogical forms of anxiety, and whereas Thérèse was terrified of the influence of dissolute companions and sexual depravities, she seemed oblivious to the far greater dangers of tuberculosis or other contagious diseases of the sick children in the Rothschild-sponsored hospital at Berck. Henri and his sister had to spend all their holidays at Berck, and were forbidden to explore the delightful surrounding countryside. Instead they were compelled to play sand-castles in the dunes with the inmates of the hospital. This switched Henri's scientific interest from natural history towards the human animal, and he decided to read medicine. His mother — despite her fear of the vicious habits of medical students — was delighted, for her ardent desire to see her son contribute actively to the welfare of the human race and care for the poor and needy was almost a pathological obsession with her.[8] In due course Henri passed his examinations and became

---

* Jeanne, deprived of friends, fell in love with the coachman and was nicknamed "La Vierge du Cocher", while the German Fraulein seduced Henri herself.

a respected pediatrician, with 127 original publications dealing with child nutrition to his credit. Like Walter, he was a creative fellow — the design and building of the Théâtre Pigalle, the invention of jam in tubes, and the Unic Taxi-Cab, etc., etc. — but he lacked Walter's drive and involvement. Henri was interested in his little patients and his autographs (and a keen and most successful collector from the time his tutor banished the schoolroom menagerie), and, later on in his fabulous assemblage of *"têtes-de-mort"*, but his sentiments could not be compared with Walter's passionate and obsessive love of Giant Tortoises. In Henri's childhood reminiscences[7] he mentions two weeks' holiday in England, but if he saw Walter's collection it left no impression. Late in life — already suffering from heart disease — he paid a visit to Emma at Tring. The cousins seemed to find nothing to say to each other and it is not on record whether Henri was marched off after lunch to see My Museum. Walter's nieces remarked that both men had tiny piano-castor feet, far too small for their bodies. Henri's feet positively twinkled. But he was fat and lacked Walter's imposing burly figure — more like a dancing master....

Colette, who wrote the preface to Henri's cruise round his memories, described how she had witnessed, unobserved, a meeting in the Rue Pigalle between the author and his mother. Although she almost certainly did not know of Emma's existence, her description somehow conjures up the iron apron strings. Henri, like the King, and unlike Walter, made good his escape by virtue of an early marriage. But Colette, with her enviable light touch and penetrating insight, drew the conclusion from the grave and fervently respectful manner in which he kissed that white, despotic, wrinkled hand, that whatever he might choose to say in his memoirs, his childhood had not been an altogether unhappy one. There seems little doubt that the cousins[9] inherited their serious scientific bent from Emma and Thérèse, and only the influence of a snuffling tutor had changed a French version of My Museum into the Théâtre Pigalle.

Natty, it should be noted, was never satisfied with merely enjoying his farm and garden. For him, improvement and excellence was the salt of life — a fact which was evident even in the letters he wrote to his parents from Cambridge. Perhaps like the Aga Khan,[10] he felt that improving stock had a truly creative element about it. In the early days, a country estate with a home farm, heated greenhouses, a splendid equipage, and gold and silver plate had been symbols of worldly success and the aristocratic life-style — so assiduously imitated by the previous generation of Rothschilds. In this respect they were highly competitive, especially within the family. But as far as Natty was concerned all that belonged to the past. He was involved in developing new agricultural practices and world-wide improvement of stock.[11] He and King Edward VII[12] sent foundation shire horses to America to stimulate interest in the breed; he sent Jersey cattle to Cecil Rhodes[13] to form a foundation herd in South Africa, and stallions to Japan[14] to help in the improvement of their National Stud. *Country Life*[15] records that "Tring Park was included in most of the itineraries drawn up for the guidance of foreign visitors on account of the high reputation for excellence of general management.... the greatest distinction was in breeding pure-bred stock".

Natty was president of the Jersey Cattle Society in 1897 and his agent, Richardson Carr attained this distinction in 1918 — the first agent ever to do so. Natty's Jersey herd was described as the best in England and "phenomenally successful at all leading shows both in inspection classes and in butter and milk trials". When *The Jersey Cow* commenced publication five years after his death, Twylish 11th, one of Natty's most famous gold medal winners, figured on the frontispiece. The dispersal sale when he died was described as "a tragedy for the breed" and the shorthorn sale as "one of the greatest events that has ever occurred among Dairy Shorthorns". *The Field*[16] pointed out (in 1906) that minutely careful and elaborate tables of statistics had been published annually for 200 herd of cattle. Not a pint of milk was wasted, and that year only 44 gallons out of a total of 33,404.20 gallons were unaccounted for during handling.

The cows and sheep perhaps attracted most attention, but Natty's shire horses were more famous. His interest began when he bought a massive fen stallion "Thorney Tom" for 500 guineas. The sole purpose was to loan the animal as a sire to tenants and local farmers for a reduced fee of £1 (£4 was the usual sum) in order to improve their stock — a practice Natty continued for thirty years. But five years later the shires proved irresistible and he began to build up his own stud which, according to the recent *History of the Breed* (1976), proved the most successful, financially, of all time and Natty himself the most successful exhibitor. Thus at one moment he owned eight different winners of nine championships in nine years, while his stallion Champion "Goalkeeper" fetched a then all time record price of 4100 guineas. Chivers believes that most of this phenomenal success was thanks to his stud manager, Tom Fowler, who was a unique spotter of future champions while they were still foals. "By some sort of sheer genius he selected just those colts which lived to become not only great sires but even more than that, those which were to become the chief fathers of all succeeding generations. And those ancestors of the modern Shire were all at Tring Park together. This is undoubtedly the most astonishing achievement in the whole history of the breed...."[12]

It was fashionable to attribute Natty's flair to Tom Fowler plus a big bank balance, and his success with the cattle to his herdsman, Wilkins, and Richardson Carr,[17] just as Walter's museum and his publications were — apart from his financial backing which was always recognised and overestimated — said to be the result of the fortuitous selection of Jordan and Hartert as curators. Like Walter, Natty certainly had a great gift for picking the right man, and then somehow developing his talent to the full and increasing his potential. But there is more in successful breeding — just as there is more in creating a great museum — than wealth, drive, an excellent staff and a bit of luck. You have to understand the animals you breed and collect. You have to experience a certain affinity....

Walter himself was not enthralled by farm stock and although Charles wrote enthusiastically from school about the Jersey Sale — an annual event — both brothers were bewitched by wild nature, not by agriculture or horticulture. Walter, however, was sufficiently excited by animals of large size to keep some shire horse bones for his museum, and he was so successful at propagating and cultivating orchids that the Royal Horticultural Society made him one of the original holders of their Victorian Medal of Honour. A glorious *Vanda* was named after him.[18] He also financed the sheep dog trials at the annual Tring Agricultural Show, which was held in the Park and attracted 25,000 visitors. But it is certain that, long after Alston Rose and Twylish 11th and Blue Peter and King Tom are forgotten, Walter's animals will be remembered.

"We have much pleasure," wrote Oldfield Thomas,[19] "in naming this very handsome little animal after Lord Rothschild by whom and to whose generosity the Natural History Museum owes a series of all the species obtained including the types of *Poeclictus rothschildi*." The "handsome little animal" was a Zoril with a skull only 55 mm long, a black tip to its tail and an insufferable smell. But this was a relatively unobtrusive *rothschildi*, for there was a Rothschild's Giraffe, a Rothschild's Elephant, porcupine, rock wallaby, hare, fish, lizard, cassowary, rhea, bird of paradise, humming bird, swallow, Galapagos finch, an exquisite snow-white grackle with a Cambridge blue patch of wattle round each eye ("Rothschild with blue eyes", ran a Swiss newspaper headline in 1963)[20] and an improbable fly of which the female sex carries its eyes on the ends of huge stalks. As for the moths and butterflies — they were legion: hawkmoths, a Morpho, tiger moths, silk moths, swallow-tails, a jewel-like bird-wing.... Altogether some 250 species and subspecies of animals were named in his honour[21] (See Appendix 1) ranging from monkeys to a humble intestinal worm. There is nothing more delightful and flattering in the world than to know your name is linked for ever to some dazzling little bird, darting among the trumpet-shaped flowers, dangling from a towering, tropical tree. A bird no-one has ever seen before.... The fairest epitaph the heart could desire, and Walter so ardently desired immortality linked to his animals! *

It was another of his weird contradictions that, although he loved animals so profoundly, he still banged away at wild duck for sport — well into his sixties. Günther,[23] who had not been conditioned by the traditional upper-class attitude to shooting, marvelled how men like Walter and Lord Walsingham could revel in the huge bags they obtained, and boast to each other of the number of driven birds killed on their country estates.[24] Shooting for a collection was a different matter, especially a hundred years ago, when the fauna was still an unknown quantity and tigers were fierce and dangerous animals, not vanishing marvels of creation in need of care and protection. Is collecting merely an extension of the hunting and gathering instincts, fixed in our chromosomes sometime in the dim and distant past? A gene for loving and understanding animals must surely have possessed considerable survival value at a period when dogs had to be tamed and trained not only to protect their owners' territory, but to pull down game in the open. Professor Lindemann[25] once suggested that the adolescent girl's love of horses — which he noticed with surprise among his friends' children — was an overt expression of the fact that for centuries the well-to-do had depended for survival on a good relationship with their horse. "A horse, a horse, my kingdom for a horse", was a cry echoing down the ages, and not relevant to Bosworth Field alone.

Jews in the ghetto, virtually isolated from wild life for a thousand years and more, would have had little use for this particular gene. Walter and Charles, with their intense love of nature, of wild animals and plants in uncultivated surroundings, away from farm and garden, seemed rather odd mutations — or perhaps the expression of a recessive gene due to the double first cousin marriage[26] of their parents, but a glance at the family tree on end paper must convince the most sceptical environmentalist that an interest in animals and plants was probably an hereditary character of the descendants of Mayer Amschel, Walter's great great grandfather (the founder of the Rothschild Bank in Frankfurt) and could not be written off as a normal reaction to sudden emancipation, and a desire of the *parvenu* to mimic the aristocracy. Jordan,

---

* If, out of idle curiosity, you look up "Rothschild" in any encyclopaedia, whether French or English, you will search in vain for Rothschild's porcupine or Rothschild's purple *Vanda*. Nor is there a shadow of doubt or dissent among the compilers of these learned tomes: A Rothschild (as opposed to *the* Rothschilds, who are designated bankers originating from Frankfurt) is one thing only: an un-upholstered armchair with a stark horizontal head rest — one of the hard facts of life.[22] In some cook books (Constance Spry, f.e.) there is a recipe for a rather dull form of pudding — vanilla with chunks of crystallised and preserved fruits added — called Soufflé Rothschild, and some restaurants in Argentina serve Scrambled Rothschild, in which most *unkosher* shrimps are recumbent in a thick egg sauce.

who Charles maintained was "the cleverest man I have ever known"[27] was probably right when he recognised the brothers' considerable contribution to natural science, but insisted their approach was essentially emotional not intellectual. Perhaps individuals who are extremely shy and inhibited find communication with animals both much simpler and more satisfactory. "For the animal," said Henry Beston[28] "....shall not be measured by man. In a world older and more complete than ours they move finished and complete, gifted with extensions of the senses we have lost or never attained, living by voices we shall never hear. They are not brethren, they are not underlings; they are other nations caught with ourselves in the net of life and time, fellow prisoners of the splendour and travail of the earth." Walter and his brother Charles would have agreed.

# CHAPTER 25

# The Great Row

AS JON KARLSSON[1] remarked (see p. 122), a disconcerting characteristic of the very gifted is their ability to tolerate ambiguities. One of the oddest traits in Walter's character was his combination of extreme indiscretion and extreme secrecy. In fact it was his habit of sudden explosive indiscretion, as well as his acute lack of business acumen that influenced his father at last to accede to his son's unspoken desire to leave New Court. He never, for instance, ceased to blurt out to his mother things which in old age she could well have been spared. Thus on one occasion a madman flung himself in front of a car driven by one of her grandsons — then aged 17 — and was killed instantly. Walter rushed up to his mother, who at the time was in her eightieth year, brandishing an evening newspaper and exclaiming in great agitation: "Mama! Mama! Have you heard this dreadful thing?" "No," said his mother, alarmed, "what dreadful thing?" "Well, then, perhaps," said Walter, pausing like a guilty schoolboy, "perhaps I had better not tell you."

Yet he kept the shattering blow of the future sale of his bird collection so secret that despite the team of technicians who had come over from America to pack the birds, and who were billeted in Tring,[2] his mother and the other members of the family heard of the tragedy for the first time, as we have seen, from articles in the daily newspapers. This was a most unfortunate revelation from Walter's point of view, for by this means the blackmailing peeress, not to mention his two friends Marie and Lizzie, were informed that he must have some ready cash at his disposal — a most unusual event.

The psychologists tell us that normal sexual relations are made difficult for a man who is deeply attached to and dependent upon his mother. Two different reactions are most commonly noted, for he is pushed either into homosexuality, or into amorous relationships with women who in every sense are diametrically opposite to his maternal parent — physically, psychologically and socially. Homosexuality was not Walter's scene, although the emotional outlet it provides would have proved in many ways an easier solution — even in Victorian times. For, after all, Cyril Flower and his circle lived just over the hill, and Natty and Emma were constant visitors to Aston Clinton. The graphologist who studied Walter's calligraphy says that in his twenties his mother's stern influence was still apparent in his handwriting. His painful shyness made relationships with Evelina's friends well nigh impossible. Lou-lou Harcourt's diary,[3] which is full of delectable gossip, says (quoting Henry Calcroft) that when Walter was 24 and just about to open My Museum, "the Rothschild family were delighted because Walter has at last attached himself to a prostitute — and immediately went and told his mother."

In view of the intense secrecy observed by Walter over his love affairs, and the reaction such an admission would have called forth from Emma, the story is clearly a pure invention. But it is amusing, because it shows that Walter's childish attachment to his mother was obvious to all and sundry and not merely a talking point in the family circle.

It was in November 1905, twelve years later, that beautiful Mary Anderson,[4] who was a good friend of Walter's, gave one of her rather diverting parties for the King. Among the guests was a fairy-like, 16-year-old stage-struck girl, Marie Fredensen, in the care of a *louche* friend of her mother's, Lizzie Ritchie. The friend pointed out the various celebrities present including Walter. "There," said Lizzie, "Is the man who, if he so chose, could put you behind the footlights and set you on the road to fame."

About a year later Mary Anderson died, and to Walter's surprise he received a letter of condolence from an unknown female who signed herself something-or-other-could-be-Fredensen. At the end of the letter she begged him to launch her upon a successful stage career.

The Rothschild family received a never-ending stream of letters from complete strangers, all of which were answered conscientiously,[5] most of them asking for financial assistance, but there were also many bizarre offers and requests, ranging from a presentation bottle of rejuvenating medicament to the free stuffing of a baby seal, and, needless to say, in those addressed to My Museum, scores of pleas for funds to support collecting expeditions. Walter immediately realized that Marie Fredensen's letter contained an only thinly veiled suggestion of another kind, unconnected with her stage career. He nevertheless took quite a lot of trouble concerning that aspect of her request. He asked his uncle Alfred's secretary for an introduction to George Edwards. (Alfred had many connections with the stage, whereas he himself had none, despite his great friendship with Mary Anderson.) But at the same time he did not think he was expected to beat about the bush. He suggested that he might call and make Marie's acquaintance (Walter spelt her name incorrectly), and if the visit led to a more intimate relationship in the future — so much the better. If, on the other hand, she would not agree to a liaison, he would still help her with her stage career.[6] Marie Fredensen said yes at once, for a wealthy patron was what she needed more than anything in the world, and then, much to her surprise, Walter suggested a meeting at Marylebone station at 5.30 pm. She was even more astonished when, after twenty minutes conversation, mostly about her forthcoming interview with George Edwards and his friend Malone, Walter — true to his time table and his internal clock — announced that he must leave immediately for a public dinner at which he was taking the chair. Without more ado he drove off hastily in his carriage, slamming the door so violently that the vehicle swayed on its wheels, and leaving her standing alone on the pavement. She might have been Sokolow!* Not for a moment conscious of the fact that he had affronted Marie, Walter knew, as his horse clattered along the streets — for he insisted, as usual, on driving at a furious pace — something extraordinary had happened to him. He had fallen desperately in love at first sight with the fairy from the top of the Christmas tree.

If the psychologists had set out to construct a blue-print of the antithesis of Emma — someone to release Walter from the octopus toils of his oedipus complex — they would have invented Marie Fredensen. As pretty as an Edwardian birthday greeting card, sweet, pert, cuddly, kittenish, simple, dependent, wheedling, adoring and light as a feather.... A girl less than half Walter's age, who looked upon him as an intellectual giant, a wonderful lover, her lucky, lucky, lucky star. He was bowled over.

Walter wrote to Marie again when he got home:[6]

> *"I cannot refrain from writing to tell you how agreeably surprised I was on seeing you as I had imagined you the very opposite to what you are. I can assure you I fell in love with you at first sight."*

Walter ended this letter in a somewhat surprising manner, considering his acquaintance with Marie consisted of a twenty-minute conversation on Marylebone Station and two brief letters about her stage prospects.

> *"I write this to put in writing that I[6] promise that should any unfortunate accident arise to separate us in the future I will arrange that you get such an allowance of money per week that you will never want."*

Tenuous threads of coincidence linked Walter and Marie's father. Fredensen was a Polish Jewish doctor and, like Natty and Emma, had taken a profound interest in the Dreyfus case.** He had, in fact, donated his life savings

---

* Nahum Sokolow, a leading Zionist, see Chapters 27-28.

** Emma was always grateful to Balfour for stating that Dreyfus was in his opinion, innocent, and that furthermore he considered his conviction illegal.[7] In response to a letter from Dreyfus's wife, Walter's grandmother made a personal appeal to Bismarck. "I hope," wrote Emma to Baroness Lionel, "That her efforts will be crowned with success — it is a good thing to be able to appeal to the fountain head."[8]

to the fund created to finance the unfortunate man's defence, and was himself ruined in the process. Fredensen contracted a mortal illness and died. His widow was plunged into acute financial difficulties, and his daughter never wholly recovered from this period of great poverty and constant anxiety. All her life Marie remained obsessed with real or imaginary worries about money, which turned her into a tenacious gold digger and effectively threw a permanent shadow over her long association with Walter. She never realized how profoundly her obsession worried and depressed Walter, for not unnaturally Marie believed he was rich and powerful without a care in the world. But it was only during the last year of his life, when he had left Tring Park after his mother's death and was enjoying the relaxation of the very first home of his own, that he sheltered behind the ban on Tring imposed by Charles's settlement, and Sister Claire's moral support (see p.307) and doggedly refused to see either Marie or her arch enemy Lizzie.

In 1905, however, Walter was so much in love that he had only one objective, which was to install Marie with the utmost secrecy as his mistress in a flatlet in London. He felt, not without justification, that this was precisely what she had wanted — a rich patron who could make her independent and help launch her on a stage career. Meanwhile, however, Marie had had second thoughts. She was by now crazily in love with Walter, and she no longer fancied either the footlights or the role she had originally planned. She now had more ambitious plans — simply she wanted to become Walter's wife. Such an idea had never, of course, occurred to him. It was impossible — unthinkable — something both parties had known from the first exchange of letters. The chasm which separated his mother, Tring, his life and home from his love for Marie, and the sort of deliciously free life he could lead with her, was unbridgeable. Walter's mind boggled at the mere thought. But he did not consider it for a moment — it spelled ruin. When he received a letter from Marie refusing cohabitation and saying it was better for them to part forthwith, he was deeply upset, astonished, baffled and unhappy. He could not understand what had gone wrong. He wrote her a naive, stammering, but rather charming letter,[6] which ended thus:

> *I still consider that the matter is not decided and that if you think over matters you will change your mind. It does not matter if it is to-day a month or a year hence I shall always remain anxious to receive a letter from you telling me you have changed your mind....*

But neither then, nor in the 33 years which followed, did he ever mention or consider marriage, although Marie herself — tenaciously — never gave up hope.

Soon after this he received another communication which greatly surprised him. It was from Lizzie Ritchie, the friend who had originally pointed him out to Marie at Mary Anderson's party, and to whom he had once been introduced. "I have always been desperately in love with you," wrote Lizzie, "for years ago I fell in love with you at first sight at one of the Anderson evenings. If only you had made the offer to me which Marie has so coldly turned down — for she has confided in me — you would have received a very different answer."[9]

Walter, depressed and baffled by Marie's *volte face*, drowned his sorrows in a binge with Lizzie. He was just ripe for a party or two. He found Lizzie as unlike the naive Marie as chalk and cheese; she was a bit crazy but knew how to buoy up his spirits and restore his self-confidence. Furthermore she was quite intelligent, and very easy to talk to, and avidly sought information on all subjects including the Stock Exchange — and what a good listener! Before you could say Jack Robinson she had procured a flat, offered Walter what he most craved in London — a cosy little room far from the chilly marble pillars of 148 Piccadilly, genuine adulation, an uncomplicated, satisfying sex life and sincere gratitude.

Despite the cloak of secrecy with which he enveloped the affair, rumours of his association with another lady somehow reached Marie. She was frantic. She sat down and wrote a love letter to Walter. Yes, indeed she had changed her mind as he had prophesied.... She could not bear another moment of separation.... She could only be happy living with him on any terms....

Walter was over the moon. He sent his fairy a ring, and told her it was like a sort of engagement ring....[6] and he would make their *ménage* "as happy as the world could be." Henceforth Marie called herself Mrs. Lionel Walters. All ideas of a stage career were abandoned, and she became a lady of leisure. In December 1906, ten months after Marie had changed her mind, her daughter Olga was born.

Since Walter had a pathological horror of scenes and could not bear reproaches and criticisms — it was too near the bone — he studiously avoided any mention of Marie's change of heart to Lizzie. He sensed that she was emotionally involved too, not merely a jolly *fille de joie*, so he continued to support her as he had begun to do after Marie's initial change of heart, and kept his own counsel, perhaps hoping the affair would peter out.

Throughout this time Walter — who was now 38 — led two completely separate lives, for he lived at Tring and 148 Piccadilly with his mother and, strangely enough, except during his summer collecting trips in Europe never spent a night away from her protective roof. Not a soul in his *entourage* suspected the existence of either Marie or Lizzie.

The year 1906 must have been a hectic one. For Walter fought and won an arduous election campaign in January[10] making innumerable speeches from one end of the constituency to the other — and he was becoming more active in the House of Commons, assisting at the Committee stage with half a dozen bills. There were ten days vacation, collecting hard in the Alps, and a week in Florence and Paris examining and comparing zoological type material. But all his mornings from 9 am until about lunch time, when he left — ostensibly for the House of Commons, but in reality for South Kensington Department of Entomology — he spent at his desk at New Court. During every weekend he worked at My Museum from 9 am far into the night.

Furthermore he and Jordan published in *Novitates* one of their classic revisions, this time of the American Papilios.[11] It contained 333 pages of text, and six plates. The small-type references for one species alone occupied two whole pages — over 200 citations. In addition Walter published fifteen shorter papers, and he had begun work on ten others which appeared in print in 1907.

Despite increasing weight he still enjoyed a fast gallop with hounds, and found it the one relaxation which "cleared his brain". (Up to a few years before his death Walter rode in Rotten Row before breakfast whenever he spent a night at 148 Piccadilly). But at this period he was already a very worried man, for the blackmailing peeress was now pressing him hard for a very large sum, Marie and Lizzie both expected to live in reasonably grand style, and the network of collectors was proving expensive. Walter's furious bursts of activity were sometimes followed by a menacing sense of confusion and disintegration when he felt that everything had got on top of him and there was no light at the end of the tunnel. It was in this mood that he sent for the first of the linen baskets and began to consign his post, unopened, to its capacious oil-cloth lined interior.

Simultaneously — although he did not dare aggregate his various debts — he decided to follow the blackmailing couple's advice and take a wild gamble in several companies in which they were interested — companies not listed on the Stock Exchange. Provided he had the capital — so they asserted — he was bound to make a killing.... This would satisfy the lady for the time being and pay some of his own debts too. Walter plunged.

Gradually the situation became more and more perilous. It was the faithful chief clerk, Joseph Nauheim, at New Court who finally decided to break the news to Natty and show him the famous list of Walter's extraordinary holdings.

Then, during the same week, a new and totally unexpected disaster occurred. Lizzie discovered the truth about Marie Fredensen. In Edwardian times it was by no means unusual for a man to have two mistresses, and in view of the manner of Lizzie's more than casual self-introduction to Walter and their equally casual relationship, the discovery should not have evoked such an alarming reaction. In fact these were the first early signs of a psychosis, but such signs are rarely recognised by those who know the sufferer well, for they merely appear as an extension of fairly familiar but hitherto subdued traits in the person's make-up, and Walter was therefore taken completely by surprise. He had merely expected an almighty scene — not a raving virago. He was terrified. After threatening Marie with personal violence Lizzie arrived in Tring, bought an empty house 200 yards from the Museum, and then stormed round to see Emma. This she attempted to do on several occasions, but failed, so she resorted to the telephone: "Do you know," said Lizzie, yelling into the instrument, "that your son has contracted a morganatic marriage with a whore and is the father of her child?"

We have no knowledge of how Emma reacted to this announcement. The secret was apparently well kept. Not even Rozsika knew.

In 1914 the troops produced a ditty in which we were all enjoined to pack up our troubles in our old kitbag and smile, smile, smile. Walter seemed to possess this ability to an extraordinary degree, for during the two years before these events had occurred, he had symbolised his ability to insulate and isolate himself from his worries by pitching his unopened mail into the linen baskets and had then gone on just as usual. At no moment did his staff at the Museum, or his family, or the wedding guests at Charles's marriage in Vienna, or any of his zoological contacts suspect that he had a care in the world. He attended various scientific meetings and congresses, smiling benignly — his elephant eyes twinkling — and as boyishly enthusiastic as ever. Even after the *débâcle* Hartert never realized the extent of Walter's financial difficulties, and wrote to Charles thanking him for the time he had spent on straightening out his affairs. On Christmas Day Charles wrote back, highly amused: "It has taken me rather more than *nine* hours to try and arrange his things!"[12]

Natty refused to discuss anything with Walter, neither his debts nor his amorous adventures. He asked Charles to assess the situation and report back to him. Emma was thankful for this small grain of comfort, for she dreaded above everything a confrontation between Natty and Walter. Walter himself was infinitely relieved. It was not until he handed over the keys of the linen baskets to Charles and told him the story of Marie and Lizzie that he realized how worried he had really been. It was like awakening from a terrible dream. But he made one fatal mistake for his future peace of mind: he retained one of the locked baskets. He did not mention the blackmailing couple, for he felt that his mother must be shielded from this particular infamy. A sordid escapade with two would-be actresses was one thing, but a scandalous divorce involving a society belle and a vindictive husband would prove a totally different kettle of fish. An unbearable thought. However trustworthy Charles appeared to be — it was safer to say nothing. When his brother suggested that it would be easier to deal with everything in his absence, Walter was even more relieved. His spirits began to rise and, like an excited schoolboy, he made preparations for a trip to the Sahara. He told Marie that under the circumstances they must part for the time being. Marie had to agree.

Meanwhile Charles re-organised the finances of the Museum, which was now to be severed from the Tring Park Estate Office and set up as a separate enterprise. After paying his debts and the mortgage on My Museum, Natty provided Walter with a not ungenerous settlement of which the capital was safely tied up. Charles then decided to make settlements on both Lizzie and Marie. They were each to receive a house and a sum equivalent to about £10,000 p.a., but only on condition that they never set foot in Tring or communicated personally or through a third party with any member of Walter's family. (Lizzie's house in Tring stood empty until after Walter's death.) They agreed to these conditions but quite independently inserted one proviso of their own: Walter must undertake in the one case never again to see Lizzie Ritchie and in the other never again to see Marie Fredensen. Meanwhile Richardson Carr wrote apologies for absence to the House of Commons (for Walter's name was again down for a dozen committees in 1908), the Tring District Council, the Board of Guardians, the British Museum Trustees, etc., etc.

On February 12th 1908 Walter and Hartert sailed for North Africa. They reached Marseilles on "a cold and frosty morning out of which a bright sun rose in an immaculate blue sky". Walter had really — at last — left New Court for ever! Travelling on to Biskra he saw from the train, across a flat expanse of water, hundreds and hundreds of flamingoes.... It was going to be a marvellous expedition. Before plunging into the desert he would book the entire hotel and arrange a tremendous party. History does not reveal Hartert's thoughts on the subject.

Both Charles and Emma were profoundly sorry for Walter, and knew how to express their silent sympathy. They consulted Hartert and Jordan and discovered that a separate wing was urgently needed at the Museum in which to house the now enormous insect collections — thus at the same time providing the birds with more space. When Walter returned to Tring after an absence of eight months he found the insect building almost completed. "Do not think I am not grateful for what my mother and brother have done"[13] he wrote to Hartert from Marseilles on the 25th June, "but we shall not have a perfect establishment until all the iron buildings are disappeared...." He was already optimistically full of plans for the future.

Meanwhile Lizzie had found a husband — her name was now Tenderson — and she vanished to Poland. Legend has it that after a brief but stormy period of matrimony she divorced and entered a convent. It is more probable, judging from her handwriting and the sheets of paper about two foot square on which she wrote her letters, that she had suffered a spell in a mental hospital. Walter, needless to say, paid scant attention to the clause in Lizzie's settlement which sought to deprive him of Marie's company, for he was very fond of Marie. Once he had returned from exile and settled down under his mother's roof, he again began leading his strange double life, but now everything was in a minor key. Walter felt middle aged — he was forty. Then suddenly Lizzie reappeared. She had been made independent but she didn't want independence — all she wanted was Walter. After the havoc she had wrought it is astonishing that he could bear the sight of her. But like the terrorists to-day, she was very difficult to deal with. Also she was paranoid. So Walter weakly forgave her. Perhaps he thought that an occasional visit and lots of letters from Tring provided the simplest solution. Despite his sluggish pen he was a tireless correspondent. At home, it seemed to his nieces and nephews that Walter was only rarely aware of their existence, but Lizzie's replies revealed that he not only noticed everything about the household and its inmates, but passed many of his observations on to her. It did not take long for Marie to hear of the wanderer's return, and a sense of insecurity again overwhelmed her. There must, she thought, be some reason for Walter's toleration of "that woman" after her insane and criminal behaviour — perhaps secretly, he had married Lizzie — out of fear, for he was, like all men, a moral coward. "Is she your WIFE," demanded Marie over and over again in an agony of apprehension, for she still truly loved Walter and despite everything continued to do so for the rest of her life.[14] But she needed constant reassurances on the score of his affection — tangible evidence in the shape of generous gifts of money. If he failed to respond she bought a cottage in the country, or a fur coat, or a piano, or a four weeks holiday for Olga and then wrote a despairing letter, swearing that this was the very, very last time.... "But you know how very unstable I am about money.... I try to do better though I feel it is too late." Once she remarked: "If I really didn't ask you for money you wouldn't believe it was a letter from me, would you?"

Walter rarely got angry and he realized that Marie, like his collectors, thought he was Croesus — a mean Croesus. But he began putting her letters away unopened. When he died there was a drawer full of these melancholy reminders of Oscar Wilde's best known poem.[15]

# CHAPTER 26

# Catherine wheels

JORDAN COMMENTED that few scientists were aware that there was "another active side to Walter's life...".[1] He did not add that this activity, as far as he could make out, consisted of a series of flying leaps from one sort of frying pan into yet another fire — whether the conflagration concerned hunting, politics or women.

Walter — when he was in his twenties — was dragged from his horse and manhandled by the "snobs" at a meet of the Rothschild Staghounds on the outskirts of Tring town. The "snobs" was a nickname given to the six hundred employees of a Tring Boot factory. The business had fallen on evil days, the management was inefficient, the economic climate was adverse, and the factory was about to close and throw the men out of work. They took a poor view of the frivolous rich and the indifferent local farming community who could hunt a carted deer for fun at such a moment of anxiety and distress. Natty was shaken by the episode, and the Staghounds never met in Tring town again.[2] He bought one of the disused factory buildings and helped organise free meals for the children of the ex-employees. Emma took a keen interest in the enterprise and supported it with endless gifts and gallons of hot soup.[3] Somehow this story is symbolical of Walter's whole life. He seemed to generate absurd excitement, whatever he did or wherever he went. Like old jokes in Punch, the rows in which he was involved have lost their significance for posterity, and to-day we puzzle over their meaning. Thus Walter walked into the House of Commons in a so-called white top hat and caused a minor sensation. The incident was reported in virtually every newspaper.[4] In view of the casual clothes worn in the House to-day, this seems incomprehensible to us, but at that time it constituted an irreverent outrage. To us it is also strange that political speeches delivered at some obscure county gathering by the local MP should elicit an immediate response from members of the Cabinet. Or did the name of Rothschild in 1905 — willy nilly — command attention?

Walter, despite his *avant garde* ideas about zoological nomenclature, was by conviction profoundly conservative — not only in his politics but in his tastes and his attitude to change. As far as he was concerned, legislation never changed for the better. He was also fearful of sweeping revolutionary movements of the proletariat, which he felt would not only destroy the old order but annihilate culture, science and learning. In fact he envisaged a modern Gengis Khan, the Khmer Rouge or something akin to the Chinese Cultural Revolution destroying Europe. This became a profound and gloomy conviction in old age.

Walter's great grandfather, Nathan Mayer, was essentially a man of finance, rather than a public or political figure. Baron Lionel, in the next generation, became a respected personality on the outer fringe of aristocratic society — the spokesman for the municipal middle class, still struggling for complete emancipation from the hierarchical system of the first half of the century, and a representative of liberal non-conformists as well as Jewry. Natty, Walter's father, was accepted in the governing aristocratic circles and sat in the House of Commons as a Liberal MP from the age of 25 until he was elevated to the peerage. In middle age he threw in his lot with the Liberal Unionists — Home Rule being a policy he could not accept from the Gladstone Liberals, but he never liked Tariff Reform and, despite his allegiance to Balfour and the latter's particular brand of conservatism, he retained at heart many of his Victorian liberal ideals. His conflict with Lloyd George left a false impression — namely that Natty had abandoned the beliefs he held in his youth, and was now a dyed-in-the-wool conservative.

The record books tell us seven[5] Rothschilds have served as MPs, five of whom were elected before 1910. These five all sat as Liberals,* until in 1886, when Gladstone split the party over Home Rule, they joined the Liberal Unionists. Ferdinand, Natty's brother-in-law and cousin, was more liberal than he, and believed in the emancipation of women, whereas Natty himself subscribed to the anti-suffragette league.[7] Walter also had no use for women in politics, and threw up his hands in horror when women County Councillors were approved — after a very noisy debate in the House. What next! Charles, on the other hand, was openly left wing, and while he was an undergraduate at Cambridge described himself as a radical socialist tinged with conservative individualistic views: "....glad if socialism gained ground, but aware that individualism must have its place".[8] "Participation" was one of his hobby horses.

Walter was really not particularly interested in national issues: he preferred local politics and specialist subjects such as the protection of wild life and

---

* The political parties in Britain at this period were all too confusing for continental writers. Poliakoff,[6] in his history of anti-semitism, designates all Rothschild MPs as Conservatives in Disraeli's party — even Baron Lionel!

inoculation against typhoid. He was elected unopposed at a by-election in 1899, after his uncle Ferdinand's death, and again at the election in 1900, and contested and won the seat in 1906. He was rather popular in the House: apparently Walter had no enemies in any walk of life. He made only two speeches as an MP, one in support of the Sea Fisheries Bill[9] and the other to oppose the Rickmansworth and Uxbridge Valley Water Bill[10] but he was a tireless speaker at local gatherings, averaging about two political meetings per month. In 1905 he addressed the annual dinner of the Chesham Conservative and Unionist Association,[11] and his speech elicited telegrams of protest from both Mr. W.E. Gladstone (son of the late Prime Minister) and Lord Balfour. He certainly had the knack of putting his foot in it. In the first instance Walter was championing the cause of the refugees from the Russian pogroms* and criticising the Aliens Act which would hinder the easy entry of refugees into England. Those unfortunates who deserted from the Czar's army rather than shoot down their co-religionists, and were hoping to find a refuge here, were caught on the German frontier and sent back, presumably to a dreadful fate in Russia. Owing to his vocal impediment, Walter "made a blunder in the heat of speaking"[12] and skipped a few phrases and described how the men were *shot on the frontier.* Natty had apparently given him private information concerning some unfortunate refugees who, having been turned back from English ports and elsewhere, had then disappeared.... Deserters were always shot. "I will try," wrote Walter desperately in his reply to Gladstone's furious telegram (March 17th) "to find out on Monday particularly from where he [Natty] got this information and if I am able to do so I will tell you on Tuesday, but anyhow it will have to be absolutely privately, as if once news of that kind is published i.e. the source of news, it puts a lot of people in danger." Then he was writing again next day[13] to emphasise the point that, whereas bankers on the continent had loaned money to the Russians, his family in the UK, owing to the disgraceful treatment of the Jews, had absolutely refused to do so. At the time it was common knowledge that the firm of N.M. Rothschild & Sons had foregone very large profits by refusing to make this loan.[14]

Balfour's displeasure sprang from an entirely different passage in the speech. An acrimonious debate had occurred on the previous day in the House, marked by an outburst of ill temper on the part of the Prime Minister (Campbell-Bannerman). Was Walter present? — We do not know — but dur-

---

* Both Walter and his father were almost obsessively concerned with the fate of the Russian Jews (see p. 33-36).

ing the Chesham speech he tackled the vexed question of Tariff Reform and a proposed tax on imported grain and foodstuffs. Walter was one of the Unionist Free Traders, a minority in the party. Like his father, he believed in Free Trade, and so did many of his constituents. The *Bucks Herald* quoted him as follows: "Mr. Balfour (whom without exception every Unionist in Britain and Ireland acknowledges as their leader) with a skill his followers were unable to fathom found that that line of argument would have been unfruitful in results.... and so instead of making his speech as leader of the unionist party he confined himself to the narrow limits of the leader of the opposition...."

The new Liberal government, with its huge majority, had simply wanted to place on record its economic views, and to Balfour's annoyance it simply laughed off his attempt to get a debate going at a serious level. Quite apart from Balfour's annoyance at not getting an answer, and his hopelessly weak support, he had to face accusations from his own followers that he was not giving a strong lead which would unite his party. The charge, coming as it did from those who were causing the disunity, exasperated Balfour and stung him into writing Walter an unusually long letter.[15] He began by suggesting that "the public" were suffering under a serious misconception, and he ended by claiming that the event to which Walter had drawn attention was "unexampled in parliamentary history". Alas, we neither know how Walter appeased Balfour, for his reply is lost, nor do we know how Natty reacted. Perhaps on this occasion he secretly agreed with Balaam's Ass.

When Walter first decided to enter Parliament he was still deeply involved in his love affair with the giant tortoises and he feared further encroachment on the meagre time he could allocate to his important scientific work. He soon discovered, however, that his seat in the House afforded him a golden excuse to leave New Court immediately before lunch, ostensibly for Westminster, and after a token visit to the House of Commons, proceeded at top speed to the Natural History Museum where he spent an enjoyable and profitable afternoon. Unfortunately one fine day his father was anxious to know how the House had voted in a particular division, and rang Walter to enquire. The official at the end of the line had obviously received no instructions for dealing with this eventuality — another example of Walter's carelessness — and replied immediately: "We haven't seen Mr. Walter here for months, my Lord." Natty was not pleased and made no secret of it.*

---

\* His sister-in-law also considered this a grave dereliction of duty. To-day the choice seems to us to have been rather a good one. But Rozsika's censoriousness sprang from a feeling of resentment that Walter should desert his post and leave the work and responsibility to Charles — for he, too, would have preferred an afternoon at South Kensington.

However Walter did not confine his disputes to extra-museum activities. Many of the wrangles which arose with his fellow scientists concerned the slights he received, real rather than imaginary, from certain professionals who resented him both as a successful amateur and also who greatly disliked the streak of charging rogue elephant in his make-up. On one occasion Walter submitted a paper for publication in the *Ibis*. The artist he had employed to illustrate the birds was Frohawk, and despite Walter telling the editor, Sclater, that he himself had worked out the descriptions, Sclater had said to Frohawk, "We cannot accept Mr. Rothschild's or your papers on birds as being new. That is not enough for this society until you bring me *Sharpe's* word for its being new".[16] Walter was very angry. He asked Günther to intervene on his behalf:[17] "I should be much obliged to you if you will either write or tell *Dr. Sclater* from me that I feel *very hurt* at the way he treated my paper on the new Rail. I have worked out all the birds I have received from the *Sandwich* and adjacent *islands* with Wilson and coming from you it will have more weight. I also wish you to tell him that I am not an ignorant amateur but that I am so far acquainted with zoology that I am quite able to find out for myself when a bird or an insect is new...."

Bowdler Sharpe, the leading ornithologist of the day, seems to have put in a good word for Walter, since the description was duly published in the *Ibis*.[18] Furthermore a few years later[19] he gave a boost to his wilting ego in an article written at the invitation of *The North London Illustrated*.

> *Mr. Walter Rothschild has won for himself an immortal name as a naturalist.... not only one of the keenest students living, but no man has a wider knowledge of natural history. As a child his interest in all classes of animals was wonderful, but he has been, from his childhood days, gifted with a phenomenal memory. The Museum at Tring in which he has placed his collections is simply a wonderful tribute to any man's energy. There is nothing like it in the world, and great as the name of Rothschild is in this country, there is no member of the family more remarkable than the Hon. Walter.... to those who consider the conditions of life in the present day it must be a source of congratulation to study the career of this young naturalist who makes such a noble use of wealth and opportunity.*

Walter was appointed a Trustee of the Natural History Museum in 1899[20] and, at the age of 31, was the youngest member of the Board[21] and served on it doggedly for 30 years — the longest period anyone had acted in this capacity. He was a regular attendant at quarterly meetings, and a most generous benefactor of all the departments — his gifts ranging from fossil plants to monkeys. He really cared about the B.M. He also held strong views concerning the arrangement of the collections, and here he fell foul of one of the Directors,

Ray Lankester. The few surviving letters[22] addressed by him to Walter are really surprisingly disrespectful in tone: for an employee to write thus to a Trustee was extraordinary. There is no doubt that Walter's "benevolent despot" manner, coupled with his stammering hesitancy and his wincing sensibility, encouraged a discreditable bullying tone in quick-witted men like Ray Lankester and Alfred Newton. This particular dispute (but there were many) concerned a "new" subspecies of giraffe with five horns instead of four, which Walter presented to South Kensington. He had financed the expedition which discovered it. He was so furious with the director for refusing to display the animal in the prominent position he had selected for it, that he promptly withdrew the gift[23] — to the great embarrassment of Ray Lankester who now had to offer some explanation to the Board of Trustees. Eventually Walter got his way — as he usually did — and the Museum got its giraffe. There arose an apocryphal tale that the two men, thereafter, had to take simultaneous vacations. For, in the director's absence, the animal was promptly moved downstairs, but if Walter left the metropolis it was swiftly hauled up to the top floor. What added to Ray Lankester's discomfiture was the fact that, in the meantime, the beautiful creature had been named Rothschild's Giraffe — in Walter's honour. (See. Appendix 1).

Walter was treated more leniently by the press than most members of his family, probably because no one understood what he was doing anyway — for the black beards of mountain gorillas, the wing of the flightless cormorant or even the fifth horn of Rothschild's Giraffe are not matters of burning interest to any but a handful of specialists. But it is an extraordinary fact that, at the time of the Great Row, when the rumour that Walter was "nearly bankrupt" was going the rounds of the clubs and House of Commons[24] — it was completely ignored by all the newspapers of the day. This is all the more astonishing in view of the popular interest shown about this period in Walter's father, "the real ruler of England" as the *Daily News* described him in 1908.[25] In fact the omission was the most sincere tribute ever paid by the press to Natty.

Some time after Walter's death an explanation of the Great Row and his departure from N.M. Rothschild & Sons appeared in Cecil Roth's book *The Magnificent Rothschilds*.[26] Cecil Roth chooses not to give references for the various statements he makes, on the grounds that to detail them "would be wearisome". He claims no original research, but his chief sources, so he says, were "memoirs of Victorian and Edwardian times". His style is modelled on Lloyd George's Limehouse speech — no doubt as a sales gimmick — but he has not got the Welshman's gift for sparkling political invective, and strikes a dreary note. His paragraph about Walter's *débâcle* runs as follows:

*His [Walter's] collection was by now becoming a terribly expensive luxury. There came a time when he made some speculations on a very large scale, which turned out unfortunately. It was not the sort of thing that his father would appreciate, and another way out of the trouble had to be found. In order to balance his accounts, he raised (it is said) a large sum on the security of his expectations, and insured his father's life for £200,000 so as to be certain to be able to repay when the time came. But one thing upset his plan. Large commitments of that sort are often divided by insurance companies with one another, so as to minimize the prospective risk. One day, Lord Rothschild was paying his customary weekly visit to the Alliance Assurance Company which had been founded by his grandfather and of which he was himself — as it were by hereditary right — Chairman. According to his usual practice, he asked to be shown a list of the major risks contracted during the week. To his amazement, at the head of it stood an insurance policy on his own life, taken out by his own son. The truth of the entire affair came out: and from that time, the heir to the title ceased to take any part in the work at New Court.*

Like the tale of the battle of Waterloo and the "seduction of the House of Lords" by Natty over the Licensing Bill, there were elements of truth in the story which encouraged a familiar "fungoid growth".[27] Walter was fairly heavily in debt,[28] although the grand total of the various sums involved was known to no-one but his father and his brother Charles. Even more amazing, Walter himself had never computed the details of his various purchases — not even the taxidermist Doggett's account — since for two years he had studiously ignored his mail. As we know, it took Charles, with four clerks, six weeks to open and sort his letters. The list which had stunned his father (in the Roth myth this became the Alliance Assurance list of weekly policies) was in fact the sorry inventory of Walter's worthless investments, which a worried Joseph Nauheim (chief clerk in the private department at the Bank) had, in desperation, screwed up his courage and shown to Natty. What Walter effectively concealed was the fact that these were investments he had made with his own funds at the request of the blackmailing peeress' husband. Most of the companies he sponsored were not listed on the Stock Exchange. One of his man's recommendations bore the improbable name of the Nonpoisonous Strike Anywhere Match Company. Michael Bucks[29] told me that this fantastic list was still a topic of conversation among the clerks at New Court fifteen years later when he first arrived on the scene, and it was no doubt the strange nature of Walter's speculations — rather than the size of the sum involved, which was never revealed — that resulted in the leaking of the story to the outside world and made his father finally agree with Walter that the city was definitely not his *forte*. Like his cousin Henri, he displayed a positive anti-talent for finance. Bucks, who was a shrewd man, thought that some outside party must have had a hand in it all....

When Lord Balcarres (a Unionist M.P. whose father (Lord Crawford) was a co-trustee of Walter at the British Museum) recorded the story in his diary, it had already burgeoned and Walter was reported to have borrowed money in anticipation of both his parents' deaths.[24] The Austrian lady mentioned below is also an invention — but at least there was a lady!

> "....what is really more important (than John Burns' views) at any rate more piquant, is that Walter Rothschild is on the verge of bankruptcy. Papa has already paid his debts once or twice; now he has speculated, and he has expended huge sums on a rather indifferent book about extinct birds [A copy of this book fetched £700 at Sotheby's in 1980] and they say that a lady friend has absorbed many shekels. Anyhow poor fat Walter has raised money on the post-obits of Papa and Mama. The former is furious: most of all that for the first time in history a Rothschild has speculated unsuccessfully. It is a great blow to the acumen of the family. They say that a meeting of the Tribe will be summoned at Frankfort or Vienna or wherever the financial headquarters are — so that Walter may be tied up more severely in the future. We don't want him to resign his seat tho' I fancy it is pretty safe. Personally I rather like him. He has certainly this much which is interesting — namely a clumsiness of person, voice and gesture which is quite unique.
>
> 9 Feb. 08. They say that Walter's mucker amounts to a sum between £750,000 and a million. It is gorgeous. But Ishmael has paid up, and there is to be no by-election. The lady's Austrian and the percent speculations dealt with Americana. It would be amusing if Lord Rothschild had to sell up half a dozen portraits of his gentiles.

The gleeful tone of Lord Balcarres' comments suggests that the tale would lose nothing in the telling — but his overt rejoicing is not directed at Walter, whom he considered a likeable figure of fun, but towards his formidable father. One can also detect the traditional note of anti-semitism already referred to as part of the conventional education of the aristocracy at this period.

It was a short step to the next embellishment — namely that Walter had secretly insured his father's life, although those who are familiar with the business of life insurance pointed out that the story is nonsense. In 1906-08 both parties would have been expected to sign the policy, and, furthermore, a medical certificate would have been required for Walter's father, who was sixty-eight years old at the time. As for the suggestion that the autocratic chairman of the Alliance would not have been informed instantly if any other company had attempted to "spread the risk" on such a policy — this was equally absurd. What Walter in fact did was something rather sensible: to avoid the possibility of landing his father with a large posthumous debt he insured his own life. If he died before his father the latter would receive £11,000. On his father's death the policy would lapse.

231

It is most unlikely that this distortion of the facts ever reached Walter's ears — certainly his sister-in-law had not heard the tale until, several years after his death, she read it in Roth — faithfully copied in due course by Morton[30] and Cowles.[31]

In a relatively recent publication it was suggested that Walter had willed a "niggardly" million to his niece.[32] Perhaps the author had confused the number of set butterflies in the Tring collection with the legacy, since both pieces of information had been given to her simultaneously — and which happened to be one of the most original and delightful bequests of all time: 140 mother-of-pearl handled fish knives (the forks were missing), a gold repeater watch, 600 copies of old sporting prints, a Pyreneean Hound, 500 live parakeets, two live Bruijn's Echidna and of course the browsing Galapagos tortoises.

British & Foreign Life & Fire Assurance Company,

No. 4, New Court, St. Swithin's Lane.

# CHAPTER 27

# If His Majesty's Government will send me a message....

IN MARCH 1914 Walter departed for six months collecting in Algeria, but he became apprehensive about the political situation in Europe and thought it prudent to return in June. He and Hartert collected birds together for the first two months. When the Lepidoptera season began in earnest his curator pushed on into the desert and Jordan joined Walter at Biskra: travelling eastwards they began collecting butterflies and moths, and at Hammam-Meskoutine were fortunate in capturing some new species of fleas for Charles.

This joint exploration of the Algerian fauna, coupled with Faroult's breeding enterprise, was enormously successful, and all three men produced good descriptions of the large collections they had amassed. Walter published no fewer than three dozen papers during 1914/15, but World War I put an end to his second great burst of energy, which under more propitious circumstances would have resulted in another big revision which he was turning over in his mind — the silk moths — temporarily abandoned following the Great Row.

Charles was less pessimistic about the imminence of war and decided to take his holiday in Hungary with his wife and two elder children. He was collecting butterflies at the end of July when he received a telegram from his father instructing him to return immediately. He only got home by the skin of his teeth — after an adventurous journey, partly on foot carrying the little girls — which lasted over a week. The station master at Dover was "mighty glad to see them walk off the boat", for Natty had been giving him a gruelling time.

The outbreak of hostilities was a shattering blow to the brothers. Their interests, whether business or zoological, were essentially international. Walter's enterprise was virtually brought to a standstill. For Charles the war added what proved to be an unbearable load of responsibility,* coupled as it was by intense worry concerning the greatly loved family in Hungary and Rozsika's distress on their behalf, which he could not alleviate. Charles always found the suffering of others hard to endure. War was anathema to him. He lacked the

---

* The New Court partners were all over 70 and Alfred was ill. Leopold's sons were away on active service and furthermore Charles gave much of his time to the Ministry of Munitions.

robust streak so characteristic of his father who — despite his age and the painful illness which, within seven months, was to prove mortal — immediately rose to the occasion and had the satisfaction of knowing that he had collected over a million pounds in his first big drive for the Red Cross.

Natty succumbed to a surgical operation in March 1915, and eighteen months later Charles left for Switzerland on account of his severe depressive illness. Walter had suddenly become his mother's sole companion. His life had turned full circle for it reminded him in a nostalgic ripple of the good old days at Tring, when he was a child and his father spent the week in London working at the Bank. He and his mother had then watched the hay-making in the Park during endless summer afternoons, while spotted woodpeckers drummed among the beech trees.

All his life Walter was a short-term optimist — almost an optimist of the moment, for a baddlyng of baby ducks or a freshly emerged hawk moth could instantly dispel his gloom — but an unshakeable, long-term pessimist. At this point in time he became hopelessly despondent about the future of the Jews in Europe. Not that he envisaged systematic extermination in German concentration camps, but he believed that financial chaos and revolution would follow in the wake of hostilities. The Jews, caught in this vicious pincer movement, the Red Peril on the one hand — a threat which constantly preoccupied Walter — with its many sinister side effects, and the destruction of the world trade on the other, would succumb to new waves of violent persecution and terror. With regard to the magnitude of his fears for the Jewish people in Europe, he was some twenty-five years ahead of events. One day during lunch at his home, one of his nephews, then a child of eight, asked in a high piping voice what would happen after the war. Walter fixed the boy with narrowed elephant eyes and paused grimly: "You will be put up against a wall and shot."

The children shrieked with laughter. Uncle Walter said such funny things.... In truth, he believed that this would happen.

Contemporary writers have been determined to assign to Walter a completely fortuitous role as the addressee of the Balfour Declaration. They view him like some sort of vast barrage balloon, which drifted into the picture as if by accident — a useful object which kept the enemy from bombing vital targets — attached to the ground by a light wire hawser bearing the tag "politically innocuous, communal, notable".[1] He is further described as a "tepid and belated convert to the cause" of Zionism.* Cecil Roth,[2] who was at

---

* Presumably Schama arrived at this conclusion by inference as he offers no evidence to support this belief. Meinertzhagen, who was a fanatical Zionist, remarks in his diaries that he and Walter had both ornithology and Zionism in common.[3] Had he found Walter "tepid" he would have certainly mentioned it, since he was brutally frank in the diaries (see also footnote on p. 241).

times spitefully anti-Rothschild, but especially anti-Natty, described the choice of the recipient of the Declaration as "incongruous". Walter, like Edmond de Rothschild,[4] did not commit his hopes and fears readily to paper, and unlike his more articulate French cousin (See Appendix 3). He was unable to talk of them freely to anyone.

Nevertheless his letter to Weizmann[5] on April 10th 1917 telling him he had arranged a meeting with Balfour — leaves no doubt about his own sentiments. "I fully realize the great importance of doing everything to further the Zionist cause with the Government in view of the persistent and puerile opposition carried out by Lucien Wolf and the C.C. (Conjoint Committee). Apart from the first and foremost great national aims of our people which are strikingly and consistently being urged now in every country, there is to my mind a very much greater need for establishing the real Jewish nation again in Palestine...."

About a month after Balfour[6] had invited the Zionists to submit a draft declaration, Walter wrote[7] to him as follows:

148 Piccadilly, W1
July 18th, 1917

*(Circulated to the War Cabinet)*

*Dear Mr. Balfour,*

*At last I am able to send you the formula you asked me for. If His Majesty's Government will send me a message on the lines of the formula, if they and you approve of it, I will hand it on to the Zionist Federation and also announce it at a meeting called for that purpose. I am sorry to say that our opponents commenced their campaign by a most reprehensible manoeuvre, namely to excite a disturbance by the cry of British Jews versus Foreign Jews. They commenced this last Sunday, when at the Board of Deputies they challenged the newly elected officers as to whether they were all of English birth (myself among them).*

*Yours Sincerely,*
*ROTHSCHILD*

Stein, in his brilliant book on the Balfour Declaration, discussed the question of the addressee in the following paragraph: "Why, it may be asked, was the letter to be addressed to Rothschild? Weizmann was the President of the English Zionist Federation, whereas Rothschild held no office either in the Federation or in the World Zionist Organisation, of which it was the English branch. It was, however, through Rothschild that the Zionists had submitted their formula in July. In transmitting it to Balfour he had suggested that the

Declaration he was asking for should be addressed, in the first instance, to himself. "If His Majesty's government will send me a message on the lines of the formula.... I will hand it on to the Zionist Federation". There is no difficulty in understanding why this arrangement was favoured. Sokolow, the titular representative of the World Zionist Organisation in England, was a foreigner and a member of the Executive of an international movement having its headquarters, at least nominally, in Berlin. Weizmann, though a British subject and the head of the English Zionist Federation, would have been an awkward choice, for eminent as was his personal standing, Sokolow was his senior in rank in the Zionist hierarchy. The selection of Rothschild not only avoided all these embarrassments but had the decisive advantage of associating the Declaration with the most potent name in Jewry."[8]

The arguments make good sense, but in fact none of them played any part at all in the phrasing of Walter's letter to Lord Balfour. It never entered his mind that the Declaration — if it was to be made to an individual — *could* be addressed to anyone but himself. He would have regarded any other mooted recipient as a nonsense. If his father had been alive it would have, automatically, been addressed to him, but since his father had died he was, willy nilly, their man on behalf of world Jewry. Whatever other misgivings Arthur Balfour may have entertained, he certainly had none on this score.

Weizmann, in his autobiography[9] (his memory is by no means faultless), asserts that Balfour asked him (his p. 262) to whom the letter should be addressed since he, Weizmann, was President of the English Federation of Zionists* and was the natural addressee. We can envisage the blank look of incredulity that would have come over Walter's face had he read this passage. To Walter Weizmann wrote:[10] "It will be rightly said that the name of the greatest house in Jewry was associated with the granting of the Magna Carta of Jewish liberation." Walter was a Zionist by persuasion, but before 1915 (the precise date of his "conversion" is not known) he played no active part in the emergence of Zionism either as a creed or a policy. But he had, from earliest childhood, witnessed the tireless and sustained efforts which both his father and mother had made on behalf of persecuted and impoverished Jews the

---

* Autobiographies are notoriously inaccurate and not infrequently tinged with wishful thinking. The careful, meticulous Herbert Samuel[11] provided his own rational explanation for the signal honour conferred on Lord Rothschild, asserting (see his p. 146) that the Balfour Declaration was addressed to him as "President of the English Federation of Zionists". However he gave Weizmann credit for being a man of outstanding qualities — even if he deprived him of his presidency — "Far sighted, tactful, passionately devoted to the Zionist idea, with tenacity, resourcefulness and remarkable powers of persuasion". He admitted that Zionism owed its astonishing achievements to his leadership. Curiously enough Blanche Dugdale also deprived Weizmann of that presidency and allocated it to Walter![12]

world over. If he had been at a conventional boarding school much of this would have passed him by. But very early in life he became aware of the importance of his father's personal influence in this sphere. Natty, by his rectitude, his genuine concern for the welfare of his fellow men, his zeal in helping those who sought his advice or his assistance, his keen mind and sober, well-informed judgements, coupled with a generous heart and a streak of gruff sentimentality, had moulded public and private opinion about the Jews. Those who eventually became involved politically in Great Britain with matters concerning Jewry and Palestine had known him personally. Few, if any of these politicians were moved solely by their concern for the fate of the Jews or their unhappy plight in Russia, Roumania or elsewhere. Although some were motivated by humanitarian and religious sentiments, their chief concern was, not unnaturally, with power politics and the future of the Empire. There was, however, a certain number, of which Balfour, Churchill and Lloyd George were all variations upon the same theme, in whom the two motives were inextricably mixed* — for their imagination had been fired by their childhood indoctrination which had engendered a romantic interest in the Old Testament** — but not infrequently the British political interests or military strategies played the major part in their final decisions. Robert Cecil was something of an exception, for he always doubted, gloomily, whether political or territorial advantages would accrue to Britain from espousal of the Zionist cause. He was, in fact, an ardent Zionist against his better judgement, whereas Churchill,[14] as far back as 1908, considered that "Jerusalem must be the ultimate goal.... the establishment of a strong, free Jewish state.... should be an immense advantage to the British Empire".

Walter's views, with a somewhat different emphasis, added up to much the same.*** He thought there were undeniable advantages to the British to procure a loyal bulwark in the Middle East, but as far as the Jews were

* Asquith wrote in his diary "The only partisan of this proposal (Samuel's memorandum) is Lloyd George, who I need not say does not care a damn for the Jews or their past or their future but thinks it an outrage to let the Holy Places pass into the possession or under the protectorate of "agnostic atheistic" France."[13] Asquith apparently attributed emotional or religious motives to Lloyd George rather than Imperial strategy.

** At the meeting in Gaster's house on February 7th, 1917[15] Herbert Samuel remarked "Even to-day the Bible exercises a vast influence over important classes of Englishmen and has won their desire to assist at a Jewish return to Palestine".

*** He also agreed with various British politicians, and for that matter Herzl and Brandeis too, that, hopefully, Zionism would prove "a counterpoise to the revolutionary elements in Jewry" and "a stabilizing force and a healthy outlet for revolutionary activity". It might defuse the Bela Kuhns and Szamuelis of this world, and attract Karl Marx to Tel-Aviv rather than Hampstead!

concerned Palestine and Zion was just a non-starter unless the British Army was on their side. Of all the big powers England had shown itself the most tolerant of their Jews and from whom they could expect justice, and when Robert Cecil said on a later occasion that "in supporting Zionism the country had been carrying out its true policy" he was inclined to agree. But he also believed that his father — although not pro-Herzl — had played a major role in rousing the interest and conscience of these eventually pro-Zionist statesmen, since it was under his influence that they had evolved a new image of Jews and Jewry. They had a genuine personal liking for Natty, and an easy relationship with him (Haldane,[16] for instance, had his own room at Tring, always at his disposal). He was someone who understood politics, talked their language, and with whom a reasonable discussion was possible, usually an informative one.

In his boyhood Walter sat in on many of these discussions which centred on the persecution of the Jews, especially in Eastern Europe. Joseph Chamberlain's interest in the Jewish problem — slight though it appeared to be — was due mainly to his acquaintance with Natty. He and Lansdowne stayed at Tring,[17] and Walter listened to the pros and cons of settling refugees in the Sinai peninsula (the El Arish Scheme), the British East Africa Protectorate and the Argentine. Louis Mallet, Churchill, Tyrrell and Milner* were other visitors, and they, too, were influenced by Lord Rothschild. Edward Grey — a deeply religious man at heart — who visited Tring regularly from 1905 onwards, had a double bond, since he was not only devoted to both Walter's parents, but shared with their son a great love of ornithology. After Natty's death in 1915 Emma spent thirty years in virtual seclusion, and, apart from the family, saw only a very few old friends. Edward Grey was among them, and in April 1922, a few weeks before his death, he wrote to her deploring his bad health, but saying he wished very much to visit her.[19] According to Vereté[20], Lloyd George and Balfour, with regard to their Zionist policies, were in fact followers of Grey. And Grey, although neither a Zionist nor pro-Zionist, "was willing to perform an act of Zionism policy" motivated by British political interest. But it was an act performed with sympathy and good will towards the Jews, thanks to his genuine respect and friendship for Natty and Emma. Odo (Theo) Russell, private secretary to Grey in 1916, had played his part by passing on and pressing Rozsika's views on the subject of Palestine and the Jews (see p. 246).

---

* Natty first invited Milner to Tring[18] to meet Mr. & Mrs. Chamberlain in September and November 1893. In 1924 Walter cabled Milner an offer of the vice-chairmanship of de Beers — this particular Tring contact spanned two generations.

In 1902 Walter's father had had some brief but friendly contact with Herzl[21] (the interview, Herzl noted, began in German but finished in English), but decided this idealist's dream of a return to the Holy Land was totally impracticable* — in fact a blue print for getting the settlers' throats cut by the Turks. Natty said as much to Weizmann — who had at long last achieved an interview — which infuriated the latter. Weizmann, whose jealousy of the Rothschilds was forever seeping through in his letters and reminiscences,** openly attributed Lord Rothschild's views to the fact that he was only concerned with the impregnable position he and his family had achieved in Great Britain and had neither sympathy for, nor understanding of his persecuted brethren. Natty was won over by Herbert Samuel's memorandum on "The Future of Palestine" (shown to him rather late in 1914 or early in 1915) which was also, apparently, quite well received by most members of the Cabinet. This document stressed the feasibility of a Jewish homeland, and it assumed a British Military presence in Palestine. As far as Natty was concerned this completely altered the situation for the Zionists and their aspirations — changing a wild dream into a sober and desirable possibility. Charles later pointed this out to Weizmann. "*As you know*," he wrote in a letter on June 9th, 1915[26] "my late father was strongly in favour of Mr. H. Samuel's*** scheme and I would give

---

* "It is so difficult to get Palestine" said Natty wistfully to Israel Zangwill.[22]

** Weizmann, at this stage, never missed the opportunity of a sly dig at the Rothschilds. Thus, for instance, Rozsika introduced him to Walter's French first cousin, Baron Henri, who she thought might be well disposed towards Zionism. Henri, who, as we have mentioned, had taken a medical degree and was a qualified pediatrician, during the war had organised a free distribution of milk in Paris which was still (in 1919) distributing £4000 of free milk per year, and had become a large factor in the commercial milk supply of Paris. In his letters Weizmann referred to him[23] scornfully as a "self styled scientist" and "the loafer". Henry's playboy act could not have deceived Weizmann for a moment. Emma, however, seems to have agreed in some measure with Weizmann, possibly because she disapproved so strongly of Henri's penchant for the stage and his design and building of the Théâtre Pigalle.[24] In any case, when she was nearly ninety, one of her granddaughters wrote to her and asked if there was any truth in the report in the Express that Henri had a brother who had become a priest. "Let me first answer your question whether our relative Henri had a brother by a decided negative and perhaps it is a blessing that a second edition of Henri does not exist.... The 'Daily Express' is I believe a very unreliable newspaper...." On the other hand, as a boy of ten Weizmann wrote to his teacher regarding "a place to which we (the Jew) can flee for help", and added that the thanks of the Jews must go to Sir Moses Montefiore and Baron Edmond de Rothschild for their work in Palestine.[25]

*** In April 1916, Herbert Samuel asked Charles to the Home Office for a discussion on Palestine, and later invited him to serve as a member of a committee which was dealing with some aspects of the financial and economic development of Palestine. Charles was too ill to accept, and suggested James de Rothschild should take his place.

any moral support I can to this movement.... which I cordially approve of...."
Earlier than this, in November 1914, Weizmann wrote to Dorothy de Rothschild:[27] "I was still more delighted to hear that Mr. Charles Rothschild
considers the Palestinian plan as the only possible future. I am always frightened of bringing up this problem before people who are not Zionists, lest they
should take one for a hopeless dreamer.... It was therefore more than gratify-
ing to hear from you, that an unbiased person like Mr. Rothschild looks upon
Palestine as a possible rational solution."

As we have seen, it was Natty who awakened Balfour's concern over the
plight of the Jews, although before he was a regular visitor to Tring, he had had
various discussions on Jewish matters with Sir Anthony de Rothschild and his
family at Aston Clinton.*[28] At the turn of the century he was writing to him
about the notorious Dreyfus case. Balfour believed Dreyfus was innocent and
that his conviction was illegal.** The letter was marked "private" and was sent
from North Berwick (15th September, 1899).[30]

*My dear Natty,*

> *Thanks for your note and the interesting letter from Paris which accompanies it. My
> information is to the effect that the case will be tried before the Court of "Cassation" in
> connection with certain illegalities into which the Court Martial have fallen, that the
> Court of Cassation will, on these grounds, quash the verdict, and thus, without legally
> rehabilitating Dreyfus, will put an end to all further proceedings — a lame and impo-
> tent conclusion! There seems no news of importance from South Africa....*

> Yours ever,
> ARTHUR BALFOUR

Walter and his father, as we have seen, had both been keen soldiers in their
day. The believed implicitly in the British Army, and felt that the probability of
a British victory in the Middle East added an entirely new dimension to the
Zionist dream. Now it could become a reality. In fact their attitude coincided
with that of Milner.

Against this background, which had been part of the fabric of his daily life
from boyhood to maturity, Walter regarded his meetings with Balfour on the

---

* In 1886 in a letter to her mother-in-law from Aston Clinton Emma[29] wrote, "Louise (Lady de
Rothschild) and Constance have a great gift of bringing pleasant people together. Mr. Arthur
Balfour, whom I have never met before, struck me as particularly clever and agreeable."

** Balfour's "pro-Dreyfusism" was influenced to some extent by his anti-catholicism.[31]

subject of the Declaration as an extension of the past scene — almost in the nature of a dialogue — with Sokolow and Weizmann playing a role similar to that of Hartert and Jordan at his Museum. Like his curators they were, fortunately, articulate and splendid draftsmen....* Just what was needed. Walter disliked politics and he felt ill at ease and embarrassed by the irritating currents and cross currents of Zionism,** its plots and counter plots, intrigues and machinations, its brilliant, touchy, quarrelsome, vociferous, ambitious, idealistic, zealous and tiresome supporters. Walter was a zoologist, a palaeontologist. His time scale was different from that of the politicians although Rozsika and Weizmann had fortunately galvanised him into activity and crystallised his long term cogitations into short term views. He looked beyond the turmoil of pro- and anti-Zionism which, he felt, just obscured the current scene.... On ethnological, historical and religious grounds Palestine was for the Jewish people. But the State — the Jewish State — must have the support and protection of the British Army. Walter had a rare genius for simplification. It was a miracle that, in his letter to Balfour, he did not write: "....I will hand it [the declaration] on to the Zionists". Perhaps his sister-in-law added the word "Federation".*** The original draft is lost, but the published version is sufficiently Walterian in style and approach to guarantee that the original lacked all punctuation. Again Rozsika may have added a few appropriate full stops. She certainly corrected one of Weizmann's own drafts.

The historians are right in one particular detail: if indeed Walter Rothschild had had, to quote Schama,[34] "the attention of posterity summarily thrust upon him", it was through the good offices of his sister-in-law, Rozsika, for whom he had both respect and deep admiration. The realization that it would be extremely helpful, perhaps essential, for Walter to publicise his latent views on Zionism, only dawned gradually as the dream began to crystallize and became first a possibility and then — unbelievably — a reality. Balfour had met Weizmann on several occasions between 1905 and 1915**** and had established an

* Walter checked the various drafts and on one occasion[32] did not insert Sokolow's suggested alterations but sent them in a separate letter to Balfour, the more to emphasise the point (see also footnote on p. 265).

** In one letter which Walter wrote to Weizmann on August 22nd[33] 1917 about a visit he had arranged to Lord Derby, he said: "I assure you there have been so many various statements and hole and corner intrigues I hardly know what I am favouring or what objecting to all I know is that I wish strongly to support the Zionist cause." This does not suggest a "tepid convert".

*** The term Zionist Federation nevertheless puzzled Stein.

**** It is enlightening to note in "Palestine and the Balfour Declaration" a minute prepared by Ormsby Gore, stating that[35] when the matter was first broached in 1916 by Sir Mark Sykes, "Dr. Weizmann was then unknown".

excellent *rapport* with him — a somewhat ambivalent relationship developed later which is described with insight by Stein. But the historians of the sixties and seventies fail to point out that among the crowd of agitated and agitating British-born Zionists, tepid, warm and red hot, upper, middle or working class, only Walter was on easy personal terms with Balfour. Officially it was "Dear Lord Rothschild", but privately his letters were addressed to "Dear Walter". (Walter's father usually signed himself "Affectionately" to A.B.). Particularly at this stage it was a help, for it made those in authority more accessible. More to the point it is impossible for the historian of to-day to envisage the quality of the aura which surrounded Walter's late father, and which still clung to the bearer of the magical title. It was a unique phenomenon. For the large mass of Jews in the United Kingdom, but especially for those in the East End of London, if Lord Rothschild decided to back something, be it a horse or Zionism, it was OK by them. They need not consider the pros and cons — they were ready to cheer.

It was Dorothy de Rothschild who suggested that Rozsika might coax Walter out of his familiar role of silent observer or at least persuade him to commit his views to paper. With this in mind James de Rothschild arranged a meeting between Rozsika and Weizmann to discuss the possibilities and put her fully in the picture. Weizmann, in his autobiography, states he had "to explain our viewpoint, our philosophy, our hopes in the most elementary terms". No doubt he did so! The more they thought about it, the more indispensable Walter appeared to be. However, it was obvious to his sister-in-law that he was in a very depressed state of mind and profoundly worried by a host of personal problems, although she was at that time totally unaware of the black secrets of the linen basket. This was July 1915 and Walter was still stunned by his father's death which had occurred four months previously. Although his relations with his father had been ambivalent he was, in a sense, deeply dependent on him, and he now felt as if a wrathful, all-powerful yet personal God had been removed, leaving a total void. He had a premonition of inevitable doom. New responsibilities and new financial difficulties assailed him. Although the publication of his father's will must have been a sore disappointment to the blackmailing couple baying at his heels, it nevertheless revealed that he had inherited an additional annuity of £3000 and they were determined to wrest it from him forthwith. He was deeply distressed at the mere possibility of inflicting upon his sorrowing mother the anguish of threatened scandal, and weakly gave way to their threats. Then it was obvious to him that his brother, following a virulent attack of influenza, was very ill indeed, and was suffering from a pathological depression.[36] Would it respond to treatment? Charles had only recently — after considerable difficulties with his uncles — succeeded his father as senior partner in N.M. Rothschild & Sons, and Walter's mind boggled at the thought of what would happen if, through his brother's enforced absence for any length of time, the bank was left in the hands of his two uncles.* He was also finding it difficult to organise his museum with half his staff in the forces (he continued to pay their salaries), and both his curators were worried by the war and their unenviable position as foreigners in a small, hostile market town. To crown it all, his own health was now giving rise to considerable secret anxiety: he experienced disquieting neurological symptoms and was assailed by intermittent bouts of intense fatigue, which in fact never altogether left him.** It was necessary to curtail his normal sixteen hours working day. Finally, he forced himself to tackle his own accounts and after temporarily assuaging the sordid blackmailing peeress, he was left with a yearly income of £10,000 with no capital to fall back on. How could he possibly maintain the Museum and pay his staff with this relatively paltry sum?

Rozsika was herself beset with terrible personal anxieties. She, too, realized that Charles was a desperately sick man, and although he participated in all the early talks with Weizmann and supported his wife in all her activities in this area, his depression was suicidal in intensity. Both she and his close friend Theo Russell feared for his life.*** She realized that it was essential that he should leave the Ministry of Munitions — where he was acting as their chief financial adviser — and seek treatment in Switzerland, but as a Hungarian by birth she would not dare accompany him, for she then rated her chances of returning before the end of hostilities as fairly remote. Furthermore Charles would certainly wish her to remain behind to look after the four children (all under ten years of age), his widowed mother and his affairs at N.M. Rothschild & Sons. She quailed at the thought of a long — how long? — separation. Then, for the first time in her life she was completely severed from her father and family — to whom in peace time she wrote several letters a day! Her father was

---

* Arthur Balfour hinted that he, too, did not altogether relish this idea, for he wrote to Walter's mother in 1915 saying how delighted he was that Charles had become a partner in the historic firm: "No step so far as an outsider can judge was more desirable — *or even necessary.*"

** He wrote to Sokolow about his ill health — as always minute-conscious — on July 3rd 1918[37] "I have been very unwell lately.... I may have a relapse but I hope not.... I will take the Chair subject to the two conditions that I am well and that I can catch the 6.30 train i.e. I must leave at 5.45."

*** Rozsika wrote to Weizmann on June 29th, 1915 "My husband often says he is sorry to have lived to see, or more, to know all this, and sometimes I think I agree with him."

over 70, and she feared with justification that she would never see him again. Her brother was fighting in the Austro-Hungarian cavalry, her 17-year-old nephew had lied about his age and had been killed in action, with the result that his mother had suffered a mental breakdown. Her other sisters were scattered in various military hospitals nursing the wounded. Finally the climate in England during the first world war was intensely hostile to foreigners such as herself — there was no distinction between, for instance, pro- or anti-Imperial Germany as there was later between anti- or pro-Nazi — they were all lumped together as enemies. Those of us who experienced only the climate of World War II, when the country was full of free French, free Poles, displaced persons, refugee commando troops and the Ordnance Corps manned by anti-Nazi Europeans, can hardly imagine the bitter hostility shown in 1914-18 to the unfortunate people who were either naturalised or married to Englishmen. Hartert's only son was born in England, fought in the British Army and was killed in action — yet the day his mother received this news she was stoned in Tring High Street. The soldiers in the hospital where Ada Jordan, aged 17, worked as a VAD, refused to be nursed by her when they discovered by chance that her father was a German by birth though English by nationality. And, most amazing of all, when Walter was called to London on urgent business in the middle of a meeting of the Tring Council — on which he had served as a founder member and which had accepted his father's benevolence for 15 years — a member proposed that all aliens or persons born alien (this presumably included his mother and Rozsika) should be imprisoned or deported. In the absence of the Chairman, Richardson Carr, the motion was carried! On his return from London Walter resigned on the spot.

Despite this background of personal anxiety Rozsika immediately put herself at James's disposal and agreed unhesitatingly to tackle Walter. Her experience in Hungary as a girl left her in no doubt on one score — anti-semitism was not going to evaporate after the war was concluded, no matter who was victorious. Her relations in England, blinded by the personal regard in which her father-in-law had been held, and his magical influence — were sadly misled if they believed that in Europe as a whole their co-religionists could look forward to better days. Anti-semitism might well decline in good times, but it would rise sharply again the moment the economic situation worsened. On a simple level it would never really change because Governments as well as office boys would always need a cat to kick downstairs.

Rozsika had accepted this state of affairs in Hungary — *numerus clausus*[38] and all — as a matter of course, as if it were part of some inviolable hierarchy, one of the disagreeable facts of life, comparable in its inevitability to the English climate, or the pains of childbirth, or the illogical fashion that men wore trousers and women skirts. She shared one quality with Walter: she was marvellously unselfconscious about her Jewishness, and like her brother-in-law she had never experienced any sense of inferiority about her race or religion. Because she accepted the situation as a fact of life she was completely free of the quirks of personality characteristic of persecuted minorities the world over. Her attitude towards her fellow men and women was one of interest and sympathy — and a lively desire to help all those in need, irrespective of creed, colour or race. During her pioneering campaign for the Save the Children Fund* in Hungary in December 1918, an admiring journalist said to her: "You are the only person I know who has succeeded in being at one and the same time a good Jewess and a good Hungarian." One of her most endearing qualities was the exuberant zest and determination with which, all her life, she tackled every problem, great and small — whether it was a new loan to the Hungarian Government from a pusillanimous N.M. Rothschild & Sons, or a bust bodice to restrain her adolescent daughter's unruly bosom. Her meeting with Weizmann was an unqualified success. The books which touch on this encounter tell us that Rozsika was "captured" by Weizmann's charisma and charm, but for once it was Weizmann who was swept off his feet. He thought Rozsika was one of the most beautiful and intelligent women he had ever met. On her side she was impressed by his political acumen — and how delightful it was to talk freely once again with a continental who knew the geography of Europe! She could never get used to the fact that Lloyd George, for instance, had only the haziest ideas of the national boundaries of the countries of central Europe — and although Weizmann shared her unbounded intellectual admiration for the British, he understood and endorsed her not inconsiderable emotional exasperation with the Island Breed! He felt that Rozsika not only possessed integrity and political sense, but was well informed, and determined. She was as brave as a lion. He could come straight to the point, and talk to her man to man. She not only served as a go-between with her tongue-tied brother-in-law, but obtained for Weizmann, in 1915, interviews with Lady Crewe, Lord Haldane, Lord Robert Cecil,** Theo Russell, and later with General Allenby, Walter's mother, Anthony de Rothschild and many others who she thought might prove useful to him, and even tackled

---

* Eventually founded by Eglantine Jebb in April 1919. In 1918 Rozsika gave an interview to the press and described the plight of the population of Hungary and the pathetic state of the children there. Banner headlines: "A Rothschild's sister starving" attracted a lot of useful publicity.

** After this meeting with Weizmann he became, in his own words, "A Zionist by passionate conviction".[39]

Edwin Montagu,[40] but with no success. She herself discussed their Zionist problems at great length with Theo Russell who was an intimate friend of Charles and herself. In fact during this period (1915-17) the Russells and their three children were living at Tring and spending the summer months with Rozsika at Ashton. Russell was assistant secretary to Balfour and Tyrell, and later to Grey, and it was probably through his influence that Grey in 1916 put forward the proposal for an autonomous Jewish commonwealth in Palestine. Theo Russell had a German wife and he understood the quality of continental anti-semitism better than most Englishmen. After the Declaration was published, Weizmann wrote to thank Rozsika for what she had done, and on November 11th 1917[41] she replied from Tring: "Very many thanks for your kind letter and most kind words which I am afraid I do not deserve — I have done so little, but that little with real belief in the cause and that was why I was able to influence some persons who have helped us. I need not say how happy I am and you must feel proud indeed of all you have achieved."

Rozsika considered the Declaration the most improbable happening of all time — certainly the most unlikely event she had witnessed during her life time — and she felt that Weizmann's role had not only been decisive but that he had somehow defied all natural laws in attaining his objective. She was not at all surprised when, in May 1917, Walter's suggestion that Lloyd George should receive Weizmann and himself together was ignored.[42] It was not only the memory of the many occasions when Natty had come off best in an argument and Lloyd George's desire, in consequence, to snub Walter,* but all ministers, including Churchill, were afraid of Weizmann — for they began an interview having decided to refuse some request, and finished by granting it.

It was strange, however, that no misunderstanding arose between Weizmann and Walter during the negotiations of the Balfour Declaration, for the circumstances were such that this could have easily occurred. There were a great many people in their entourage who would have liked nothing better than to engender a rift between them. In fact at Christmas 1917 some mischief was afoot and an unknown troublemaker at work, but fortunately Weizmann consulted Rozsika at an early stage; she quickly poured oil on the waters before the storm could break: "As regards my brother-in-law,"[43] (for there had also been a *mal entendu* with James de Rothschild) she wrote to Chaim, "you need not apprehend that any misunderstanding *could* arise regarding yourself — he

---

* Walter sent a letter by messenger on May 13th and was nonplussed because the man waited all day for an answer but none was forthcoming. C.P. Scott subsequently arranged a breakfast session between Weizmann and Lloyd George.

knows too well that the last thing you would do is to speak against him, quite apart from his absolute trust in your loyalty, he is so convinced that he has done all according to his lights in order to help the cause, that he never could believe that those with whom he worked should abuse him in any way.... I believe everyone's nerves are on edge through the terrible stress of present times...." She then went on to deplore the fact that Weizmann's holiday had to be — in view of his work and worry — so brief, and reminded him that Walter's mother hoped to see him before he left. "My little girl is still very, very weak and looks so frail and it will be a long time before I can be happy about her. I try to be cheerful and forget that this is the first Xmas since 11 years which Charles and I spend far away from one another." Perhaps Weizmann's conscience pricked him, for Rozsika had her worries too, and nothing more was heard about this particular "misunderstanding". Walter himself may never have known of its existence.

Twenty-five years later it so happened that the three of us — Chaim Weizmann, David Bergmann and I — were trapped together for four hours in an air-raid shelter. Weizmann was still talking of the meeting with Rozsika. "Your mother," he insisted, "was of a totally different calibre to any of the other Rothschilds, and for that matter your father, despite his ill health, shone with integrity and sincerity. But your mother was the most remarkable of all the Rothschilds. James," he snapped his fingers, "basically a lightweight — a devious, unpredictable fellow. Dollie, very, very sweet, simple, charming, delightful.... But what more can I say...." I interrupted his sentimental paean of praise about Rozsika* with a query. "Walter?...." he murmured, while the muffled sound of the Battle of Britain rattled and droned far overhead. "Walter?...." he paused with his head held slightly on one side, and, in the half light, a melancholy hooded expression drifted over his face. "I confess I never understood Lord Rothschild."

One of the most contradictory facets of Walter's enigmatical character was the fact he was a great gossip *manqué*. There was nothing he enjoyed more than

---

* Weizmann incidentally failed to correct the proofs of his autobiography, in which she is referred to as Jessica![44] Poor proof reading, however, could not account for the nonsensical letter he[45] wrote to his wife Vera in 1921 describing Rozsika's despair because Charles (recuperating in Switzerland and shortly to return home) had left her! Charles had changed his views about the desirability or need for a national home for the Jews in post war Europe following what he hoped would be a "just peace". Was Weizmann hurt or astonished because — at what he considered the whim of a sick husband — Rozsika had decided she must give up working actively for him and Zionism? Did he feel he required some explanation for Vera and posterity and "guessed" this was the real answer for Charles's long absence abroad? Or was this a moment of wishful thinking? In fact it is impossible to find a reasonable explanation for this letter.

imparting lively confidential bits of news, and his prodigious, freakish memory was in this respect most useful, since he remembered verbatim — with bland, ponderous, irritating accuracy — every conversation he had ever heard. But an unkind fate decreed that he could listen, but only rarely participate. You cannot gossip confidentially if you are apt to break out, suddenly and unexpectedly, into a loud bellow.

"Weizmann," he once remarked, with not a little insight, "is not the lofty detached chemist you believe him to be — he is bitterly jealous of Jimmy.* Up to the time Jimmy got himself seconded to the 39th Fusiliers and arrived in Palestine he enjoyed considerable personal acclaim, but once Jimmy was there he became a nonentity — thrust aside, forgotten, *ignored*. Weizmann could never forgive him. I know," he added in a triumphant, sepulchral whisper, "because Harry Dalmeny was on the spot and he told me". And his kindly elephant eyes twinkled.

* James de Rothschild

# CHAPTER 28

# Dear Lord Rothschild

IN HIS PRIVATE life Walter Rothschild was, as we have seen, a painfully shy, nervous, hesitant man. He was obsessed by the fear of upsetting or angering his parents — like an eternally guilty schoolboy — easily cowed and confused at home by quick repartee or a sharp word* — and bullied, albeit most affectionately, by his mother to the end of her life. He also worried about his health and John Foster,** a friend of the family, once jokingly remarked — with almost second sight — that he was the type of man "blackmailers would pray for". But he was full of tenacious courage when impersonal issues were involved. And he enjoyed a good fight. Rather unexpectedly, as we have seen, in his early thirties he took up boxing in an unsuccessful effort to reduce his weight. He saw everything as dead black or pure white. The world was peopled with the good and the bad, and he found no difficulty in deciding to whom he owed allegiance. Consequently, although a hopeless politician, he was a very successful partisan — a fact appreciated by the Aylesbury electorate who had returned him four times to the House of Commons,[1] unopposed on two occasions (1899 and 1900). One facet of his public support of the Zionist cause cannot be underestimated, namely his implicit, unquestioning confidence in his father's judgement. He would never have succumbed to Weizmann's persuasiveness, which had converted Balfour into an ardent Zionist unless he had known that in the last months of his life his father had seen and approved Herbert Samuel's memorandum and modified his own views accordingly[2]. Walter believed that British military power in the Middle East was a crucial factor, and shed an entirely different light on what, in 1902, had seemed like the crack-pot scheme of some half-crazed continental idealists, a view held to the bitter end by Edwin Montagu and certain other Anglo-Jews, and also sub-

---

* It was manifestly unfair to argue with Walter, since he could never find the words with which to reply in time, and he never in his life had the best of an exchange with his decisive sister-in-law.

** Sir John Foster, QC, MP for the Northwich Division of Chester (from 1945-74), Secretary of State for the Commonwealth Office, and a tireless worker on behalf of persecuted Jews all over the world.

scribed to by Asquith. In order to discuss this issue his mother organised a Zionist lunch at 148 Piccadilly at which Herbert Samuel,[3] James de Rothschild, Charles and Rozsika were also present. Unless she had known of Natty's views, it is unlikely that Emma would have acted as hostess to such a gathering, for Zionism had split the family in two. Walter was greatly fortified by the knowledge that at long last he had parental approval for something he believed in — for apart from his sister-in-law and his brother, there was only James in the family who had not lined up solidly and vociferously against him. Charles was now, in fact, sunk in deep melancholia, and James, despite his brilliant brain,* could not be considered a very serious character. Walter therefore found himself somewhat isolated. But although badly handicapped by his speech problems, he was not altogether averse to plunging into the limelight, for despite his perennial doubts about his own inherent ability, he was at times convinced that he had considerable powers of leadership which could somehow never be realized — certainly not while his father was alive.** How serious his speech problems could prove in an ordinary discussion was well illustrated by an entry in Richard Meinertzhagen's[4] diary. The latter had aroused Walter's wrath by putting the following question to Balfour during a lunch party given by Peggy Crewe: "Is this [the Balfour Declaration] a reward or a bribe to the Jews for past services and given in the hope of full support during the war?" Meinertzhagen summed up the discussion which followed in these words: "Walter Rothschild was quite incoherent."

As we have seen, Walter's chief concern throughout the period prior to the Declaration was linking the concept of a future "home" for the Jews with the

---

* "Never," said Walter's mother, anticipating in essence a *bon mot* of Churchill, "Has a member of this family done so little with so much." Lady Rothschild herself had no patience with any of her men folk, however brilliant their intellect, who put in less than a regular twelve hours work a day at their desk — nor did she approve of gambling, and it was no secret that James betted heavily on horses.... But Emma at this time was seventy and was inclined to live in the past. She did not realize that James had long since sown his wild oats and that, in fact, he played a dominant role in founding the State of Israel. James was actually more of a dyed-in-the-wool Zionist than his father Edmond. Edmond[5*6] — like Baron Hirsch — was first and foremost a founder of colonies, but, from religious conviction, linked to courage and vision he selected Palestine instead of the Argentine for the experiment. He was, in effect, a pre-Herzl Zionist. James joined Weizmann's statemanship and political acumen to his father's practical genius and vitality in the realization of an apparently impossible dream.

** The graphologist who analysed his handwriting independently came to this conclusion: "Inwardly he was liable to feel different from other men in a way that made him aware of being especially chosen, almost as a man of destiny". Walter had a streak of megalomania which showed itself in his love of outsize specimens for his museum. He felt outsize animals were magnificent in a special way.

British Empire. He felt that sooner or later those in authority must see the light, and, like Lloyd George[7] at a later date (April 2-3, 1917), would realize that the Russian and French claims were irrelevancies which could be swept aside. A Jewish Palestine without the British Army was a nonsense, while a well-disposed, loyal, Jewish population in that area would justify its existence by greatly adding to the strategic advantages which the Empire would thereby gain. Walter was quite consistent in pushing this point throughout the preparatory period. A conference was due to take place at Moses Gaster's[8] house on February 7th (1917) with Mark Sykes in the chair, to consider the document drafted by Sokolow on behalf of the "Declaration coalition" a revised version of which (see Stein's footnote, p. 369) had been sent to Sykes the previous week. Sokolow, when drafting this document, had explained to Brandeis that it must be acceptable, not only to the English Zionists, but also to the Rothschild family, adding, "I need not dwell on the value of their support."[9] The need for a document setting out briefly and clearly the Zionist programme was again strongly urged by James de Rothschild at the lunch at 148 Piccadilly on November 15th, 1916. Rozsika[10] passed this information back to Weizmann and told him that such a memorandum should be submitted immediately. The third and final draft was completed by November 25th, but it does not seem to have been finally presented to Walter until January 26th. His mother, who was appalled at bad English, made two "linguistic corrections" which were duly passed on to Weizmann by Rozsika,[11] together with some comments about the necessity of stressing the advantages to Britain of the Palestine scheme. At the eleventh hour Walter was afraid — unnecessarily as it turned out — that he would not be able to attend the conference at Moses Gaster's house, and so in great haste he dashed off one of his characteristic letters to Weizmann setting out a few points to which he wanted to draw Mark Sykes's attention[12]. Walter's speech problems and hesitations were surprisingly evident in his letters. He left out words, letters and punctuation. He could not spell and mixed up his tenses. On this occasion he inadvertently dated this communication February 8th instead of the 6th, and wrote "condominiun" for "condominium".

*Tring Park*
*8th February*[12]

*Dear Professor Weizmann*

*I am writing this in case I am prevented at the last moment from coming to the conference; and I should like you to lay it before Sir Mark Sykes. It is my earnest opinion that if the Palestinian scheme is to be a success that a Condominiun of France and England is out of the question and that England must have sole control.*

251

We do not know how Weizmann reacted to this letter, since Walter, after all, attended the conference, so presumably a verbal answer sufficed. However, Mark Sykes and the other members of the conference were apparently of the same opinion and "strongly opposed to an Anglo-French condominium".[13] By early April Lloyd George had reached the same conclusion and told Weizmann that he was "altogether opposed to a condominium with France". Like Walter, he took the robust line that might, in the guise of the British Army, was right. "We shall be there by conquest and shall remain."[14]

Walter's letter went on to deal with a number of subsidiary points:

> *I am fully in sympathy with and uphold thoroughly the appeal of the Jews, that under the scheme they should have full autonomy in regard to their private and personal concerns, such as education, religion and trading locally. Further I consider it necessary financially to establish a "Development Company" to make full use of the resources of the country. If it should be objected that a Development Company is synonymous with a Chartered Company and that the latter are not just now in favour, I should urge that this should be a Company really for the purposes of Development and not a profit making machine. It should be international with a Board of Directors consisting of Jews of all nations and if it proves possible to make its dividends free of taxation these should be limited to 3% any surplus being divided between the British local administration and an Irrigation authority. This company under the tutelage and control of the British administration, should carry on the general administration of the country other than in those matters the Jewish colonists wish to control. Then I should like to see as many of the British Governors administrators as possible to be Jews provided due care is taken that no anti-Zionists are thus appointed....[15]*

At 11.30 am Moses Gaster opened the proceedings with a statement, in general terms, of the aims of the Zionists. Apart from Gaster and Lord Rothschild, the following were present: Herbert Samuel, Mark Sykes, James de Rothschild, Nahum Sokolow, Chaim Weizmann, Joseph Cowen, Herbert Bentwich and Harry Sacher.

Walter was then called on to speak,[16] but as we have seen he had already handed a letter to Sir Mark Sykes. He stressed that he sympathised fully with the development of a Jewish state in Palestine under the British Crown, but was irrevocably opposed to a condominium. He made only a few points[17] which had not been set out in his letter. He felt that priority should be given to immigrants from oppressed communities — notably Russia and Roumania. He had no objection to the control of the Holy Places passing to other nations, Russia excepted, if this step was considered necessary. It would not be fitting that officials of a nation which was oppressing the Jews should exercise authority in Palestine.

Herbert Samuel, who spoke next, agreed with Gaster and Walter. In fact there was no dissenting voice. Sacher, however, stressed the difference between a nation and a state — the latter concept involved political obligations, whereas a nation was a spiritual entity — Jews outside Palestine would be members of the nation, and this implied no political involvement. Samuel drew a rather dubious parallel between this situation and the Roman Catholics and Rome. The military and strategic importance of Palestine to Britain was mentioned, the constitution of the Chartered Company, the boundaries and frontiers, the Arab National Movement, French and Italian aspirations, etc., etc., were discussed briefly.

The draft under revision, oulining a programme for the Jewish Settlement of Palestine had already been handed to Sir Mark Sykes. It was confirmed at the meeting and summarised thus:

### SUMMARY

*Palestine* to be recognised *as the Jewish National Home, with liberty of immigration to Jews of all countries, who are to enjoy full national, political, and civic rights; a Charter to be granted to a Jewish Company; Local Government* to be accorded to the Jewish population; *and the Hebrew language to be officially recognised.*

In April Weizmann again became alarmed by rumours of France's ambitions in relation to Palestine, and wrote to Walter describing the situation:[18]

*If I may express an opinion I would like to say that the most important point at present is to strengthen the hands of the British Government in these particular negotiations as much as possible. There can be no doubt at all that the French have no claim, and it is incredible that they should press in such an aggressive form a case for which there is no slightest justification, but of course I realize how difficult the position might become and I would be so grateful to hear your kind advice on this point. Would it not be advisable that you yourself together with Mr. Samuel should take the opportunity of seeing Mr. Balfour?*

This Walter arranged to do, but Rozsika only a few days later wrote to Weizmann saying that Balfour had left secretly[19] for the United States, so the meeting had been cancelled.

At this period Walter had his sights firmly fixed on one objective — a Jewish state[20] — but he was now to be lumbered with what seemed to him an unnecessary and particularly irritating fight, right in the foreground of the picture. But he plunged into the fray with great gusto and drive.

The schism in the Rothschild family was reflected in the whole of the Jewish community in Britain, and also for that matter, with different emphasis, in those of France, Germany and the United States. Probably every Jew and Jewess resolves the vexed question of so-called Jewish identity in a slightly different way, but broadly speaking the pro- and anti-Zionists of the day fell into two categories — those who believed the Jews were a religious community, and those who believed they were not only a religious community, but also a dispossessed nation. Both views were upheld staunchly with a sense of emotional and moral righteousness, tempered in each case with particular individual fears and anxieties, the outcome of centuries of persecution and harassment. It depended largely on personal experience, education and orientation whether the ordinary Jew felt he was of a different race, a different people, a different religion or constituted a combination of all three — but to be Jewish was certainly to be different — for there was no situation comparable to that of the Diaspora anywhere in history from which a lesson could be learned. The anti-Zionists were almost all "assimilationists", believing that for themselves and their co-religionists everywhere, the real solution to anti-semitism and persecution lay along the lines that had occurred in Britain — the winning of full rights of citizenship in the country of one's own choice, linked to a mutual respect and public-spirited co-operation, and the freedom of religious worship. The extremists like Edwin Montagu and Marie Perugia (Mrs. Leopold de Rothschild) considered the Zionists' creed a blueprint for anti-semitism. Montagu[21] thought that a "National Home" would turn all Jews outside Palestine into aliens, encourage their expulsion, change Palestine into a huge ghetto glorified by the cynical name of "home" and give every country a valid excuse to be rid of its chosen few.

Marie had a more feminine approach. She was an Austrian by birth and had identified herself completely with the country of her adoption. She loved England and its elegant life-style, and the mere thought of being compelled to emigrate to a hot, sandy Eastern country like Palestine filled her with absolute horror. She was vituperative about Walter's shameful and deceptive behaviour. In her opinion he was a traitor, who had disgraced and betrayed the memory of his father and grandfather, who, by their sagacity and lives of impeccable public service had led them all out of the shadow of the ghetto into the freedom and sunlight of the Vale of Aylesbury.[22] The thought of Walter — a butterfly buffoon — endangering her precious and hard earned niche in a corner of the British aristocratic scene — for among other things she could now claim a close and dear friendship with Alice Derby — filled her with spitting fury. Rozsika, who was becoming almost English in her understatements, mentioned to Weizmann that she had enjoyed "quite a breezy exchange of letters" with Marie. And then there was the hideous question of divided loyalties which had once again raised its ugly snout. Suppose for a moment there was a clash of interests between Great Britain and the envisaged Jewish Homeland? What then? Walter, who was all too familiar with the problems of his two German-born curators in wartime, thought this was sheer rubbish and said so bluntly. Rozsika remarked to him privately that Marie was the only woman alive who knew both Burke's Peerage and the Almanach de Gotha by heart.... whereas Richard Meinertzhagen told her to her face that she was a worshipper of the Golden Calf.[23]

Lucien Wolf, the Secretary of the Conjoint Foreign Committee, although not such a serious or powerful menace as Edwin Montagu, nevertheless, in one sense, caused Walter a lot more trouble. Curiously enough Lucien Wolf[24] was so nearly a Zionist himself that, sixty years on, the student of this period may well be astonished at the inability of the two sides to come together. For in April 1917* at the height of the pro- and anti-Zionist controversy, he wrote an article in the *Edinburgh Review* in which he recognised the existence of a new "Jewish Nationality". But he declared that while in Russia this new national movement offered a complete and practical solution for their Jewish problem "yet the idea that this movement could eventually embrace the whole Jewish people is only a dream, for in Western Europe so far as human prescience can tell, such a development of Jewish unity is impossible".

Lucien Wolf was the chief representative of the Conjoint Foreign Committee which was openly anti-Zionist. The committee was originally appointed by the Board of Deputies[25] to sustain and assist the cause of Jewish communities abroad, especially the persecuted minorities in Russia and Roumania. An offshoot of this organisation was the so-called Special Branch, of which Lucien Wolf was director, and which aimed at keeping a watchful eye over the fate of the Jews in the countries involved in the war. In fact Wolf gave up his journalism and devoted all his time to the cause. He was both a sincere and talented man. Furthermore he believed optimistically that a new liberalisation after the end of hostilities would make Jewish nationalism anachronistic — and Zionism a retrograde step. (Charles Rothschild later also inclined to this view (see p. 249)[7]. But he never subscribed to Marie's fears that Jewish nationalism was incompatible with British patriotism. There he agreed with Walter.

The two joint presidents of the Conjoint Committee were Claude Montefiore and D.L. Alexander. To add to the confusion the latter was, in addition,

---

* In the early days of Herzl's campaign, Lucien Wolf was frankly pro-Zionist, but later he changed his views (see Friedman, 1973 and the references cited by him). Vereté (1970) believed that despite his dislike of Zionism in its 1916 form, Wolf had a major share in the suggested pro-Zionist formula which was included in Grey's despatch of 11.3.1916 to Paris and Petrograd.

president of the Board of Deputies, while the former was also Chairman of the Anglo-Jewish Association.[26] Claude Montefiore was a sincerely religious man and he felt that the Jewish faith (as he understood it) was the rock on which the future of the Jewish people was built. They had emerged from the confines of the ghetto into world-wide freedom — something BEYOND nationalism. The Conjoint Committee was a powerful unit, for it represented the long established *élite* of Anglo-Jewry, and through Lucien Wolf had to some extent gained the ear of the Foreign Office. Stein, in his excellent book, suggests that the anti-Zionist attitude of the Conjoint Committee and the pro-Zionist attitude of the Board of Deputies reflected in addition a subsidiary conflict between the old order and the new — and was "mixed up with a struggle for power in the internal politics of Anglo-Jewry". Such a struggle may well have been in progress but if it was, it certainly passed Walter by.

Matters were brought to a head in May 1917 by an injudicious and violent protest on the part of Lucien Wolf and the Conjoint Committee, who decided they were being misled by the alleged assurances of H.M.G. to the Zionists, and must forthwith publish their views on the Palestine question, or they would be forestalled by Weizmann, Sokolow and the English Zionist Federation. A letter[27] signed by Claude Montefiore and D.L. Alexander was therefore published in *The Times* and was introduced as follows:

*THE FUTURE OF THE JEWS; PALESTINE AND ZIONISM; VIEWS OF ANGLO-JEWRY*

*In view of the statements and discussions lately published in the newspapers relative to a protected Jewish re-settlement in Palestine on a National basis, the Conjoint Foreign Committee of the Board of Deputies of British Jews and the Anglo-Jewish Association deem it necessary to place on record the views they hold on this important question.*

The letter included major sections dealing with "The Cultural Policy", "Nationality and Religion" and "Undesirable privileges".

Walter could hardly believe his eyes when he opened *The Times* on Thursday morning. He went charging along to his mother, waving the newspaper. First of all he knew that the Conjoint Committee had been told by the Foreign Office that a public discussion at this moment in time should be avoided. Secondly the Conjoint Committee had, in his view, no right to publish the Manifesto without the knowledge and the approval of the parent body, and was thus deliberately flouting the authority of the Board of Deputies. What was particularly galling to Walter was the certainty that had his father been alive this could never have happened, since Natty would have cut the capers of the Conjoint and seen to it that Lucien Wolf, an unqualified admirer of the old man, toed the F.O. line. The manifesto was therefore a sort of hidden insult to him personally. Thirdly, it was clearly creating a schism in British Jewry and at a most ill-chosen moment and, moreover, conveyed an entirely false — subversive — impression to the public, namely that British Jews were on the whole opposed to a National Home in Palestine. Walter was furious at this gratuitous act of sabotage and immediately dashed off an answer to *The Times*.[28]

> *Tring Park,*
> *May 25th.*

*Sir,*

*In your issue of the 24th inst. appears a long letter signed on behalf of the Conjoint Committee by Messrs. Alexander and Montefiore and entitled "The Future of the Jews". As a sincere believer both in the justice and benefits likely to accrue from the Zionist cause and aspirations, I trust you will allow me to reply to this letter. I consider it most unfortunate that this controversy should be raised at the present time, and the members of the Zionist organisation are the last people desirous of raising it. Our opponents, although a mere fraction of the Jewish opinion in the world, seek to interfere in the wishes and aspirations of by far the larger mass of the Jewish people. We Zionists cannot see how the establishment of an autonomous Jewish State under the aegis and protection of one of the Allied Powers can be considered for a moment to be in any way subversive to the position of loyalty of the very large part of the Jewish people who have identified themselves thoroughly with the citizenship of the countries in which they live. Our idea from the beginning has been to establish an autonomous centre, both spiritual and ethical, for all those members of the Jewish faith who felt drawn irresistibly to the ancient home of their faith and nationality in Palestine.*

*In the letter you have published the question also is raised of a chartered company. We Zionists have always felt that if Palestine is to be colonised by the Jews some machinery must be set up to receive the immigrants, settle them on the land, and to develop the land and to be generally a directing agency. I can only again emphasize that we Zionists have no wish for privileges at the expense of other nationalities, but only desire to be allowed to work out our destinies side by side with other nationalities in an autonomous State under the suverainty of one of the Allied Powers.*

> *Yours faithfully,*
> *ROTHSCHILD*

Walter was, in fact the right man to deal with this particular situation. He was, as we have seen, an isolated figure in British Jewry. Although he made not the slightest claim to be succeeding his father as lay-head of the Jewish community, the aura of the dead man nevertheless hung protectively about him. He was transparently, almost naively, sincere on the issue of Zionism, and had, at that period, neither political nor commercial axe to grind. But by far his greatest asset was his complete freedom from the fears and prejudices and the distinctive qualities, good and bad, of the average so-called assimilated Anglo-Jew. First and foremost he was proud and glad to be Jewish. Then his interests and ambitions lay in quite other fields. His outlook was international rather than parochial, yet owing to his life at home he had been kept in touch with the distinguished political figures of the day. He was eminently suitable for leading the new order emerging from the ranks of the Board of Deputies and directing its efforts towards the Balfour Declaration.

On July 20th the Deputies acknowledged this spontaneously and gave him a great ovation when he was eventually elected vice-president of their newly constituted board.* [29]

Before the board could meet to discuss the Conjoint Manifesto, the Anglo-Jewish Association also had to consider a resolution by Dr. Gaster, which amounted to a vote of censure. Much to Walter's dismay, Dr. Gaster withdrew his motion — "a huge mistake" he wrote to Weizmann — as a mark of personal respect for Claude Montefiore. As Walter feared, Wolf tried to use this fact at the Foreign Office as propaganda on behalf of the Conjoint's views. He then set to work to prepare the ground for the following Sunday's meeting of the Board, and, after a meeting at the Council of the United Synagogue, he asked Weizmann to come to 148 Piccadilly to discuss the campaign. He set out in a letter[30] one of the problems he had encountered while canvassing certain secular Deputies:

> *There are a number of Jews (including Mr. Emanuel) who, while disapproving of the action of the C.C. are held back from open support of us by a point which astonished me greatly. They say the C.C. are quite wrong in saying Judaism is a Creed and not a nationality but that we are equally wrong in saying Judaism is a Nationality and not a Creed. That the essence of Jewish Nationality is the Jewish Creed and that their fear is, that if the Nationality is put as the foundation of the Zionist cause without the creed being equally emphasised, there is every likelihood and possibility of the various countries of the World saying* AFTER *the establishment of the Jewish Nation in Palestine, to those who remain you must decide whether you are Jews or citizens of say England or France. Should this occur Mr. Emanuel and his friends are afraid many Jews who wish to remain citizens of their adopted countries will not only abjure their Jewish Nationality but also their creed and that it is just the Jewish Creed which has maintained the Jewish nationality as a living Nationality.*

---

* Herbert Samuel's brother Stuart, a man with essentially neutral views, veering towards Zionism was elected president. Later both Walter and Weizmann found him difficult and un-cooperative. The latter described him as "all grievances and pretentions".

On Sunday, June 17th 114 Deputies gathered at the Jews' College to consider the Conjoint Manifesto.[31] Mr. Elsley Zeitlyn (Dublin) moved the following motion which stood in his name:

*That this Board, having considered the views of the Conjoint Committee as promulgated in the communication published in the Times of the 24th May 1917, expresses profound disapproval of such views and dissatisfaction at the publication thereof, and declares that the Conjoint Committee has lost the confidence of the Board and calls upon the representatives of the Conjoint Committee to resign their appointments forthwith.*

Walter had only intended to speak very briefly, but he was so incensed by the sarcastic utterances of Mr. J. Prag — one of the anti-Zionist Deputies — that he decided to reply at rather greater length:

*Mr. Prag's claim that the views he has expounded are those of the National Zionists [is] quite erroneous. It is well known that so long as there was any question of Palestine as a national home for the Jews under Turkish dominion, my late father was most strongly opposed to it. But I was with him during the last six months of his life almost continuously, and I know, as I have read his letters on this subject and heard his statements, that when it was a question of a National Home for the Jews under one of the Allied Powers, he was prepared to throw himself whole-heartedly into the question. But I have always thought that such a Home was only meant for those people who could not or did not desire to consider themselves citizens of the country in which they lived, and I can truly say that the National Zionists have done nothing, and would never do anything, inconsistent with the status of the true British citizen, of which I am proud to be one, just as proud as I am of being a Jew. I am supporting the resolution that has been moved because I consider the statement issued by the Conjoint Committee un-Jewish and its manner of publication with the misrepresentations it contained un-English. As to the question raised about the emancipation of seven million Russian Jews, I know their views because I have seen a telegram from the Executive of the large meeting of Russian Jews the other day. It was very different in terms from the garbled message that has appeared in the papers. They are in favour of a National Home and they are going to strive for it. It was said at the meeting that they would not all go to Palestine, but those who remained would be zealous and good citizens of the Russian Republic. I have been asked by my constituents (the Manchester Great Synagogue) to come here and protest against the publication of the document, and in doing so I associate myself wholeheartedly with them because I feel the publication of the document to be a weapon which has brought about the very worst schism in the Jewish ranks. I consider that the Conjoint Committee and those connected with it had no right to speak in the way they did, because although everybody has a right to his own opinion, they gave in that document the impression to the English people and to the world at large, that they were speaking,*

259

*if not for a majority, then for a very large section of the Jewish people, while it is known that by far the largest number of the Jewish people are in favour of a National Home in Palestine.*

He rounded on Lucien Wolf and accused him of quibbling, and added that he did not wish his words to be turned and twisted and words to be put into his mouth which he (Walter) had never spoken....

Walter was delighted with the day's work, and wrote immediately to Weizmann — the opening sentence of the letter read something like his nephew's account of his latest house match.

*148 Piccadilly,*[32]
*17.6.1917*

*Dear Professor Weizmann,*

*I write to tell you that we beat them by 56-51 and Mr. Alexander, Mr. Henriques and the rest have all resigned. I have written to Mr. Balfour asking for an interview for yourself and me for Tuesday or Wednesday and I shall be able to prove to him that the majority of Jews are in favour of Zionism as we have forced the authors of the Manifesto to resign.*

*Yours sincerely,*
*ROTHSCHILD*

The Goodies had won.

During the ensuing critical months in the summer of 1917, Walter, — in the absence of most of his staff in the Army, — was working exceptionally hard in the Museum and struggling to fit in various important meetings with the Zionists and Ministers. His telegrams and letters to Weizmann were always polite, though only somewhat less peremptory than those he was wont to send to his curators. Between June and November he summoned him abruptly to Tring and 148 Piccadilly about ten or twelve times: "Could you come out by train 5.45 from Euston Monday or Tuesday and dine and sleep or else if you prefer it you could go back by train after dinner".*[33]

---

* He saw Sokolow[34] more often, especially when their final preparation of the Declaration was in progress. Thus Sokolow went to 148 Piccadilly four times between July 2-18 (1917). On one occasion Walter suggested meeting him on Paddington station at 10.15 am for a 15 minute discussion. Sokolow was not alone in receiving invitations to meet at stations (see p. 217).

Two days after the favourable outcome of the Board of Deputies' meeting on June 17th and Walter's letter to Balfour, he[35] and Weizmann were received by the Foreign Secretary. Walter had already informed him that they could now claim that the majority of British Jews were in favour of a National Home, and Weizmann urged him for a public declaration of support on the part of H.M.G. Balfour in turn asked for a draft to be submitted for ultimate consideration by the War Cabinet. Weizmann then departed for Gibraltar[36] and Sokolow took charge of the drafting of the proposed declaration. He was rather slow and it was not until July 13th that he sent a fair copy to Walter for his approval. Walter did not really like this draft, but he held what seems to be a rather curious view, namely that the precise wording was fairly unimportant, since it was better to get *something out* rather than delay matters by quibbling. He told Rozsika he would infinitely have preferred Sacher's* term of "Jewish State" to "Jewish Home" since "home" in zoological terms was a *nomen nudum*.[37] He received the second draft on July 18th and wrote on the same day[38] to Balfour enclosing the new draft.**

Draft Declaration

*1. His Majesty's Government accepts the principle that Palestine should be reconstituted as the National Home of the Jewish people.*

*2. His Majesty's Government will use its best endeavours to secure the achievement of this object and will discuss the necessary methods and means with the Zionist Organisation.*

An official acknowledgement came back on the 19th July, but under separate cover Balfour sent Walter a handwritten note marked "Private",[41] the opening paragraph of which filled him with deepest misgiving.

---

* At the meeting at Gaster's house on February 7th Walter had stated that he "sympathised fully with the development of a Jewish State under the British Crown". The formula Jewish National Home was doubtless intended to be sufficiently vague to be capable of more than one interpretation.

** The loss of documents is a phenomenon not confined to the Rothschild family for the original of this draft has also been lost, although in Harold Nicolson's own words: "I was only attached to Sir Mark Sykes to see that he did not lose documents".[39] This draft and the covering letter from Lord Rothschild to Balfour, and the latter's acknowledgement constituted the total "correspondence which had passed between the Secretary of State for foreign affairs and Lord Rothschild on the question of the policy to be adopted to the Zionist movement" which the War Cabinet had under consideration on September 3rd and which Lloyd George mentions in the *Truth about the Peace Treaties.*[40]

*My dear Walter,*

*Many thanks for your letter of July 18th. I will have the formula which you sent me carefully considered, but the matter is of course of the highest importance and I fear it may be necessary to refer it to the Cabinet. I shall not therefore be able to let you have an answer as soon as I should otherwise have wished to do.*

Rozsika came to lunch at 148 Piccadilly and found Walter steeped in gloom. There was nothing he feared now as much as delay, for in mid July a great disaster had occurred.... Edwin Montagu, Enemy No. 1, had been appointed a Minister of Cabinet Rank.*

Walter's fears concerning Montagu's appointment were in part due to a curious episode which had nothing to do with Palestine or the Jews. He was, of course, fully aware of Montagu's rabid anti-Zionist views, but he knew him better in quite another sphere, for he was, like himself, a keen ornithologist. Not so long ago a fine collection of bird skins had come on the market which contained specimens required by the British Museum as well as by Walter himself and Montagu. He discussed this problem with his brother Charles, who tackled Edwin, whom he also knew very well. In fact Montagu had given him, as a token of regard, the excellent zoological drawings he had made as a student at Cambridge. It was agreed among the three of them that in the interest of the British Museum the two private collectors would abstain from bidding. Much to the brothers' surprise, the British Museum failed to buy the birds in question — some totally unknown collector having paid an exorbitant price for the lot. In zoological circles it is very difficult to keep such matters quiet, and to Walter's rage and Charles's intense chagrin — for he had up to that moment thought of Edwin as a true friend and public-spirited man — it was presently revealed that Montagu had purchased the birds for himself under a pseudonym, or *"nom-de-plume"* as someone suggested. As far as Walter was concerned, this act put him beyond the pale. Moreover he now knew that the Zionists were confronted with an unscrupulous and ruthless opponent, who would clearly stop at nothing.** He told Weizmann as much without revealing his reasons for thinking so. Walter never offered explanations, but he expressed the gloomy view that Montagu's appointment sounded

* Walter's alarm was justified, for although his fears that no declaration was now going to take place were exaggerated, the emasculating of the declaration was certainly the outcome of Montagu's appointment coupled with Lucien Wolf's shadowy influence.

** It is a curious fact that Edwin Montagu and Edward Grey, whose views on Palestine were diametrically opposed, were associated in forming a bird reserve.

the death knell of the Declaration. In fact, as late as October 13th (1917) he was writing to Sokolow in a mood of deep discouragement,[42] saying that he now feared the Government would decide against a Declaration. About the middle of September Walter heard that at a Cabinet meeting Montagu had indeed succeeded in vetoing consideration of the draft Declaration. He was immediately seized with panic. He scrawled a note to Weizmann, complete with the inevitable spelling mistake, short of punctuation and with his characteristic double hyphens.[43]

> *Tring Park*
> *Tring*
> *18 9 1917*

*Dear Dr. Weizmann,*

> *I have written to Mr. Balfour for an inter=*
> *=view Thursday or Friday and have asked him to*
> *telegraph as soon as I get the wire I will arrange*
> *for us to meet beforehand. Do you remember I said*
> *to you in London as soon as I saw the announce=*
> *=ment in the paper of Montague's app=*
> *=pointment that I was afraid we were done.*

> *Yours sincerely,*
> *ROTHSCHILD*

Three days later, on September 21st, Walter obtained his interview. Much relieved at his friendly reception, he immediately dashed off another of his inimitable letters[44] to Weizmann, whom Balfour had already seen.

> *Tring Park*
> *Tring*
> *Sept 21st 1917*
> *6.30 p.m.*

*Dear Dr. Weizmann,*

> *I have just got back. The interview was most cordial. Mr. Balfour began before I could open my lips, by saying he had seen you and that he had told you that in his, and the Prime Minister's absence, the cabinet had discussed the matter and had concluded the moment was not opportune for a declaration. I then proceeded to say that I knew this, but that both in America and Russia there was much impatience and it was felt that as France and Italy had declared their policy England ought to do so to; I then added the financial figures, and told him of my aunt's libel and about Colonel Patterson. He then*

---

\* There is quite a hopeful note in this letter which Stein (see his p. 510) failed to appreciate.

*said I will try and reopen the question with the Prime Minister. I then asked him to try and arrange an interview with himself the Prime Minister and ourselves and he promised to do so. I then said I had evidence that I member of the Cabinet was working hard against us. He hastily said "he is not a member of the War Cabinet only of the Government — I think his views are quite mistaken" He then began talking about Brandeis and said we ought to urge him and the leading members of the American Zionists to bring pressure to bear on President Wilson to urge the necessity of the President pushing the matter as urgent with our Government. As he said this I thought it advisable to say nothing about telegraphing the interview but it is for you to wire Brandeis that you and I have reason to believe that if he Brandeis pushes the President and urges him to say something in favour of the Zionist movement to our Foreign Office (Foreign Minister) it might be of great use. But do not mention the interview or the actual source of our belief.*

*Yours sincerely,*
*ROTHSCHILD*

Then he wrote a postscript.

*After posting my letter, I suddenly remembered that, unlike to France, letters and wires to America are censured [sic] both sides. It therefore is advisable, whether you cable or write, to only say to Judge Brandeis that you and I "Have reason to think it of supreme importance for more energetic propaganda to be made in highest influential quarters" on the other side of the Herring Pond. I mean to use my words in " " and not to mention either the President's name or our Foreign Office.*

The panic was over, at least for the time being.

But the following day, on September 22nd, Walter decided to write another letter to Balfour for he felt that it was necessary to stress the fact that the Germans not only had their eye on Palestine, but were also touting for Zionist support at home.[45]

*Tring Park*
*Tring*
*Sept. 22nd 1917*

*Dear Mr. Balfour*

*There was one point I forgot to mention on Friday and I think you might draw the Prime Minister's attention to this; during the last few weeks the official and semi-*

*official German newspapers have been making many statements, all to the effect that in the Peace Negotiations the* Central Powers *must make a condition for Palestine to be a Jewish settlement under German protection. I therefore think it important that the British declaration should forestall any such move. If you, as you promised, can arrange the interview I suggested please let Dr. Weizmann know as I am going away for a few days on some special business and Dr. Weizmann can get at me quicker than if the message is sent to me direct as there will be no responsible person at Tring as my mother is away also.*

> *Yours sincerely,*
> *ROTHSCHILD*

It so happened that a few days later (October 2nd) a despatch from the British Consul at Berne strongly corroborated Walter's suggestion that the Germans were endeavouring to make use of Zionist aspirations for their own aims. The question of the Declaration came before the War Cabinet on October 4th, 1917, and in presenting the case Balfour relied upon three main arguments, the first of which was Walter's point that the German Government were making great efforts to capture the sympathy of the Zionist movement.

Nevertheless Walter, with good reason, was still obsessed by the Montagu menace and he and Weizmann thought it necessary to despatch another memorandum[46] to the Foreign Office (on October 3rd) voicing their fears. Further delay, they felt, might be disastrous and the Declaration permanently shelved. "We cannot ignore rumours that.... the anti Zionist view will be urged at the meeting of the War Cabinet by a prominent Englishman of the Jewish faith who is not a member of the War Cabinet." We do not know who drafted that memo, but this phrase has a characteristic Walterian ring tactfully watered down by one of his henchmen. One is pretty confident that this time Emma did not correct the English.

It was over three months after sending off the first draft[47] of the Declaration to Balfour that Walter received the amended text in its final form.* A message

---

\* Sokolow had sent Walter several drafts[48] for his consideration on July 4th, and altogether seven different versions were produced. Later, during Weizmann's absence in Egypt, Sokolow suggested that he and Walter should meet at least once a week — either at Tring or in London — to keep him informed with regard to the latest developments and to "Have the benefit of your valuable opinion with regard to them".[49]

to the effect that the Cabinet had at long last approved the Declaration must have reached him at Tring on October 31st, for the following evening[50] — after waiting all day for a letter to arrive — he wrote to Weizmann congratulating him and Sokolow "on the very considerable step forward in our prospects". Balfour's own communication did not reach Walter until the evening of November 2nd.* Did he open it at his desk at the Museum?** We do not know. At the time he was hard at work on the birds of Rossel Island and the Apollo butterflies — during 1917 he had completed six papers of his own and got *Novitates Zoologicae* off to press as usual. The coloured plates depicting the heads of Moorhens had proved very troublesome....

His brother Charles, like his mother, slit open envelopes with a paper knife, so neatly that they still appeared intact: Walter, like his father, completely destroyed them in the process. On this occasion he was a little more careful. When he eventually removed the heavily sealed envelope, marked "Secret", and realized that the scruffy bit of typescript was in fact the Declaration, did he shout: "Hartert! Hartert! Schauen sie! Schauen sie! SEE WHAT WE'VE GOT!'"? It is more probable that he gazed at it in silent disbelief.

On December 2nd a Great Thanksgiving[52] Meeting was held at the London Opera House. In order that Walter could preside, the meeting had been postponed for a week, for on the 19th November the news of Evelyn de Rothschild's and Neil Primrose's death in action in Palestine had reached the family.*** Balfour unfortunately was prevented from attending, but Robert

---

* Walter sent Weizmann Balfour's personal covering letter when he wrote to him on November 7th/8th (?) telling him to come and see him at 148 Piccadilly that day. He wrote again on November 11th asking Weizmann to return this letter as he wished "to retain it among other similar documents".[51] But, like so much of the Rothschild correspondence with Balfour — whether from Natty or Walter — it has vanished. Dr. Nehama Chalom has confirmed that this letter is not at the Chaim Weizmann Archives at Rehovot.

** Walter's envelope was addressed to 148 Piccadilly and taken by messenger to Tring. Sir Ronald Graham wrote to Weizmann on the 1st November 1917 informing him that the Declaration had been dispatched to Lord Rothschild by Balfour.

*** It was constantly reiterated — in order to demonstrate Marie Perugia's violent anti-Zionism — that when her son was killed fighting in Palestine with the British Army[53] she wrote to Weizmann ordering him not to make pro-Zionist propaganda out of this melancholy event. It is a dreadful irony that Evelyn was killed fighting in the Holy Land. But in fact it was Rozsika[54] who wrote to Weizmann begging him to be careful on this score, for Marie had already received some extremely tactlessly worded condolences from his fellow Zionists, and Rozsika feared that the bitterness and ill feeling between the pro- and anti-Zionists would thus be exacerbated. Marie had, in fact, suffered agonies during the war, for she had various members of her family in Austria, and when the second world war threatened, she announced she could never bear to live through another.

Cecil, Mark Sykes, Herbert Samuel and Chaim Weizmann were on the plat-form and of course James de Rothschild and his sister-in-law Rozsika.* The so-called Anglo-Jews were conspicuous by their absence, but the hall was packed to overflowing, in fact many people failed to gain admittance.

Walter was in high spirits. He began his address by pointing out that this was the most momentous occasion in Jewish History for 1800 years.... Apart from touching briefly upon the more obvious and critical topics such as the escape from persecution and oppression, the need to respect the rights and privileges of non-Jews in Palestine, and the compatibility of patriotism and Zionism, he pressed the necessity for cultivation of the land and a good agricultural policy in the future. Walter had always taken an ecological view of Palestine. He then moved the resolution.[52]

> *That this mass meeting, representing all sections of the Jewish Community in the United Kingdom, conveys to His Majesty's Government an expression of heartfelt gratitude for their Declaration in favour of the establishment in Palestine of a national home for the Jewish people. It assures His Majesty's Government that their historic action in support of the national aspirations of the Jewish people has evoked among Jews the most profound sentiments of joy. This meeting further pledges its utmost endeavours to give its whole-hearted support to the Zionist cause.*

Walter was delighted with Robert Cecil's speech, for he tactfully bracketed the two draftsmen Sokolow and Weizmann with himself, which was exactly as it should be, and then he went on to exclaim: "Our wish is that Arabian countries shall be for the Arabs, Armenia for the Armenians and *Judea for the Jews*". Walter had not expected the Under Secretary to have gone as far as that! He looked at him thoughtfully with his characteristic half-lowered elephant eyes. Was it possible that dry, cool, Cecil had been slightly carried away? Herbert Samuel and Walter had long ago given up the unequal struggle of understanding one another and, having agreed that each was incomprehens-ible, got on rather well. But again Walter was astonished that for once the stony Samuel, after a measured, sober introduction, became emotional — almost rabble-rousing, in his sincerity. He finished his speech: "....we shall be able to say, not as a pious and distant wish, but as a near and confident hope, 'Next year in Jerusalem'!" Loud and prolonged cheers followed. Samuel was accorded a mini ovation.

---

* Also Dorothy de Rothschild, Israel Zangwill, Sokolow, the Chief Rabbi, the Haham, Moses Gaster. Walter had refused to allow Dr. Tschlenow to speak unless he did so in French, Polish or Russian.[55] He felt that a speech in German would create a bad impression.

Until this moment the crowd packed into the Opera House had listened in silent attention to the speeches from the platform. But now a wave of restlessness and pent-up emotion passed through the hall — people had had enough of this sort of thing. The resolution had been seconded — they wanted to celebrate. Without the benefit of loudspeakers the crowd of well wishers outside, who had failed to gain admittance, felt cheated — they wanted to join in too. What was the cheering for? What was going on inside anyway? — and they began to jostle and shout. Cat calls and hullos echoed round the hall, scuffles broke out near the entrance doors, and a missile flew dangerously past Sokolow's head. In a few seconds pandemonium broke out, and Rozsika realized with horror that the situation was virtually out of control. The scene was set for total disaster. Walter, after a moment of stunned surprise — as if awakening from a dream — suddenly lumbered to his feet, and faced the turbulent well of the hall — six foot three tall, and weighing 22 stone, with an enormous bald head. It was obvious that he was very angry. There was an awkward pause. Then he roared:

## SILENCE!

Later, everyone agreed that, unquestionably, this was the loudest voice ever to roll round the Opera House. Almost immediately the noise subsided — like air escaping from a pricked balloon. People again sat down, sheepishly, in their allotted places. The audience was suddenly aware of the aura and authority of Walter's late father — this was the man to whom the Balfour Declaration was addressed. Rozsika said to herself that for the first and probably the last time in her life she had actually witnessed a miracle.

Walter swept the silent hall with a half-lowered gaze, like a trumpeting bull elephant about to charge. Someone on the platform half expected his ears to be raised.... Not a sound rose from the auditorium. In a frozen hush he sat down.

At that moment, the stunned committee sitting on the platform, saw the chair splinter like matchwood beneath his weight, and collapse pancake-wise with a resounding crash onto the floor boards, each leg directed starkly to another point of the compass. Chaim Weizmann and Sokolow helped Walter to his feet, noting with surprise that he was remarkably agile for a man of his size. There was no murmur from the hall, no nervous giggle from the ladies on the platform. Not a pin dropped.

Walter thanked the two men in his awkwardly polite manner, and sat down ponderously on another chair. He signalled the Chief Rabbi to continue.

Rozsika eventually reached home — and threading her way through the crowd outside the Opera House proved something of an ordeal — all the children were long since in bed and sound asleep. But at breakfast next morning her daughter, aged nine, was all agog to hear the news. What had happened at the meeting? Tell! Tell! Somehow this child, while scooping out the bottom of her boiled egg, got the impression — an impression dispelled only years later — that, after all, only ONE person had been there — Uncle Walter.

# CHAPTER 29

# What has become of two ostriches....

| | |
|---|---|
| Leopold Amery: | "The Balfour Declaration is one of the most momentous declarations that have ever been made in recent history." |
| Mark Sykes: | ".... a turning point in the history of the world." |
| Lord Meath: | ".... will be recorded as one of the great landmarks of history." |
| Lord Robert Cecil: | "One of the greatest achievements of the war...." |
| Lord Rothschild: | "The greatest event that has occurred in Jewish History for the last eighteen hundred years." |
| Lord Crewe: | ".... the partial attainment of a great ideal and the reward of an undying faith." |
| David Lloyd George: | ".... a contract with Jewry." |
| Charles Webster: | ".... the greatest act of diplomatic statesmanship of the first world war." |
| Lord Harlech: | ".... behind it all was the finger of Almighty God." |

The Declaration was received with a burst of vocal euphoria[1] although in private there were expressions of deep misgivings.

Richard Meinertzhagen, at this time Chief Political Officer of H.M.G. in Cairo — as well as author of an excellent book on the birds of Egypt — who was starkly realistic and did not keep his doubts to himself, described[2] the

Declaration as "an ambiguous document.... gives with one hand and denies with the other.... and to me it is disappointing".... and again; "The Balfour Declaration, which Weizmann regards as a great document, a charter of freedom, is in fact a paradox meaning nothing at all, like so many other things emanating from A.J.B."

Walter, although to some extent agreeing with his fellow ornithologist, still stuck stubbornly to his guns and with his sights, as usual, fixed on the bare essentials, insisted that there was only one crucial point — to have got something out in black and white which was, in a general way, positive.

In the Epilogue to his "Middle East Diary"[3] Meinertzhagen supports Walter's viewpoint and says: "The contradiction in the Balfour Declaration is obvious and is a masterpiece of rendering an important document one hundred percent nonsense.... But all the same it gave the Jews an opportunity and they seized it."

Once the Declaration had been published Walter felt his principal role was over and the professionals must now take over, but he was willing and indeed anxious to assist Weizmann by keeping an eye on the disruptive elements still lurking in the background at home. He wrote (as noted on p.266) congratulating him and Sokolow[4] on the notable progress so far achieved and suggested Weizmann should come as soon as possible to 148 Piccadilly to discuss future strategies. But just as he had once planned some zoological activities behind the lines when he volunteered for active service in the Boer War, and before that, while still in his teens, had organised the whole of the Tring Park staff into a collecting team, beavering away on his behalf — so now at the end of this memorable year he decided it was high time the estimable Sokolow and Weizmann were put to good use hunting up a few of his live ostriches, with which, during hostilities, he had unfortunately rather lost touch. Jordan, sadly enough, was still in Switzerland nursing Charles, and Hartert, of course, could not leave Tring. But these were two reasonably reliable substitutes.... Two of the ostriches about which Walter was seriously concerned had been left in the care of a Jewish naturalist/schoolmaster called Aharoni, who lived near Jaffa.* In particular he wanted to track down a chick from the Syrian desert which this man had reared in captivity, on his behalf, in Jaffa. On hearing from

---

* Although we know Nahum Sokolow had previously been involved in the Ostrich hunt[5] no details of his campaign have survived. But Walter mentioned to Weizmann that Sokolow knew all about Aharoni. Later Meinertzhagen came to Walter's aid in dealing with Aharoni, who completely lost his head and began demanding exorbitant sums of money for further specimens. Matters were smoothed over and good relations re-established and he continued collecting for several years for the Tring Museum.

Rozsika that Weizmann was shortly to leave for Palestine, Walter immediately sent him a request to come to 148 Piccadilly:[5] "I wish to see you before you go to discuss what I can do in your absence to counteract the downright lies being carried on by the British League. I also wish to bother you with a commission.... I wish to find out what has become of two ostriches...."

Weizmann, as we know from numerous sources, possessed a certain magnetic charm, but then, so did Lord R. Walter looked at him in silence with his twinkling elephant eyes and another faithful ostrich hunter was on his way....

Meanwhile a difference of opinion had arisen between the American and English Zionist organisations, and Weizmann announced apologetically that the expedition was thereby delayed. Walter assured him he need not worry, for quite obviously the Commission to Palestine must not leave before the unfortunate hitch in American affairs was cleared up. Although justifiably confident about Weizmann's ability as an ostrich hunter Walter was not entirely satisfied with his choice of the members of the Commission, for he thought a non-Zionist should have been included. Just before Weizmann left Walter warned him that Montagu[6] had obtained an interview with Allenby in Cairo on his way to India, at which he had sought to prejudice the great man against them. Allenby must be tackled. He then gave Weizmann a sixpence sent to the cause by an anonymous well-wisher....

The Zionist Commission eventually arrived in Palestine in April 1918 — World War I was still raging — but it was not until October that J. Aharoni and the birds were triumphantly located by Weizmann, but by the end of November the schoolmaster/naturalist was once again corresponding with Walter.[7] The ostriches were safe and within a year had arrived at Tring. The chick had grown up and in 1919 Walter described it as a new sub-species, *Struthio camelus syriacus* Rothschild. It was one of the skins retained at the Museum[8] when the bird collection was eventually sold to America.

In June 1918 Weizmann and the Emir Feisal were brought together at Aqaba, a meeting initiated by Allenby, and an amicable exchange of views took place. (In his autobiography Weizmann said that this first meeting in the desert laid the foundations of a life-long friendship.) Six months later Balfour strongly advised Weizmann to seek an agreement with Feisal[9] — and to facilitate renewed contact Walter gave a dinner at the Ritz* just before Christmas in

* Owing to the fact that Sunday December 21st was the only suitable date, the party had to take place at the Ritz, since Walter could not offend his mother, who kept the Sabbath religiously, and a dinner which would have involved preparatory work during that day was, of course, impossible. The Emir ate all European dishes, but like Walter himself eschewed pig, and of course did not partake of wine or any other alcoholic beverages.

honour of the Emir (who was visiting London), and invited Weizmann,[10] together with Milner, Robert Cecil, Crewe, T.E. Lawrence, Herbert Samuel and various Zionist leaders. Lawrence translated the Emir's speech, which was widely reported in the press. Walter was a trifle uneasy about some of Weizmann's comments and he may well have suggested to the *Jewish Chronicle* that it would be diplomatic to print only Feisal's speech and no other. The evening was an unqualified success and led up to an agreement between Weizmann and Feisal which was signed on January 3rd, 1919. The Emir believed that Jewish and Arab national movements could work in friendly co-operation — a belief held from the first by Walter, who said so in no uncertain terms — but Arab extremists forced him to abandon the view that peaceful co-existence could evolve and prosper. By August 1919 Curzon was informed that "the agreement was not worth the paper it is written on".

Meanwhile, trouble with the Board of Deputies was brewing yet again. This time it was Sir Stuart, Herbert Samuel's brother, who, as their chairman, was throwing his weight about. He confounded Walter by asserting that the Jews of England through the Board should "have a say in the disposition of Palestine at the Peace Conference" and was about to submit a memorandum to the government along these lines. Walter passed on this information in confidence to Weizmann and then added in his peremptory tone,[11] "You must inform the Foreign Office.... they ought to say in reply that they would let Sir Stuart know when the time comes for the Board's help." Three days later[12] he repeated these instructions: "I feel sure there is only one course to adopt and that is for you, when communicating, *as I told you to do*, with the Foreign Office...." Weizmann took the hint.

In September 1919 Walter thought it necessary — once again — to disturb Balfour's holiday.[13] He wrote enclosing a letter from Weizmann. The latter had been sent for by Meinertzhagen, who was appealing urgently for his help in Palestine where administrative difficulties were horrendous. Weizmann was most reluctant to leave England and admitted that he felt depressed and discouraged.

*Tring Park*
*Tring*
*September 26th 1919*

*Dear Mr. Balfour*

*I have been asked by Dr. Weizmann to send you the enclosed letter. He sent it to me to read because he thought I ought to know what he is writing to you.*

*I don't wish to appear importunate but I must add my appeal to Dr. Weizmann's and ask you to try and counteract the intrigues against Zionism now going 'on. Lord Rosebery told me that his son Dalmeny, who is just come back, told him that everyone out in Palestine and at home were trying hard to get the government to revoke their terms. Asking you to forgive my intrusion on your holiday.*

*I remain,*
*Yours sincerely*
*ROTHSCHILD*

Balfour replied promptly:[14]

*Private*                                                    *Whittingehame*
                                                             *September 29 1919*
*My dear Walter*

   *There is no doubt that you and Dr. Weizmann are perfectly right in saying that there is a great deal of intrigue of one sort or another going on against Zionism. Nobody, however, has suggested to me that the Government should go back upon their declared policy. I also agree with Dr. Weizmann in earnestly desiring a good understanding with Feyzal. The undoubted difficulties of the existing situation have almost entirely risen from the long delay in settling problems connected with what was the Turkish empire. For this I do not know that anybody is to blame — certainly not the British government; as clearly nothing can be done until we know what issues from the American turmoil.*
   *In the meantime I am, as you know, not the Acting[?] for Foreign Affairs and am not therefore at the moment in a position to take any steps in the matter, if steps there are which ought to be taken. My views on Zionism are, I need hardly say, quite unchanged, although difficulties surrounding them have doubtless increased.*

*Yours very sincerely,*
*A.J.B.*

*PS. I am not sure that the Zionist Jews themselves have always [been] wise. After all, it would be unreasonable to expect it under existing circumstances.*

Walter was immensely relieved by this letter. He had a curious trait of being easily, almost childishly, reassured. He sent Balfour a grateful note:[15]

32 Karl Jordan, 44 years curator at Tring Museum, laid the blame for Walter's eccentricities and Charles's anxieties fairly and squarely on their upbringing.

33 Alfred Minall "curator" and his family outside the cottage Walter had built for him – part of his coming-of-age present from Natty.

34 Edward VII at a shooting party at Tring in 1905. Natty with his dog is on the King's left – the only one not dressed for shooting. Walter is on the extreme right and Lord d'Abernon and Lord Rosebery on the left of the King.

35 Natty with Randolph Churchill and W.H. Trafford (left) at Tring in 1890, 'conducting the business of the country together'. Roy Foster describes Rothschild as 'the man to whom the Chancellor turned for everything'. (Photographer: Charles Rothschild.)

Victoria R.I.
June 22. 1897 —

(1837 – 1897)

36  Queen Victoria. Portrait presented to Natty and Emma on the occasion of the Diamond Jubilee. Emma and Victoria were faced with certain similar problems in bringing up their eldest sons.

37 Walter first harnessed his wild zebras singly to a trap. He broke them in himself.

38 Walter, after receiving his German honorary degree (at Gissen) – an honour not conferred on him by any English University. His niece, who learned her zoology degree at home in Tring Museum thirty years later, encountered less prejudice.

39 (*top left*) A caricature of Walter's father, Natty. 'The whole of the British Capital having been exported to the South Pole as a result of the Budget Revolution, Lord Rothschild flies from St. Swithin's Lane and succeeds in escaping to the Antarctic regions disguised as a Penguin'. Potted Peers, *Westminster Gazette*, 1909.

40 (*top right*) A caricature of Walter leaving Tring Park en route for a Duck Shoot on the Tring Reservoirs by his 15 year old niece 'Liberty' Rothschild.

41 (*left*) Walter in the white top hat which outraged the House of Commons. A caricature by Spy in *Vanity Fair*, September 1900.

42 Ernst Hartert when he took over the curatorship from Minall.

43 Walter at work in his room at Tring Museum. It was from this desk that all the ornaments and personal knick-knacks disappeared mysteriously after his death.

44 Walter at the Agricultural Show with Tring Park in the background. He was one of the promoters of the early sheep-dog trials, for which he gave the prizes.

45  A visit of the Ornithological Congress to Tring. Walter is standing between Hartert
and Jordan: he welcomed the delegates in a speech in French, German and English.

46 A butterfly hunt at Ashton. Walter as usual with downcast eyes is standing between Neville Chamberlain (later Prime Minister) and the ornithologist Bowdler Sharpe. Charles was again the photographer.

47 King George V at the Natural History Museum. Walter is standing on the left – note the relatively enormous size of his head.

48  The crowds outside 148 Piccadilly at Natty's funeral.

49  The crowds lining the route to Willesden at Natty's funeral, March 1915 – a token of spontaneous regret.

Foreign Office,

November 2nd, 1917.

Dear Lord Rothschild,

I have much pleasure in conveying to you, on behalf of His Majesty's Government, the following declaration of sympathy with Jewish Zionist aspirations which has been submitted to, and approved by, the Cabinet

"His Majesty's Government view with favour the establishment in Palestine of a national home for the Jewish people, and will use their best endeavours to facilitate the achievement of this object, it being clearly understood that nothing shall be done which may prejudice the civil and religious rights of existing non-Jewish communities in Palestine, or the rights and political status enjoyed by Jews in any other country".

I should be grateful if you would bring this declaration to the knowledge of the Zionist Federation.

50  The Balfour Declaration. His Majesty's Government approves a National Home for the Jews in Palestine.

51 Walter at a public dinner sitting between Einstein and George Bernard Shaw. He had had his beard trimmed for the occasion.

52 Walter with 'Foxy' Ferdinand, King of Bulgaria, two great bird collectors who enjoyed the same jokes.

53 Walter aged 64. He is wearing his scarlet tie "ready for the revolution".

54 Walter's mother aged 91. Her reign in Fairyland had lasted over 60 years.

55 One of the Tring Museum ledgers which accidentally escaped the Museum bonfire. Note the brass lock has been prised off.

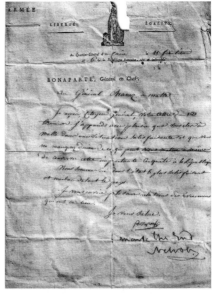

56 The luck of the draw. The only known letter signed by both Napoleon and Nelson, won by Walter in the lottery organized by Emma via her Will.

57 (*top left*) Walter's granddaughter, Olga Miriam Rose.

58 (*top right*) Marie Fredensen aged 20. The fairy off the top of the Christmas tree ...

59 (*left*) Rozsika in 1935 – with the inevitable cigarette.

60 Walter in his 69th year, dwarfed by his Pyrenaean hound, Monné.

*Tring Park*
*Tring*
*October 3rd 1919*

*Dear Mr. Balfour*

*I write to thank you very much for so kindly answering my letter. I quite agree with you that a number of extremists among the Jews have been most unwise. I see this morning that the Emir Fiesul is making difficulties and asserts that nothing will satisfy the Arabs but undisputed sway and empire over Mesopotamia, Syria and Palestine.*

*Yours sincerely*
*ROTHSCHILD*

Meanwhile the trouble with Sir Stuart seemed to have been dealt with satisfactorily along the lines suggested, but another more serious problem was already looming on the horizon. This concerned the vexed question of the boundaries of Palestine. Although Walter did not in fact take any active steps in this difficult terrain for another year it had been in everyone's mind ever since the Declaration. Furthermore, despite Lloyd George's biblical definition of Palestine as "Dan to Beersheba" no one really knew exactly what the War Cabinet meant by "Palestine", or how they envisaged its precise boundaries. Walter had always maintained an oecological, rather than a political approach, to the boundaries and extent of the Holy Land, since he believed that the agricultural development of the country was the key to its future viability and prosperity. Richard Meinertzhagen,[16] however, had already given considerable thought to the thorny problem; he wrote to Lord Curzon and put the case very well and succinctly. He stressed the importance of the boundary running to the north of the Litani River, following up the right bank and crossing it from east to west above the Litani Gorges. This would include the Hermon waters which flow into the Litani and Jordan Basins and, unlike the provisional boundary line based on the Sykes-Picot agreement, would not deprive Israel of the vitally important catchment area of Jordan. The new boundary was known as the "Meinertzhagen Line" but it was a dream boundary which was destroyed by future events.

Richard Meinertzhagen became a true friend of the Jews and Judaism from the day on which he personally witnessed a terrible pogrom in Odessa,[17] which he visited in 1910. He was appalled when brought face to face with the horror and cruelty of violent, senseless persecution and destruction, coupled with the callous indifference of the onlookers. "I have been shocked beyond belief....

275

overcome with anger and compassion.... There is no word in the English language to describe such vile and bestial behaviour." Weizmann, whose memory was not always very accurate, attributed this fierce championing of the Jews and Zionism to the personal influence of Aaron Aaronsohn* — whom Meinertzhagen both liked and admired, and who served with him in the Secret Service in Egypt. But his friendship with this gifted man occurred after his "conversion" to the cause of the Jews, and merely confirmed his views that they possessed both courage and integrity.** Perhaps his belief in Zionism was largely due to Aaronsohn's influence, but it was not the original spark igniting the torch which he bore with such persistence and determination throughout his life. Meinertzhagen was a wonderfully attractive but very strange man. As a collector Walter considered him most eccentric — a bit mad — and, on his side, he most probably wrote Walter off in some ways as *non compos* — but at this period of their lives the two men met on very amicable terms as ornithologists, and great specialists and lovers of the Palaearctic birds.

Weizmann and Meinertzhagen respected and liked one another but one senses some intuitive reservation on Weizmann's part, for he described him rather flatly and conventionally as an observer rather than as a close and intimate friend: "He was a man of lion-hearted courage and had fought on almost every front. He was repeatedly wounded and sent home.... With Aaronsohn, Meinertzhagen had many talks about Palestine and was so impressed by him that he completely changed his mind and became an ardent Zionist — which he has remained to this day. And that not merely in words. Whenever he can perform a service for the Jews or Palestine he will go out of his way to do so."[18]

Meinertzhagen in a less objective, more warmhearted vein, wrote: "Chaim was a great chemist, a great Jew, a great man and a Prince of Israel. He and Smuts are the two outstanding figures of my generation.... As a man I had no better or loyal friend than Chaim Weizmann. I remember spending a weekend at Helouan with him in 1918; it was then I got to know him and love him...."[19]

A shaft of light is thrown on Meinertzhagen's character by his reaction to T.E. Lawrence's virulent description of him in the *Seven Pillars of Wisdom*, the manuscript of which was shown to him by the author! Meinertzhagen, who

---

* Weizmann was probably confusing Meinertzhagen with Sir Mark Sykes whose conversion to Zionism was completed by Aaron Aaronsohn.

** Meinertzhagen, in a confidential letter to the Prime Minister, described the Jews as "virile, brave, determined and intelligent".

enjoyed a love/hate relationship with Lawrence, remarked there were no secrets between them. "I believe I was the only one of his friends to whom he confided he was a complete fraud.... I was devoted to him.... to me he spoke the naked truth, his whole soul uncovered; and in response I opened my heart to him."[20] The controversial paragraph ran as follows:[21]

*"....a student of migrating birds drifted into soldiering.... Meinertzhagen knew no half measures. He was logical, an idealist of the deepest, and so possessed by his convictions that he was willing to harness evil to the chariot of good. He was a strategist, a geographer, and a silent laughing masterful man; who took as blithe a pleasure in deceiving his enemy (or his friend) by some unscrupulous jest, as in spattering the brains of a cornered mob of Germans one by one with his African knob-kerri. His instincts were abetted by an immensely powerful body and a savage brain, which chose the best way to its purpose, unhampered by doubt or habit."*

Meinertzhagen commented mildly: ".... almost amounting to libel."

It is highly improbable that Walter and Meinertzhagen at this time had any direct communication or consultation concerning the northern boundaries of Palestine although their views exactly coincided. Weizmann was the intermediary and he suggested drafting a joint memorandum on the question which Walter would sign and send to the Prime Minister. The matter appeared to Walter so urgent that he pursued the ex Foreign Secretary during his Christmas holiday at Whittingehame:[22]

> *Tring Park*
> *Tring*
> *December 30th 1919*

*Dear Mr. Balfour*

    *I am enclosing the copy of a letter the Zionist[23] Organisation is sending to the Prime Minister. I venture to hope when you have read it you will champion our cause with the Prime Minister. I would especially point out and ask you to urge upon Mr. Lloyd George that whereas the Litany water supply is all important for Palestine it cannot affect the country north of it in any way and therefore the French only want to ruin Zionism and detract from the value of the country under British Mandate* [the next sentence has been cut out by an autograph hunter who sliced out the signature on the reverse side of the page] *also by insisting on leaving this country in the French area they seek to deprive the Jews of some of the best of the already established Zionist colonies.*

> *Yours sincerely with best wishes for the New Year*
> *ROTHSCHILD*

Balfour responded immediately to this *cri de coeur*:[24]

*Whittingehame*
*January 2nd 1920.*

*Private*

*My dear Walter*

*I am very anxious to see the frontiers of Palestine so drawn as to give a full economic chance to Zionism; and I will do what I can to further these views, though, of course, I have not exactly the same official position in relation to them which I had while Foreign Secretary. I believe the Prime Minister to be most sympathetic.*

*With every good wish for the New Year*
*Believe me, Yours sincerely,*
*ARTHUR JAMES BALFOUR*

He followed this up with a letter to Philip Kerr[25] (later Lord Lothian) enclosing the memo from Walter and voicing the opinion that "nothing more preposterous was ever suggested even by the Sykes-Picot Agreement than the attempt to fix the Northern boundaries of Palestine in such a fashion that some of its important water supply was handed over to Syria...." Philip Kerr replied that as far as he knew, the Prime Minister's views were as strongly sympathetic on this subject as they ever were.[26] Six weeks later Balfour was again writing to Philip Kerr[27] on the question of the "Meinertzhagen Line".[28]

*Private*

*February 23rd 1920*

*My dear Philip*

*I have this instant got the enclosed letter from Weizmann. I do not know how the matter stands, but I entirely agree with him that, if we are to make a success of the Jewish Home in Palestine, it is of enormous importance that the water power should be secured to Palestine which naturally belongs to it, and which can be of no possible use to Syria or the French.*

*I bored you about this subject at Christmas time: but if there is any doubt in the minds of the Conference, please bring it again before the attention of the PM. It is a subject on which I feel most strongly.*

*Yours ever*
*ARTHUR JAMES BALFOUR*

Throughout this period, indeed from the moment of the publication of the Declaration, Walter was experiencing difficulties engendered by the activities of the League of British Jews, a strongly anti-Zionist organisation of which his cousin Lionel de Rothschild and Sir Charles Henry were the moving spirits. It was common knowledge that they aimed at persuading the government to rescind the Declaration. Friedman fancied[29] that Lord Beaverbrook, newly appointed to the Ministry of Information, had fallen under Lionel de Rothschild's spell, and it was he who inspired his marked anti-Zionist attitude. But this is not correct. Beaverbrook had quarrelled violently with James de Rothschild, at one time a close friend, and he was prepared to go to any lengths* to thwart and annoy James. He therefore jumped at the opportunity of anti-Zionist propaganda, knowing the cause was dear to James's heart. Lionel himself was a sincere assimilationist, but added to this was a subconscious element of family jealousy, for he resented the fact that Walter, the inarticulate butterfly buffoon, who was not a partner in N.M. Rothschild & Sons and stood aloof from business and politics, had been automatically acknowledged as the spokesman for the family. He thought this ridiculous, grotesque, unfair, and he seized the opportunity presented by the League of becoming leader of at least one section of British Jewry. Walter and Rozsika, by their joint efforts, eventually persuaded both Lionel and his younger brother Anthony to modify their attitude and learn to live with the Balfour Declaration, while Ormsby Gore successfully tackled Beaverbrook. Furthermore Walter told Weizmann that he must be less openly aggressive towards the assimilationists — whatever he might feel in private — and he insisted that the invitation to attend the International Zionist Conference planned for the summer must be sent out and signed by the leaders of all sections[30] — otherwise it would fail in its objectives. He then arranged a meeting[31] (June 20th) between Herbert Samuel, Stuart Samuel, Weizmann and himself to discuss the means by which this could be brought about. Simultaneously he was arranging yet another consultation,** this time between Weizmann and certain other members of the Zionist Federation, with Samborn, Caultie and Alcock of the Tropical Disease Prevention

---

* His nephew Bill Aitken, who worked on the *Evening Standard* told me that they had blanket orders to "knock" James de Rothschild in the paper whenever the opportunity presented itself, and to make opportunities if none occurred spontaneously.

** Walter was always ready to help. In May[32] he accompanied Sokolow to the Italian Embassy (where he had arranged an interview with the Ambassador concerning Italy's alignment with the British declaration in favour of Zionism) and again to lunch at the Ritz with Mark Sykes.[33] Walter telegraphed: "I will be there". On another occasion he corrected drawings of the desert locust for the Palestine Office of the Zionist Organisation.

Society (of which he was a member) to discuss the prevention of disease in Palestine before such an emergency might arise.[34] He wrote to tell Charles that in this particular area he was, indeed, sorely missed (see letters of June 27th and 29th).

Two months later the San Remo Conference of the League of Nations was upon them, and, once again, the Balfour Declaration and the British Mandate hung in the balance. A letter was addressed to the Prime Minister who, with Curzon, led the British delegation at San Remo. Walter headed the list of thirteen signatories which included Robert Cecil, Ormsby Gore, George Barnes and Josiah Wedgwood.

> *"There are strong rumours here that the British Government are hesitating to accept the British Mandate for Palestine. We venture to submit that Zionist opinion throughout the world regards a British Mandate as essential to the effective fulfilment of the Balfour letter and we agree. We also urge that the early settlement of this mandate question is vital to the pacification of Palestine."*[35]

On this occasion the fates were kind to the Zionists, for the Declaration was confirmed and the Mandate awarded to Britain;* but only a year later everything once again seemed in jeopardy. Meinertzhagen visited Tring in April[36] and told Walter that the whole of the staff in Egypt from Allenby downwards consistently worked against Zionism. Furthermore the appointment of Haj Amin al Husseini (by Herbert Samuel, then High Commissioner**) as Grand Mufti of Jerusalem was sheer insanity. It placed a man, who hated Britain and the Jews, in a position in which he could do untold harm to both. Meinertzha-

---

* In December 1920 Walter was the first signatory of the appeal for the Keren Hayesod (Palestine Foundation Fund) which was to provide the financial resources for the upbuilding of the Jewish National Home in Palestine. Although in a greatly modified form, this fulfilled some of the suggestions he had made to Mark Sykes in 1917.[37] Thus[38] an Economic Council had been formed, composed of men of affairs of high standing in the commercial and financial world; it would examine projected undertakings "which can be regarded as, in the stricter sense, reproductive of executing such of them as are approved, and further of assisting the Board of Directors with expert advice in the general administration of the Fund." He and Mond also signed letters to the leading Jews of Iraq.

** When Sir Herbert Samuel became High Commissioner of Palestine after the Balfour Declaration he "leaned over backwards" in his efforts to be scrupulously fair to all parties concerned. It was in this honourable but misguided spirit that he reinstated Husseini as Grand Mufti of Jerusalem — an appointment which justifiably appalled Richard Meinertzhagen.[39] Natty in his day had never been the victim of doubts and fears of this sort. He was candid, confidently upright, and he was not afraid of criticism and paid no heed to "what people might say" or think. Walter lacked judgement and confidence — he was too nervous for candour — but, like his father, not a leaner.

gen added that the French intended to go flat out for a French condominium rather than a British Mandate at the League of Nations. Walter was pressed by Weizmann and Alfred Mond to pass on this information to the Prime Minister, but he was unwilling to do this since he had been give the facts under seal of secrecy.

By December the situation had worsened and Walter concocted one of his panicky telegrams to Balfour[40] — albeit at Weizmann's suggestion:

*Right Honble A.J. Balfour*            *13 December 1921*
*British Embassy*
*Washington*


*Situation most critical Balfour Declaration is threatened with reversal and Zionist cause desperate strongly support Weizmann's appeal to you to interfere.*

*ROTHSCHILD*

Stein, with Walter's approval, toned down the wording before despatching it the following day.

On July 5th 1921, six months before these disquieting events, the World Zionist Conference had opened in London. Walter presided at the meeting convened at the Memorial Hall to consider and approve the recommendations of the various committees appointed earlier that week. The meeting was marred by violent arguments and considerable disorder. In particular the minority report which called for complete nationalization[41] of land in Palestine was the signal for a general uproar. After listening for half an hour to shouting and cat-calling, Walter lost his temper, announced that the meeting was closed, and stormed off the platform. Order was eventually restored, and the majority report on the land policy was adopted.[42]

The resolution expressing the deep gratitude and heartfelt thanks of the Jewish people to Britain for the acceptance of the Mandate was passed with boisterous acclamation. In contrast to this noisy and obstreperous gathering, the huge thanksgiving meeting on July 12th at the Albert Hall[43] — despite the fact that thousands who tried, failed to gain admission — was impressively well behaved and patient, for the speeches were exceedingly long-winded. Walter again presided, and with him on the platform — at long last! — was Balfour; he felt that the Mandate was what they had both fought for and believed in, even if their view-points had not exactly coincided. *What would his father have thought?* Speaker followed speaker, Balfour, Cecil, Crewe, Ormsby Gore, the Chief Rabbi, Silver, Sokolow, Weizmann, Nordau, James de Rothschild. Yes, what *would* he have thought? Eventually the chairman called on Josiah Wedgwood. "Lord Rothschild and Jews!" began Wedgwood.... The crowd caught on at once. They liked it. So did Walter.

# CHAPTER 30

# A figure in the background

THE DECLARATION was signed in 1917 and when introduced by Balfour, was ratified by the League of Nations in 1922. Yet by January 1923,[1] an unofficial note from the Foreign Office to the Cabinet declared that upon the origins of the Balfour Declaration little if anything existed in the way of official records.

> *"....Indeed, little is known of how the policy represented by the Declaration was first given form. Four, or perhaps five, men were chiefly concerned in the labour — the Earl of Balfour, the late Sir Mark Sykes, and Messrs. Weizmann and Sokolow, with perhaps Lord Rothschild as a figure in the background. Negotiations seem to have been mainly oral and by means of private notes and memoranda, of which only the scantiest records are available, even if more exist."*

The note adds that Sir Mark Sykes's official papers "have unfortunately been dispersed and that, so far, little referring to the Balfour Declaration has been found among such papers as are preserved".

Vereté[2] states that in his view "Balfour's share in the Declaration was rather small". Bolstered by his knowledge of the Foreign Secretary's friendship with his late father, and because he felt that Balfour had valued Natty's counsel, Walter played his part in bulldozing him in the right direction. Balfour was, for him, a magical personality endowed with all the virtues and the necessary influence, and willy nilly Walter was going to hitch their Zionist wagon to this particular star — since his father would have wished it. We find Balfour struggling against the tidal wave:[3] "I much appreciate the proposal that a deputation should come here to express the thanks of the Zionist Federation for the Decision taken by H.M. Government, but is not this really somewhat superfluous? While I do not wish to discourage the idea if you support it, I am afraid next week is going to be a particularly busy one as far as I am concerned."

But Walter persisted, and perhaps the letter he extracted from Balfour on November 9th[4] contained the assurance he was seeking. "As you press for a deputation I will certainly try to receive it on some date such as you suggest. I do not think you need have any fear that H.M. Government will rescind the policy which has been deliberately adopted after full consideration and discussion by the Cabinet."

The Memorandum titled "Palestine and the Balfour Declaration" (prepared early in 1923 with the object of "setting out the History", so the Duke of Devonshire informed the Cabinet)[1] was sent to Lord Balfour for his comments. Balfour was ill when the document reached him, but a reminder was sent in February and he then replied as follows:

> Whittingehame
> Prestonkirk

*My dear Eddy Marsh*[5]                    *February 19th 1923*

*Your letter just received to-day reminds me of one that you sent me when I was ill, enclosing an official document prepared in the Colonial Office with regard to the so-called Balfour Declaration.*

*It was so official that it really gave no information on the subject with which it dealt — I presume because formal documents were not available. But don't you think that something authentic but informal might with advantage be written on the subject? My memory is so defective that I am afraid my contribution to such a paper would be worthless; and Mark Sykes, who had the whole thing at his finger ends, is unhappily no longer with us. But I believe Billy Gore knows it all and could easily put it on record.*

*I have, of course, vague recollections of long conversations with Weiszman [sic], Sokolow, etc., and I wrote a preface, since republished, to the latter's book on Zionism.*

*I also remember in the middle of a discussion (I think on some wholly alien subject) which took place during the Peace Conference at the Quai d'Orsay, making my modifications in the declaration suggested by the Jewish authorities.\* I only intended these to be suggestions for consideration; but as a matter of fact they have, for the most part, been embodied in the official documents. My memory, however, of all these things is very hazy.\*\**

*I shall arrive in London almost as soon as this letter.*

> *Yours sincerely,*
> *A.J.B.*

---

\* This is most obscure, since if it refers to the Balfour Declaration of 1917 (as it seems to do) it is not clear what modifications were contemplated at this late stage. Perhaps Balfour is referring to some declaration to be issued in the name of the Peace Conference itself, or an early draft of the Mandate. Or was he thinking of "The Statement of the Zionist Organisation regarding Palestine: Proposals to be presented to the Peace Conference" (Paris, 4th February, 1919)?

\*\* One contemporary historian with whom I discussed this peculiar letter, believes that Balfour remembered perfectly, but was determined not to get involved, while another maintains that Balfour never knew the details of these discussions — simply "he was too lazy".

It takes a long time for an individual to realize that he is different from everyone else; we assume, as children, that we all see, and hear, and feel in the same way. It took Walter years to realize that he was a strange phenomenon and other people did not remember their own lives as accurately and in such detail as he did, for both his parents were also endowed with exceptionally good memories and he fancied that this was "normal". It is true that his own recollections were spotty — he was a fool at his books and could not, for instance, remember Greek verse, but on the other hand, unlike ordinary mortals, he never had to keep a diary of his engagements since he easily carried every relevant date in his head. He had a tiresome habit of quoting the exact time of all future meetings — like his recorded bird sightings — as if he was reading carefully from a timetable.

It took him even longer to realize that when people maintained they had forgotten a conversation, of which he remembered every detail, they might indeed be telling the truth. Nevertheless — making every allowance for Balfour's age, for he was now over 70, and taking into account the relative forgetfulness of most of the people he knew — Walter could not believe (when somehow the gist of the letter filtered back to him) *he could not believe* that the great man had only "vague recollections" of long conversations with Weizmann, or for that matter of their own exchange of letters and discussions, and that his memory of all "these things" — yes, *these things* — was very hazy....

Walter never wrote to Balfour again. He felt Balfour was no longer really interested — he was old and tired — and only his niece, Blanche Dugdale, from time to time prodded him into activity and pricked his conscience. Nor did Walter accept Weizmann's invitation to take the chair at the opening of the Hebrew University in 1925, which may have surprised Balfour, who went himself.[6] With a mixture of pessimistic foreboding and discouragement, yet in a sense with infinite relief, he figuratively locked the insoluble long-term problem away in the linen basket, and directed all his energies once more to an area where he not only felt at his ease, but which he loved, and to which he believed he could make a useful contribution — to the world of Tiger Moths, Swallowtails, Tree Kangaroos and the Cassowaries, which had somehow miraculously survived it all. He turned his back stubbornly on the Zionist conflict and in a curiously diminished spirit, seemed to join his mother in the mood of her subdued and orderly widowhood.

Several years later (1924) Walter presided at a dinner in London to mark the anniversary of the Declaration and to celebrate the planting of the Balfour Forest. Samuel was there and Sokolow, Mond, Reading, Wedgwood, Vansittart, Mrs. Snowden, James de Rothschild.... During a long speech by Weizmann, who was sitting beside him, Walter's attention began to wander.

Weizmann spoke well but he went on and on.... He was reminiscing in a semi-humorous, sentimental vein about the war and the Declaration, "....and the troops in Palestine were singing something like this....

    '.... And then you see they gave the holy City
      into the hands of the Zionist Committee....'"

and Walter drifted off into a peaceful sleep, his beard resting lightly on his immaculate shirt front. Balfour, whose attention was drawn to this happening, remarked that he was "scarcely surprised" and gazed at the chairman with undisguised envy.

And at the dinner party and ball given at 148 Piccadilly for Walter's favourite niece's "coming out", Balfour sat next to Emma.... But then he had never told his mother about "these things".

During 1923 — apart from his presidential address to the Entomological Society — Walter wrote and published only three papers,[7] on the birds of Yunnan and Algeria, and a few short notes including "Remarks on a Mountain Gorilla".

Then came the news that Charles had died. *Nature*[8] in its obituary notice, said that "by the death of the younger son of the first Lord Rothschild, nature in a literal sense, entomology and, it may be added, tropical medicine, have each sustained a formidable blow". Walter was stunned. He felt as if a very large piece had suddenly been knocked out of him. For the family, for the Museum, for his home, it was an immeasurable disaster. He suddenly knew with dull certainty that there was no future.

# CHAPTER 31

# You will be painted blue and yellow and exposed in the High Street

WALTER, looking vast and imposing, but hesitant as usual, took Charles's place at his nephew's Barmitzvah, a ceremony which was held a couple of months after his brother's death. A cousin tried hard to introduce some light relief into the gloomy scene by a whispered suggestion that Walter was bound to drop the Scrolls of the Law and the community would have to fast for ten — or was it eleven? — days to atone for the disaster. But the children remained, like their uncle, awkwardly silent.

Charles had died eight years after his father — a blow from which his family could never really recover. There was no-one to take his place. It is said that a man is never irreplaceable and whether he is prime minister or police constable there is always someone on whom the mantle will fall. But this was the exception which proves the rule. Nature conservation in the U.K. was virtually brought to a halt and for the next 25 years marked time; Tring Park gradually disintegrated; New Court sunk into genteel inactivity. After Charles's Will was read, Emma sold her potential dower house, Champneys on Wigginton Hill, which Natty had provided for her in case she decided to leave Tring Park to make way for the younger generation, and Walter settled down to a future of twelve years solid, conventional, pedestrian zoology at her side. Did she ever discuss his future residence with him at this stage? Or did she assume that he would eventually live amicably with her grandson in Tring Park through ripe old age? We do not know.

Walter left the running of the household entirely to his mother, with a staff of fourteen to help her, and the farms and gardens and financial problems to Rozsika. Very soon he began to depend completely on his sister-in-law for the advice and support he always needed. Where he was concerned, she was scrupulously fair, incisive and slightly disapproving. Apparently that was exactly what was required to assuage his perennial feelings of insecurity. Struggling with the disaster of the second imposition of massive estate duties within eight years of the first, and with her four children to rear, Rozsika was faced with a difficult task. Drastic economies were also necessary and these are always unpopular, but she found that administration and authority could provide a sort of salvation from overwhelming grief, and lay the necessary tramlines which would eventually make life more bearable. Fortunately for the family she had had seven years of hard training while Charles, although too ill to work himself, had nevertheless still been there to guide and advise. Glancing through his letters from Switzerland one finds him asking her to deal with the following matters: Frohawk who was to make 100 drawings of adult butter-flies, a trust deed which Lord Haldane was to look at, the purchase of land in Essex as a Nature Reserve, Sharpe who had ten daughters — and needed a job, the pension problems at New Court, the plans for making small 10oz gold bars for use in India after the war, the changes in the sale of quicksilver, the payment of Uncle Alfred's death duties, smelting of "sweep" at the Refinery, the separation of the Bullion Department from New Court, cuttings of *Dian-thus* from the weed garden at Ashton, the sale of Rio Tinto shares, purchase of the sporting rights at the Reservoir.... There were hundreds of requests and problems to tackle.

"You are a wonder," he wrote from Lausanne, "I am a very lucky man to have you as a helpmeet".

But Rozsika never recovered from his death. Despite her extraordinary tenacity and efficiency (for example, by successful investment she doubled the value of her son's trusts during his minority); despite her positive outlook and the energetic and useful life she led, she remained at heart a sad and lonely widow. Her doctor persuaded her to learn bridge for relaxation, which she quite enjoyed. She took innumerable lessons and played hundreds of hands alone at home, but remained remarkably inept at the game. However it initiated a change of scene.

While Charles was ill in Switzerland it had seemed quite natural as a temporary measure for Rozsika and her children to live with Emma at Tring, but now it was like some appalling Greek play — sitting opposite Walter at meals, sleeping in adjoining bedrooms — all boxed up together in a ponderous gilded cage. In a way life had taken on a mad Macbethian character, unimagi-nable if she thought of her carefree, impecunious girlhood in Hungary. The only meaning she could give to it was that, in her present circumstances, she could help Hungary along the road to postwar recovery and attempt to carry out Charles's wishes. And, of course, there were the children.

Looking back, it seems astonishing that at this moment in time Emma did not suggest to Walter that he move out of the bedroom he had occupied since childhood (only a small flight of stairs separated this room from his mother's apartment) into one of the twenty empty bedrooms at Tring with his own bathroom and study alongside. Or why not occupy the uninhabited so-called batchelor's wing, which was like a separate little house, and opened so conve-niently both into the garden and into the communal library where Walter and his mother sat and read before and after meals? Emma, in a sense, still lived in the past and perhaps Walter — the most doggedly conservative of all men — clung to the familiar soothing comfort of his old quarters and was indifferent to the rowdy children yelling up and down the passages, and to the slippery bathroom floor which they shared and kept permanently awash. However, in 1916 Rozsika decided that, come what may, her brother-in-law should have a luxurious bathroom of his own, fitted — for safety — with a semi-sunken gigantic tub. Despite this precaution, one fine day Walter slipped as he stepped over the edge, and in a panic pressed the alarm bell. By the time the valet had found the right disc on the bell-board and lumbered up three floors, Walter had recovered his composure. The end of the story suggests that Charles[1] invented it.

"Are you all right, my Lord?"

"Quite all right, Woods — I only lost my centre of gravity in the bath...."

"Very good, my Lord," replied Woods respectfully, "I'll look for it later, my Lord".

The children shrieked with laughter, but one of them was not sure if she had really understood the joke.

But, bathroom or no bathroom, Walter clung tenaciously to his old quarters on the nursery floor. Moreover he still slept with his door wide open and his tremendous snores rumbled along the passages, like the massive rollers on an Atlantic beach, amazing the children who could scarcely believe it possible that such awe-inspiring thunder could be engendered by a single individual. They were all too scared of Walter to do more than take a quick peep round the open bedroom door after he had gone downstairs and Woods was laying out his evening tails. The only recollection which endured was the improbable size of the four poster bed and the fact that his mirror was completely obscured by invitation cards and notices for meetings which were stuck in the looking glass frame by one corner. Clearly Walter never looked at himself in THAT mirror.

On one occasion Emma — then in her eightieth year — awoke to perceive Walter in the dim, rosy light of her Chinese *veilleuse*, standing at the foot of her bed immaculate in pure silk pyjamas. "Mama," announced Walter in great agitation, "I have been poisoned, Mama".

Emma switched on the light. She always looked rather slight and sharp in bed — without her *toupé*, without her marvellously realistic false teeth — enveloped and rather lost in a voluminous, old-fashioned flannel nightshirt.

"You look remarkably well to me, Walter — I expect you've had a nightmare."

"Not at all," whispered Walter, "One of the pheasants due at the museum went astray and I've suddenly realized what happened to it — we had pheasant for supper last evening, Mama. The preservative is most toxic — I've got a pain — I'm sure I've been poisoned."

"Then WE have been poisoned," said Emma calmly, "because I had a slice of that bird too, and I'm quite well, thank you, Walter. You will feel better in the morning."

After a long, thoughtful pause he turned on his heel and meekly went back to bed.

The children were embarrassed and uneasy in his presence, something which Charles — remembering his own boyhood and Walter's skill and enthusiasm as a guide to birds and butterflies — could not envisage. Hearing that his brother was accompanying Emma to Ashton for a summer visit, he wrote to Rozsika: "I can just imagine how the children will pepper him with questions." But they dared not ask him anything. A large meaningless void — the generation gap — yawned between them and, like Walter, they were tonguetied. Despite appearances to the contrary the nursery/schoolroom complex on the top floor at Tring was as discrete as the marzipan layer in a Christmas cake, and Emma herself remained serene and rather aloof, like the icing sugar, white and silvery on the top.

On one occasion without warning the schoolroom door suddenly swung open and Walter strode in. His two nieces were sitting demurely at their desks studying Roman history, under the eagle eye of a stout governess. He thumped over in silence to the modest butterfly cabinet in the corner, knelt down heavily on one knee, and puffing and blowing began pulling out the drawers, glancing at the contents and banging them in again. After a nod of consent from the governess one of the little girls walked over and timidly asked a question: "Are you looking for my Hungarian Purple Emperor, Uncle Walter?" she enquired — for the children had engaged in a noisy argument about this specimen the previous day over lunch. Walter did not reply. He continued crashing the drawers in and out, and the little girl — trembling for her butterflies, which she felt must be snapping on their pins — went back to her desk with burning cheeks, certain that somehow or other she had dropped the biggest brick of all time. Walter then rose clumsily to his feet, and still puffing and blowing strode out. He never offered a word of explanation.

A few years later, as a child of 13, I was waiting with my sister for the lift at Tring, on our way up to bed at 9 pm, after dining as usual with Walter and his mother. Suddenly his voice boomed out into the passage like a fog horn — although no doubt he had intended to talk in a subdued tone — and I heard the only remark about myself, good, bad or indifferent, that he ever made within earshot.

"Mama," said Walter, "isn't it strange that Miriam is completely *square*?"

It was an interested, not unkind appraisal by a morphologist — I might have been one of his giant tortoises. But we little girls neither giggled nor shrugged; nor was I mortified. It was a voice from another world.

Although the relationship was shrouded in silence, Walter liked and approved of his nieces, but he felt that his nephews secretly, and not so secretly, made fun of him. Furthermore Victor came to lunch with his hair unbrushed and with an open shirt and no tie. How could his mother put up with it? One day, when he considered that the boy was talking more nonsense than usual, Walter suddenly put down his knife and fork and addressed him in a playful bellow:

"Mister Victor, (long pause) Mister Victor, (another, longer pause) You will be painted blue and yellow* and exposed in the High Street."

Frozen silence.

Rozsika was not a little to blame for the lack of communication between Walter and the children. She was ambivalent — on the one hand she considered he was a potentially bad influence on her son and feared he might discourage Victor from working at New Court; in her heart she bitterly resented the fact that Walter had left Charles to shoulder a double load of responsibility with such disastrous results. On the other hand she now felt he was *her* personal responsibility — inherited from Charles — and a man greatly in need of care and assistance — and she rather enjoyed authority. Although she knew nothing whatsoever about the Museum or Walter's work, she spoke rather slightingly of his contribution to science. "Walter is a splitter," she commented on one occasion — that is to say a systematist who divides up one species into several different ones, sometimes unnecessarily, merely for the pleasure of naming the new forms thus created. Rozsika must have been repeating some criticism she had heard in the past — possibly from Jordan, who rightly disapproved of the criteria on which his employer created new species of cassowaries — for Walter was bewitched by their multi-coloured wattles which proved in the long run rather capricious characters. Certainly Rozsika understood absolutely nothing about the rights and wrongs of systematics. She might, with more justification, have designated Hartert a "lumper".[2] However so great was her influence that she was listened to faithfully on all subjects — like the Delphic oracle itself. Walter was a splitter — that was that. If only on some Sunday afternoon after lunch, when he departed alone for My Museum — crashing the terrace door to behind him so that the glass panel remained intact only by a miracle — she had suggested that

* The Rothschild racing colours.

the *désoeuvré* children should go along as well and be shown his treasures, how happy he would have been! And what a lot of stimulating and fascinating information they would have picked up! Instead Rozsika retired to her desk and her piles of papers, and the children to their futile but mildly amusing games on the roof at Tring Park — for eggs could be tossed down onto the lawn, and to the astonishment of the uninitiated onlookers, would bounce unbroken on the springy mossy surface below.

The generation gap did not narrow as time went on. Walter's nephew, now a Harrow schoolboy, was beginning to take an interest in Natural Science and liked to goad "the old boy" at dinner with controversial asides. One evening he began an argument about evolution with a few deliberately provocative statements. Walter fell into the trap and replied by bellowing an answer and, gradually working up steam, roared away happily and didactically. Emma, who in her eighties had become a little deaf and also absent minded during meals, suddenly looked up: "Walter," she said, quite sharply, "don't talk so much — eat up your artichokes."

There was one other topic on which Walter was quite easily drawn, and that was his *bête noire* — contemporary art. He was a very good judge of scientific illustration, both as regards accuracy and artistic merit (for he had not only bought the Audubon Birds, but also the original drawings of Moses Harris's Aurelians)[3] but he could countenance no other form of art except a sort of *trompe l'oeil* realism. One of his nieces at the age of fifteen had won a Royal Drawing Society Gold Medal for a painting — for like her grandmother and other Rothschild women, she could paint and draw. She had the temerity to defend the Impressionists. Walter for once raised his eyes from his plate and fixed her with a baleful glare: "A painting," he roared, "must resemble the object which is being painted. It must look exactly like the thing itself — not a mass of pink and blue blobs."

Afterwards she wrote in her diary: "I think Uncle W. is the stupidest man I know."

As Walter's nephews and nieces grew up they were encouraged to bring their teenage contemporaries home for weekends. Walter — who it must be admitted, greatly embarrassed the girls, for they were at an awkward age when they felt at a loss, for who could laugh off and explain his eccentricities to friends? — watched them in silence with his good-natured twinkling elephant eyes. Someone had perforce to sit next to him both at lunch and dinner, and the individual in question, whether a daughter of Winston Churchill or some reluctant local lad, enticed in to make up the cricket team, would be warned not to worry, because however hard they might try, a completely silent meal lay ahead. On one occasion, the Captain of the Harrow Cricket XI, David

Rome, had a singular success: towards the end of dinner, after the crumbs had been swept off the table cloth with a silver scoop, he switched the conversation onto dining clubs at Cambridge and foreign universities. Suddenly Walter raised his eyes off his plate and to everyone's surprise and delight broke into loud, bellowing *studenten* songs.... with pidgin English-cum-German words:

> "Here a leaf and dare a leaf,
> And hintern another leaf...."

Apparently he had a vast repertoire and, judging by the half-hidden smile on Emma's face it was not the first time she had heard them. The butler withdrew discreetly and Walter roared on happily, and without repeating himself, for the best part of an hour.

On one occasion there appeared among the girls' new acquaintances two immaculate brothers, duly presented to Walter with the other guests before dinner. He did not appear to show any interest in the young men and greeted them perfunctorily as usual, with eyes fixed firmly on the floor. Time, and the still unopened letters in the locked linen basket, revealed in due course that they were the sons of the blackmailing peeress.

Walter, like his father, never consented to become merely a figurehead on any committee, and this bewildered and irritated Jordan, who considered his frequent absences from My Museum unjustified. Walter rarely missed a meeting, whether interesting or irksome. For instance he took the chair at every monthly gathering of the Royal Entomological Society during his first year as president. The same applied to the Society for the Promotion of Nature Reserves, of which he was elected president to succeed Lord Ullswater in 1931. He lectured frequently at the Victoria Club in Aylesbury,[4] of which he was president. The list of his committees ranged from those of the British Museum Trustees (for 30 years) and Jewish Board of Guardians[5] (for 40 years) to the Jewish Peace Society and the Tring Bowling Club. He was also a keen Congressman and enjoyed the fact that the idea of the Entomological Congress originally came from Tring. He was positively fêted at the Zoological Meeting in Budapest in 1929. Reporters who crowded round the beaming giant — described by them as the "little gold uncle" — were puzzled when, true to form, he arranged to meet them on the staircase landing of the hotel at 9.15 and announced blandly, "I have one and a half minutes only...." Ninety seconds later he thumped away — with the long-suffering but resigned Jordan following, beard tilted skywards, and eyebrows raised.

During this decade Walter arranged several modest collecting expeditions, chiefly to Africa; including one for Hartert to Algeria.[6] At the last moment he decided not to go himself and sent Fred Young instead. He now averaged about ten publications per year, chiefly systematic papers on birds, butterflies and moths, but also including notes on fish, giant tortoises, a new Aardvark and Tree Kangaroo,[7] and the fertilisation of orchids.

At this time I became a student of Zoology myself, and Walter invited me to dine with him and his mother at 148 Piccadilly, and afterwards took me to a Royal Society Conversazione. Our affection was mutual but silent, for we both seemed to have lost our voices — the words frozen on our lips as if 148 Piccadilly had been the hall of the Snow Queen. At Burlington House he was obviously delighted to be able to introduce me to all the great men present, and did so with touching, awkward, almost naive pride. But he never addressed one single word to me.

"Thank you very much indeed, Uncle Walter, I really enjoyed that *enormously* — the exhibits were so interesting, especially Mr. Donisthorpe's ants.... Thank you so much," I said with genuine enthusiasm before leaving.

In the light of the streetlamps he gazed at the ground with fixed concentration — one could feel the words jostling in his brain — but he did not reply.

# CHAPTER 32

# The Primrose Way* — The truth about the Rothschilds

"....and we assert with scientific accuracy
that the Rothschilds are the most astonishing organisms the world has ever seen."

Samuel Butler (1865)[2]
*Lucubratio Ebria*

SOME TWENTY YEARS ago a paragraph appeared in a daily newspaper heralding a forthcoming book titled *Lies about the Rothschilds.* The authoress, so the notice claimed, had collected together in one massive compilation the exaggerations, misplaced eulogies, legends of untold wealth, fictitious anecdotes, discreditable falsehoods, nonsensical fantasies, libels, abuse, *bon mots* never made, beards never shaved off,[3] German accents never possessed, thousands of pounds never spent on fleas off grizzly bears[4] — the infinite inaccuracies that had appeared in print concerning Walter's family. These ranged from the citation at the head of this chapter down to the old chestnut concerning Nathan Mayer and the Battle of Waterloo[5] — carrier pigeons** and all — and the charming vignette of Walter's father by Count Corti[6] in which he is described as "the complete Englishman, quiet and of few words; he spoke no language other than English, was modest, warm-hearted and always courteous".*** This book was never published, and when the authoress was asked to

---

* "The primrose way to the everlasting bonfire...." (Macbeth, 2, 3, 22)

** One persistent story[1] suggested that an agent of Nathan Mayer sent him the news of the victory of Waterloo by pigeon post. The Rothschilds were always interested in communications, and Arthur wrote a book which ran into four editions on the history of letter post. However he apparently had not heard about the mythical carrier pigeon, although he mentions that the Persians used swallows — their feathers dyed specific colours — to send messages home.[7] The Baron Henri also wrote a small book about postage stamps.[8]

*** Virginia Cowles,[9] on the other hand, described Natty as one of the rudest men in England, but neither she nor Count Corti had had the honour of meeting the great man! She did, however, record the fact that he conversed in German with Herzl. Natty, although not a brilliant linguist like Emma, spoke German and French fluently and his Hebrew was excellent. He sometimes could not resist a sly joke in that language at the Rabbi's expense. Roth also mistakenly described him as "doggedly monolingual".[10] Recently (1981) he was provided with a heavy German accent in a television film![11]

explain why she had abandoned such a delectable project, she replied: "It was relatively easy to spot the lies, but it proved impossible to find out the truth."

This is a complaint voiced over the years by many *bona fide* research workers — there is a dearth of authentic information about the Rothschild family. "When I get to the bottom of any subject," wrote one irritable PhD student investigating the split in the Liberal party over Home Rule, "I find the first Lord Rothschild, and there the trail ends."

In Walter's own case the lack of documental evidence about his life and activities is even more marked than among the rest of the family. The fact he never married, his difficulty in communication, his isolated existence within his museum — for there he worked with acquaintances or employees, no really close friends — the disintegration of Tring Park and the destruction of 148 Piccadilly (to make way for a wider road) which scattered his parents' possessions and papers to the four winds, greatly accentuated the familiar Rothschild pattern.

Count Corti was no doubt correct in assuming that the family — especially during the years which straddled the turn of the century — were reluctant to reveal their business secrets and the extent of their interests. But there were other more personal reasons which they felt justified their mania for destroying letters and diaries. Natty, for instance, was determined that after his death the confidence placed in him would not be betrayed, misinterpreted or misrepresented by the merely curious, the indiscreet, the malicious, the gossip writers or by the calumnies of the perennial anti-semites. And his whole family concurred. Thus, for example, Disraeli's friendship with Lionel de Rothschild and his wife — though platonic — was far closer and more intimate than many biographies and books* concerned with his personal and political activities have hitherto revealed. There was a period when Disraeli visited "that admirable woman Baroness Lionel" as Gladstone called her, every day, and a handwritten note sent on ahead heralded his arrival, which read: "Is she alone?"[12] Trunks of Disraeli's letters to Baron Lionel and his son Natty were burnt by his conscientious obedient executor. In fact so conscientious was Natty that he would allow Charles to keep only an *envelope* in Dizzy's handwriting for his autograph collection. Disraeli leaned heavily on "N. Rothschild who knows everything" for information and advice other than

---

* Natty, according to Lady Derby[13] would have liked a biography of Dizzy written in accordance with his (Lord R's) views, and he suggested to Froude that he would then guarantee the financial success of the book. But Froude refused, saying that he preserved complete freedom of his pen under all circumstances.

dates in history (see p. 32). Natty's opinion was sought not only because he was unusually well informed but because he was a great financier and because N.M. Rothschild & Sons had remained the "hub of the financial universe"[14] in spite of the growing competition from the joint stock banks. It was his business acumen and remarkable comprehension of the money market and monetary transactions that encouraged all and sundry to seek his counsel and take advantage of his knowledge and objective judgement. Natty was determined their confidence would never be abused.* Furthermore Emma believed that only within a framework of strict personal privacy and discretion could harmonious relations be maintained in a large family like her own. She left instructions that all private letters — meticulously listed, packeted and dated — in her bedroom drawers were to be destroyed when she died. Charles and Rozsika modelled themselves upon Emma in this respect. Thus Walter's voluminous correspondence as a child, schoolboy and student was lost as well as Natty's letters to his wife and to Charles; the scraps of information about his childhood survived by pure accident. In 1930, when Gunnersbury House was emptied and vacated, a cache of letters addressed to Baroness Lionel, and Charlotte's own diaries about her children (originally destined for the flames, but — fortunately for us — mislaid) were discovered in a box in the cellar. The find caused Emma considerable annoyance, since even at this late hour (for she was then over 80) she found it intolerable that her sister-in-law, Marie Perugia, should read these intimate letters written to her mother-in-law in the early days of her marriage.

The destruction of Walter's own private papers was characteristic of all his activities — a mixture of cautious and deliberate policy, wild hazard and careless insouciance. At times he seemed to caricature certain trends found in the less flamboyant and less eccentric members of the family.

Thus he painstakingly tore up Marie Fredensen's letters immediately after reading them (on one occasion temporarily losing her new address in the process!)[17] and ordered her to do the same with his own, but then carelessly left large bundles of her correspondence unopened in his chest of drawers. He locked away the blackmailing peeress's letters, many unopened, in the laundry basket, but allowed the whole grim story, together with her identity, to be

---

* Although Lord Rosebery's numerous longhand letters to Natty were all destroyed by the latter's executors, a few of Natty's own letters to Archie are preserved, and these contain frank asides: "Harcourt is much on the downline. He is very pompous and boisterous and more perfidious than ever",[15] and again: "Salisbury's speech at the Mansion House was a great histrionic display but as usual he was injudicious."[16] These were meant for Rosebery's eyes alone.

revealed to Rozsika, by failing to destroy them during his life time. He frequently extracted a mass of documents from his museum files and assembled them all together in a boot box when he was, for instance, writing up some particular subject or specific group of animals. Then he would fail to refile them and the boot box in question would be casually put back on the basement shelf in horrendous confusion. The receptacle which contained Beck's diaries, Webster's letters and the various documents relating to the Galapagos expedition[18] (see also p. 185), also contained a long detailed eye-witness account of copulation between elephants, a letter from Julian Huxley seeking permission to capture, castrate and release a pair of Great Crested Grebe on the Tring Reservoir, a *cri de coeur* from the father of one of his collectors whose library roof had sprung a leak, and a despairing note from Faroult, whose eldest son had been killed in the war and his youngest son grievously wounded.

All the records of Walter's orchid hot houses (including the catalogue of the species he had collected and grown with such spectacular results, and the methods of cultivation), and the breeding records of live animals he had reared successfully in the old days in the Park were lost when the Tring Park and Ashton Estate Office papers were destroyed.

This occurred during World War II. Victor Rothschild was absent on active service and his agent, Major Fellows, faced with the problem of evacuating several tons of paper from Tring Park to Rushbrooke with very little petrol at his disposal, decided that the most practical solution was to burn the lot. An ex-employee, living in a house overlooking the garden of the Estate Office telephoned frantically to Ashton Wold[19] to announce that "a madman is burning the Tring Park ledgers". But the call failed to elicit any response, for the message was simply not understood. It did not seem to make sense. Thus the detailed history of the wonderfully successful mini-welfare state with all its ramifications, pioneer health and fire services, unemployment and apprentice and pension schemes, water and electricity supply, comprehensive milk recording projects, stock breeding and poultry fattening programmes, herd books, conservation, sylviculture, game management and game books, agricultural shows, apple orchards, sheep dog trials, aviary, reading room, allotment schemes, holiday camps in the Park for East End children, parties and Christmas hampers for all and sundry, even the details of the dogs' cemetery and its gravestones, was lost. Fortunately Walter's botanical library had been moved from the Gardens to the Cottage at the Museum, and thus escaped the Fellows bonfire.

Even more disastrous was the destruction of Alfred Minall's and Fred Young's diaries.[20] They consisted of daily records which had been kept faithfully and continuously from the day the Museum was installed until Fred Young retired. Apart from their value as log books, they also contained complete lists of the visitors, descriptions of conferences, meetings of the various societies which visited Tring, interesting anecdotes and happenings and so forth. The diaries were stored in cupboards in the basement of the Museum. When it was decided in the late 1960s to pull down the insect building and erect the new ornithological department of the British Museum on the site, an order was given to clear the basement in preparation for this event. Apparently no inventory had ever been made at Tring, and the contents of these cupboards had not been investigated after the take-over. In Walter's Will he designated the papers and boxes of photographs stored in the basements as the Museum Archives, but he did not indicate where they were located. The Works Progress Officer, full of commendable zeal, and determined to fulfil his instructions to get rid of the rubbish, decided, like Major Fellows, that his task would be considerably simplified if he, too, burnt the lot. Leonard Rance, who was caretaker at the time, protested loudly. So did Frank Smith, who, much against his will was put in charge of the Museum bonfire and ordered to load the truck with the ill-fated ledgers, letters and diaries, boot boxes full of papers and boxes of photographs — in fact the Museum Archives. Rance was particularly upset as Fred Young had placed the diaries in his personal care and he had been told that they were valuable historical documents. But their pleas were ignored. One box toppled off the truck on its way to the bonfire and rolled away. By a strange coincidence it was found and salvaged by some unknown person and returned to me, since it contained a few letters which were my property, in addition to Beck's diaries written on the Galapagos expedition, Webster's letters and so forth. By this rather circuitous route they are now finding their way back from the Pacific via Tring and Ashton to the British Museum of Natural History — one hopes to the run of shelves outside the rare book room.

Along with the contents of the basement cupboards, the leather-bound account ledgers — each weighing around 7 lbs — were also consigned to the flames. In his Will Lord Rothschild gave his trustees power to destroy any ledgers, account books and similar documents "as they thought fit". The trustees, of whom Rozsika was one, discussed this matter with Jordan, but they felt that part of the history of the Museum was enshrined in the ledgers — which recorded every purchase from pins to pythons — and therefore must be left in the Museum. The ledgers, into the bargain, were fastened with heavy brass locks of which Miss Thomas had the only key in her safe keeping. However at some moment in time brass became quite a valuable commodity and all the locks were duly prised off by some person or persons unknown. In this rather naked condition the ledgers must have proved an embarrassment to the Works Progress Officer, and their fate was sealed. Again by rather a strange coincidence, years after these weighty tomes had been forgotten, a sharp-eyed visitor emerging from the ladies wash-room recognised with astonishment three large books lying on their sides at the back of an empty shelf. These were the ledgers Nos. 7, 8 and 10. Their brass locks were missing, but otherwise they were in good condition.

A most fortunate accident saved the voluminous correspondence of some 80,000 letters,[21] all addressed to Walter's curators (including about 300 he had himself written concerning the day to day management of My Museum). I was thinking of publishing an article about Walter's early days at the Museum and was given permission to borrow the wooden crate containing the letters. Leonard Rance sent it to me at Oxford and the contents thus escaped the Works Progress Officer's bonfire and were in due course returned to Tring and stored in another part of the basement. Eventually "discovered" there by Maldwyn Rowlands, Chief Librarian at the Natural History Museum, they were rescued, sorted, filed and indexed. Leonard Rance — outraged — noticed that between the day I returned these letters from Oxford to Tring and their rediscovery by Rowlands, some perspicatious person with access to the basement had removed the old stamps from the envelopes. But all the letters addressed to Walter personally have vanished. According to the note he sent to Günther's grandson,[22] thousands of these letters were accidentally destroyed after the first linen basket episode in 1908. And the rest, which he had stored in boot boxes in the basement cupboards, along with the diaries, have now also gone. After Walter's death Rozsika and Sister Claire, following his instructions, sorted and destroyed his opened and unopened mail in drawers, boxes, trunks and a linen basket at Home Farm — a year's post, plus the huge accumulation from his bedroom on the nursery floor at Tring Park. But there certainly seems to have been a sort of hoodoo on Walter's letters, for even in the well preserved and catalogued collection of Chaim Weizmann (in the Rehovot and Jerusalem Archives and among the twenty volumes of his published letters) several of those we know Walter wrote (and which were answered) are now missing. As we have seen (p. 266) he forwarded Weizmann the personal covering letter sent by Balfour with the Declaration. There is (in the Rehovot Archives) the note Walter enclosed with it,[23] and then a further letter asking for its return[24] "since I want to retain it among other similar documents". But the Archivist can find no trace of Balfour's covering letter. Nor did Walter present it to the British Museum with the Declaration itself. What did Balfour say in this private communication to Walter? For the abrupt end to their intercourse seems strange, like so much connected with Walter.

And where are the letters Sokolow and Weizmann wrote to him during the critical period 1916-1923? All his own correspondence dealing with these seven difficult years, and which we know he preserved, is now missing — one of the reasons why his contribution to the genesis of the Declaration has been persistently underestimated. It was certainly not at Home Farm where he died, for Sister Claire examined every scrap of paper on the premises. It may well have been in the plate closet at Tring Park stored alongside his coming-of-age presents, or in one of the tin boxes now missing from New Court. Or it may have been in his desk at the Museum. But, like the various ornaments on that desk, the brass mounted compass inscribed by Winston Churchill, the ivory elephants and the student pipes from Bonn, the correspondence has vanished mysteriously without trace.

Perhaps the strangest of these missing links are the black tin boxes Walter left at New Court for safe keeping, and which contained his personal papers and some documents relating to the history of the firm. They returned safely to the bank after their war time exodus to Tring, since I checked them myself after they were brought back. Walter's name was painted on the outside in white lettering — yellowed with age — several inches tall. Yet they, too, have vanished into the blue.

This is a melancholy and tiresome tale of missing documents and the quenching of mundane facts. But the truth about the Rothschilds is that they were never greatly interested either in their own past or in the distant shape of things to come. There was a bakery in the basement at Tring Park and a huge safe, but no muniment room. Only Baron Edmond and Walter's brother Charles were endowed with vision and foresight. Members of the family, on the whole, were energetic and practical and lived essentially in the present; they were interested in everything of the moment. It is recorded that one of the daughters of a French Rothschild remarked in reply to a query: "I like everything that happens to me." Fundamentally that was Walter's philosophy. Any day one might sight an Alpine Swift or an Isabelline Warbler.

"Not that the Rothschilds are so very old," wrote B.B. in his 84th year,[25] "but they have made history as if they were centuries not generations old." If there are any *bona fide* students of this particular splinter of history, who are interested in Walter's family and the various activities of his near relatives — ranging from Natty's role in settling the great Dock Strike to his sister's astonishing talent for breeding champion cats — ten, all of different breeds — their faltering footsteps will follow a hard and stony path, leading them on inexorably to a very large bonfire.

# CHAPTER 33

# The birds cross the Atlantic

WHEN ERNST HARTERT decided to retire from the Directorship of the Tring Museum in 1930, Jordan took his place. Walter tried to engage a suitable successor for the bird room, but for the salary he could afford, coupled with an uncertain future, it was extremely difficult to find a man of the right calibre — moreover the tide of his own enthusiasm was ebbing. He did not fancy the sheer slog of training an inexperienced young man, and as long as Arthur Goodson was there he felt it was almost easier to curate the birds himself. But then, within the year, Goodson was swept away by cancer; the Department of Ornithology was crumbling around him. After considerable deliberation Walter decided to offer the post to the young Ernst Mayr, who, after taking his PhD, had collected (1927-28) for the Tring Museum in New Guinea* — a wonderful, intuitive choice, for Mayr duly became the most distinguished evolutionary zoologist of the day. Then once again Walter's financial worries escalated. Perhaps it was the fear of the blackmailing peeress's revelations that stopped him from discussing his dilemma with Rozsika and his mother, but he was also engulfed in a wave of depression. He felt trapped. Secretly he decided to sell the birds.**

In October 1931 the collection, except for the 200 ostriches, rheas and cassowaries, was offered to the American Museum of Natural History for $225,000 (a little less than a dollar a piece).[1] Walter could never have parted with the cassowaries — alive, stuffed or skinned. He also tried to retain the birds of paradise[2] but failed.

Phyllis Thomas believes that, somewhat earlier, he had corresponded with the Whitney family's[3] physician on the possibility of a sale. It is extraordinary that Walter kept the whole matter quiet for six months, and that his family, amazed, first read the news when they opened *The Times* on March 11th and saw a paragraph headed "Loss to British Ornithology" (although the sale had

---

* The expedition was a great success and cost Walter only £1017.

** By a very strange coincidence it fell to Ernst Mayr[4] to curate his collection after its arrival in New York.

been announced at the Ornithological Club two days previously). Walter behaved like a guilty schoolboy — as long as he wasn't found out, no harm would come of it.... Dr. Murphy, in a letter to Frank Chapman, described the Club meeting on March 9th, when Karl Jordan made the news public, and said: "Rothschild looked a bit hang-dog at first but presently he was bellowing in his wonted manner with his table companions".[5] Dr. Murphy, who travelled over from the U.S. to pack and supervise the transport of the collection, explains that it was essential for him to list the birds, because Walter's memory was so extraordinary that a working catalogue had never proved necessary. Hartert had, of course, published lists (between 1918 and 1931) of the Tring types which numbered 2165.

Murphy thought he had succeeded in cheering up the silently grieving owner. "I told him how Sanford had fought for his interests — had demanded payment on the *gold* pound basis — and that we would make the collection serve as a sort of Rothschild Memorial much better than the British Museum would ever do.... I promised him that his signed photograph would be framed and hung forever in the Department. He was as jubilant over it as a schoolboy on the 'honor roll'. Although he always carries the front of a lord, he is also an extraordinarily simple man."[6]

The large birds were wrapped in sheets of old English newspapers which had been stored in the basement for packing purposes, and when these ran out, Dutch journals were substituted; the little birds were tucked away in rows between layers of cotton wool. By the end of four months 185 wooden cases had been neatly filled, nailed up and labelled. The list prepared by Dr. Murphy and his wife numbered 740 foolscap pages.[7] It was typed out by Phyllis Thomas who travelled to the States with the collection and helped install it in its new home. The American Museum tempted and then begged her to stay, but Phyllis Thomas was wholly loyal to Tring. When she returned, Walter climbed the staircase to her office and asked for an account of the new installation. He listened intently — as usual in silence — but halfway through the narrative he was overcome with emotion and hurriedly left the room.

Hartert wept when the news reached him in Berlin.[8] This episode illustrates the strange gulf which separated Walter from his curator, for apparently when Hartert retired he had not the temerity to broach the subject of the birds' future and there was no discussion about the work yet to be done; he had vaguely assumed the collection would be bequeathed to South Kensington, for Walter took almost as much interest in the British Museum as in Tring.

David Bannerman declared that the sale "was an epoch making event which closes what had probably been the most progressive chapter British Ornithology had ever seen." There were only a few antagonistic and critical newspaper articles[9] and letters when the sale was made public, chiefly centred on Walter's secrecy and his "questionable right" to deprive the nation of such a unique collection. Suddenly everyone discovered that the contents of the Tring Museum were of national importance. Shades of Newton and Sclater! But Walter was too depressed to notice. He had, in fact, approached the Archbishop of Canterbury and several of his co-trustees before writing to America, but the British Museum had held out no hope of purchase. Emma apparently remained silent on the subject — which puzzled and worried Rozsika — unless she had spoken to her son privately, or did not grasp the magnitude of the disaster for Walter — but one of the museum fans stoutly defended him. Quite apart from any financial consideration, she was confident he had taken a wise decision. "*I* think Lord Rothschild is absolutely right — there's bound to be a war or a revolution in Europe; at least the birds are safe from bombing in New York. It will keep the collection intact and I think the dust problem there is far better managed than in London. But in this day and age one should be *internationally* minded about zoology. The birds belong to the world and not to the UK." Then she added: "How can Lord Rothschild manage them alone — and I fancy if the truth be known, old man Jordan is at heart a bit anti-bird. He grudges the time spent on them."

Walter seemed to shrink visibly in the period following the sale.[10] His silences were now dull, not pregnant with unspoken turmoil. He felt tired and distrait, and spent only about two hours before lunch in the Museum. It was winter — the birds had flown.

304

# CHAPTER 34

# Home Farm

IN THE NIGHT of January 7th, 1935 Walter's valet Woods was awakened by the loud pealing of a bell. He searched the board with his flashlight and realized that, for the first time during the thirty years he had worked at Tring Park, Lady Rothschild had pressed the alarm. In his nightshirt the stolid, ponderous Woods raced upstairs, two steps at a time.... Then he rushed up to the nursery floor and fetched Rozsika.

Emma died next day in her 92nd year. She had never in her life before taken a pain-killing drug or been given a sedative, and now, with the help of pethadine and an oxygen cylinder — despite her laboured breathing — she slept away her last hours in peace.

We shrink from the affliction of a desolate partner who survives the greatly beloved life-long companion; there is nothing we can do; nothing of significance that we can say. How would Walter withstand the shock and finality of his mother's death and the move from Tring Park, where he had lived and worked uninterruptedly for sixty odd years? It was an extremely disquieting prospect. Rozsika, as usual, dealt with her own anxiety by a burst of efficient activity. She furnished and arranged Home Farm for her brother-in-law and persuaded Grosstephen, the chef, who really wished to retire, to cook for him. She provided six maids, an excellent gardener and fortunately both Walter's valet Woods and his driver Christopher were glad to carry on.

Home Farm had once been the residence of Richardson Carr, and it was not for nothing that the Tring Park Estate had nicknamed Natty's agent The Duke. The house was arranged with an eye to supreme comfort and convenience — and situated only 300 yards from My Museum. Furthermore the transition from the building to the garden was ideal — an open French window and a single step from library to sheltered lawn, sliced out of the foot of the Chiltern Hills.

Almost immediately after Emma died, Walter sustained a serious injury to his leg. He fell awkwardly and heavily — for he weighed over 22 stone stripped at this time — in the tunnel linking two gardens at Tring, through which he strode four times a day on his way to and from the Museum. He lay on the ground for several hours before a passer-by in the road above heard his shouts for help. "A token suicide," commented one of his nieces, "Perhaps he will never leave Tring Park after all."

At this juncture Rozsika organised the lottery which was to dispose of Emma's personal possessions, except those mentioned specifically in her Will. This was a tradition in Baron Carl's branch of the family, for both he and his father had left similar instructions;[1] they hoped this scheme would avoid all feelings of jealousy or hurt feelings among the heirs, and it might even provide some fun for the children. Walter, in complete silence, and with a look of intense concentration on his face, took his turn at drawing carefully folded slips of paper out of a silver porringer — and won every single object of outstanding interest. Among them was a framed letter — which always hung, rather lopsidedly, in the billiard room at Tring — a dispatch from Napoleon, captured by a British frigate in the Mediterranean.[2] It was something which had always helped to break the ice for shy guests! Irritated by the tone of the Emperor's letter, Nelson had scrawled "Mark the End, Nelson" below the "Bonaparte". It was claimed to be the only letter bearing both their signatures. To his nieces' intense chagrin Walter immediately gave the letter to the British Museum,[2] "Where it will never be seen or enjoyed again", they lamented in private. The fortunate man himself neither by word nor gesture, then or later, ever referred to his amazing luck in the draw.

The knee injury kept Walter in bed for five months, but it also brought Sister Claire into his life.

Sister Claire was a star hospital sister, who, at the time of the accident, worked exclusively for Dr. Robinson, a consulting physician from London who had looked after Walter's health in a desultory manner during the last 30 years — for there had never been any major problems.

The good doctor feared that pneumonia might develop, for Walter had been deeply shocked, severely chilled and gout had flared up in the foot. He needed good nursing. He got it.

Sister Claire was a fantastically gifted nurse. She was tall and thin, a very good-looking woman in her mid-thirties, with a rose-like complexion and a wasp waist — accentuated by the starched belt of her hospital uniform. She had, if you looked closely, a tendency to red eyelids, due, so she said, to diabetes.... She was very efficient, very talented and could draw butterflies to perfection, holding a brush or pencil, one in each hand, wielding them simultaneously and achieving thereby perfect symmetry. She was also hopelessly over-emotional, and over-sensitive, and very sweet. Sister Claire was determined to get Walter out of bed, out of the gloom of the deserted nursery floor at Tring Park, and into the garden at Home Farm, which was full of birds. The rough lawn sloped up gently through the beech trees of Stubbins Wood and then more steeply, presently vanishing into the unspoilt chalk escarpment. There was a tangle of untended flowering shrubs round the perimeter. It was a little paradise for wild life and an invalid naturalist — a secret garden of the heart's desire, thought Sister Claire. She asked for a bird bath and also insisted that Walter should forthwith buy the Pyreneean Hound he was so eager to possess. After the move to Home Farm a metamorphosis appeared to take place, and almost immediately the patient seemed to shed his crippling shyness and embarrassment, and ceased to struggle desperately for words. He became relaxed and enormously benign, with a permanent expression of mild amusement in his eyes. Walter appeared, despite his two sticks, his black tie and tottering steps, a contented and — yes — a happy man. Monné, the embodiment of canine satisfaction — was always lumbering along at his side.

A great love, even if it is mutual, can prove hard to bear — although the sufferers are totally oblivious of the cause of their malaise. Nor is it unusual to find a new personality emerge in a bereaved and sorrowing widow or widower. The previously meek, rather subdued and wholly dependent wife, always loving and obedient, now suddenly blossoms out, and to the amazement of her *entourage* — instead of dying of a broken heart as they feared — runs the show with dash and efficiency and no uncertain authority. Walter, by his mother's death, had simultaneously been set free from both the beloved and tyrannical iron apron strings and the power of the hated blackmailing peeress. And then he found Sister Claire's company most agreeable and comforting! The transfer was so simple and effortless.

Sister Claire herself fell desperately in love with Walter. She confessed this to Rozsika in so many words. Rozsika listened — her cigarette ash getting longer and longer and, as usual, defying the laws of gravity — and marvelled. But a year later she was faced with a very different and depressing situation for the day after Walter died, Sister Claire collapsed. The local physician supposed she had rather overdone things.... Even with two nurses to help her, Lord Rothschild must have been a heavy man to lift. They sent for Dr. Robinson, who motored down from London. A few hours later he and Dr. O'Keefe asked to see Rozsika.

"I am deeply sorry to tell you, Mrs. Rothschild, Sister Claire has had a mental breakdown," announced Dr. Robinson, ponderously, "I propose to come down with a car and remove her — by force if necessary — and temporarily certify her."

Rozsika was aghast. On this occasion she took the cigarette out of her mouth, and stared, horrified, at the two men.

"You will do nothing of the sort," she said curtly, "Sister Claire had had a great shock — she is highly strung and devoted to her patients and it is not surprising she is in a nervous state after a spell of such dreadfully exhausting work. She refused to take a single day off during the last year and she often worked nights as well as days. I am going to keep her here with me as my secretary, to help and arrange Lord Rothschild's papers. She is an excellent secretary and she will do the work to perfection," — a confidence which ultimately proved fully justified.

"Mrs. Rothschild! Sister Claire is out of her mind...."

"All she needs is a good rest," replied Rozsika, "She is staying here with me — that is final."

Dr. O'Keefe looked first at Dr. Robinson and then at Rozsika, and dropped his gaze to his boots.

"Mrs. Rothschild, I feel obliged to tell you — to warn you — to warn you *seriously* — you are taking on a great responsibility."

"Dr. O'Keefe," said Rozsika dryly, "I am accustomed to responsibility."

But for the present, the move to Home Farm had taken place smoothly, and Sister Claire was determined to get Walter back to work as quickly as possible. She produced a wheel-chair and pushed him energetically up the quarter mile of Park Road to My Museum. And now he rediscovered some of his old zest. Monné was a constant companion — like her master she hobbled because one of her legs was weak — and Dr. Jordan, who detested dogs in the Museum, had to accept her enormous, clumsy presence with good grace, for there was a certain new authority in the air. While Walter worked at his desk on a paper describing Tree Kangaroos, Sister Claire drew butterflies and chatted amiably with the setters. The change in the patient was startling and obvious to all the old hands working in the Museum. They whispered to one another and speculated....

When the British Association meeting at Blackpool came round — Walter had been president of the Zoological section in 1931 and had, with plenty of assistance from Dr. Jordan, produced an interesting address on "The Pioneer Work of the Systematist"* — Sister Claire suggested that all three of them attend. Walter, although still dependent on two sticks, thoroughly enjoyed the meeting, and in the late afternoon took long, leisurely drives in the countryside

---

* Professor Arthur Cain[3] notes "In Section D at York (1932) the then Lord Rothschild gave a good account of work on geographical variation and speciation. Speciation was a problem untouched by the *Origin* and was one of the main ingredients in the neo-Darwinian synthesis of the 1930s and 1940s. The discussions in 1931 in Section D included evolution, vertebrate embryology, population, classification with reference to phylogeny and convergence, variation and genetics." Cain considers that "Walter Rothschild as a zoologist has been most extraordinarily neglected".

with both his companions. Owing to his abnormally sharp eyesight he could see many more details out of a car or train window than most people and he found this way of scanning the natural history of a district quite delightful. After the closing dinner party he became most expansive and treated the guests to a round of his special brand of roaring *studenten* songs. He had plenty of encores.

Soon after their return Jordan determined to tackle Walter about the future of the Museum.

Walter's original Will was a weird testamentary document. He had decided that his collections should be broken up and distributed among various museums which lacked the species concerned. Only his peculiar memory made it possible for him to recall the gaps in literally scores of collections, and name his specimens to fill these lacunae. The Will consisted of enormous lists — filling page after page — of the names of animals and their ultimate and far-flung destinations.[4] But by 1915, when his father's death made it necessary for Walter to concoct a new Will, he had fortunately decided his collections were sufficiently important to be kept relatively intact. He left the Museum and the insect collection to his brother Charles — to pass on to his nephew Victor and niece Miriam for life if, at the age of 16, either of them was still interested in entomology. Both Jordan and Charles were consulted and made notes for Walter.[5] Charles proposed that he would alter his own Will so that in the event of his death the children would have funds with which to run the Museum. Unfortunately he fell ill himself in 1916, and failed to make a new Testament before he died. Walter had then to think again and produce a further version, which he signed in November 1923. This Will still contained various quaint specific legacies such as the Angolan Pallah Particoloured Bear, the Great Auk, two Oarfish, Uroplates Gaboon, Nose-horned Viper, Chinese Takin, Giant Forest Pig, Porbeagle Shark, Abnormal Roach, etc., etc., etc. But after this effort, like the proverbial ostrich, Walter put the whole business of Wills out of sight. This is not unusual, for as people grow older they become more and more disinclined to face the disagreeable fact that the sands are running out. Furthermore, the days go faster and time telescopes and it is easy to forget that it is actually twenty years and not two, since that last Will and Testament was signed.

Professor Kenneth Roeder,[6] who was a particularly gifted and articulate scientist, once remarked when he was 70, "I prefer to go on as usual and live every day as if I was immortal". But Roeder had little to lose from this approach, since creativity was, in his case, linked to highly specific personal skills and there was no Tring Museum and its future to ponder.

Karl Jordan discussed the problem first with Rozsika and then with her children. It wås agreed that the enterprise had grown too large, financially as well as scientifically, for private individuals to handle in the modern world.

There was only one aspect of the Museum about which Walter had never faltered: Its function must be the study of systematic zoology. Neither of the younger zoologists in the family could at this period have been called dedicated systematists, despite a few new species of worms and fleas which they had described. Nor had they the gift of wide horizons which was so characteristic of Walter's and Charles's outlook — they were pigmies in the wake of a giant, modest, very modest, experimental biologists. Drawer upon drawer of set butterflies, however beautiful, seemed oppressive and overwhelming and, furthermore, how easy it would be to become a mere appendage of these naphthalene scented halls! But the total lack of communication between the two generations had all along been the essential stumbling block. Walter never discussed My Museum with his nieces or nephews. He happily assumed that zoology was zoology and that was enough. They knew nothing of his aims present or future for the Museum — except in the most general terms — and he never questioned their own inclinations or aspirations.

Rozsika solved the problem caused by her brother-in-law's reluctance to contemplate his own death, by suggesting that he should ask the Trustees of the British Museum to accept Tring as a gift *inter vivos*; the owner would, of course, retain full control during his lifetime.[7] Walter was delighted with the suggestion, which for some reason had never occurred to him, and forthwith drafted a Deed of Gift. He added a codicil to his Will saying much the same thing, but to cover the eventuality of his own death before the Trustees had officially accepted the gift (or the possibility of an outright refusal) the Museum was left, lock, stock and barrel, along with all the collections, to his niece Miriam.

Jordan attended the Trustees meeting in London at which Walter's Deed of Gift was announced.[8] He then returned to Home Farm to describe the enthusiasm with which the news was received. Walter was as pleased as Punch. It was one of his oddest and most contradictory traits that he was on the one hand an unreasonably modest man (this concerned his personal achievements and endowments) and yet at the same time imperious and even autocratic; but at a different level. For this involved his special position in the community which was inherited from his parents — an immutable fact of life. Thus, when he left the gates of Number 148 on foot in the morning, and set out to cross Piccadilly, he held up his umbrella majestically and plunged into the road without troubling to raise his eyes from the ground, blindly confident that the Bobby on duty would bring the traffic grinding to a halt for his benefit.[9] Yet now, hearing that the Trustees had been pleased to accept Tring, he was like the excited schoolboy of the eighties, thrilled to the marrow because Günther — visibly impressed by the assembly of butterflies and moths he had collected so assiduously round Tring — sang his praises unreservedly. It was as if, on his death bed, he was at last receiving the sort of accolade he had naively craved all his life — the approval his father had, once upon a time. withheld. Sister Claire was equally delighted for she had been nervous that Jordan would, by his air of gloom and despondency, reveal some private misgivings about his own and the Museum's future. "How could they help being overwhelmed?" asked Sister Claire. "After all, Lord Rothschild, *this is the biggest collection ever made by one man* and it is the most important gift the museum has *ever* received. It will take generations of naturalists to work through your eggs or butterflies or moths. You know, you've carved out a niche for yourself in the hall of fame...."

"Stop flattering me!" boomed Walter delighted, and his gaze followed Sister Claire as she whisked briskly about the room.

Had Rozsika's plan for the Museum matured, and the donor survived to assist in the transition period, some of the gravest errors since perpetrated at Tring would surely have been avoided. But suddenly Walter began to complain of backache. Sister Claire was alarmed. A sixth sense told her this was nothing to do with his knee, his sticks, his gout or his lumbago. "I can't bear to see you in pain," she said, and rushed him off to London to be X-rayed. Neither Dr. Robinson nor she told him the truth: he had developed inoperable cancer of the pelvic girdle.

Walter was, in one sense, fortunate. For the disease rapidly invaded the spinal cord, and the paralysis which followed affected his legs without causing him a twinge of pain. He had no idea how ill he was. "There are some blood vessels near the base of my spine which have become twisted," he asserted quite cheerfully, "But they are slowly untwisting and I will recover. Dr. Robinson explained it all in some detail." He had such a strong heart and constitution that he took a year to die. Jordan maintained that it was devoted nursing which prolonged his life in this remarkable fashion. Walter was still keenly interested in the daily Museum routine, and his curator came in with the post and any specimens of interest. On one occasion he appeared with a clutch of Bird of Paradise eggs, which greatly excited Walter. He instructed Jordan to attend the Ornithological Club meeting and exhibit the eggs, and in due course the old man — for Jordan was seven years older than Walter — reappeared to give an account of the meeting: who had been there; who had read papers; what new species were described; what discussions he had had. Walter listened with interest, but when Jordan appeared to have finished the recital he asked very eagerly: "What did they think of my eggs — they must have caused a stir!"

Eggs? Jordan had clearly forgotten all about them. They bored him and he had probably put them out on the exhibition table on arrival, drifted away to chat, and collected them when he left.... "Eggs?" he repeated a trifle wearily, "oh yes, I've no doubt they were appreciated."

Walter's face fell like that of a disappointed little boy — the light died out of his eyes. If Sister Claire had had a gun she would gladly have shot Jordan there and then.

Presently the curator left. Sister Claire sat down on the edge of the bed. "Do you realize," she said pensively, "Dr. Jordan is jealous of you?" Walter looked at her in astonishment. "Of course you don't...." she continued after a judicious pause, "but for the last forty years he has consistently taken a back seat — day in and day out, always over-shadowed — and he feels it, without knowing, of course.... Those eggs. I am sure they completely stole the limelight.... It was so clear — one could see it by his manner."

"You know," said Walter, brightening visibly, "that is something I had never thought of before. Those eggs are unique.... Yes, I believe you could be right."

"I am sure I am right," said Sister Claire, and quietly took his hand. It was so white lying on the counterpane. He had a skin as smooth and pale as alabaster. Walter lay back on his pillows with a secretive smile on his face. Sister Claire was such a clever woman!

He died in his sleep in the early morning on the 27th August 1937. Walter had decided that he was to be buried in a lead-lined coffin alongside his mother and father. His wishes were carried out and a most appropriate phrase was carved on his tombstone: "Ask of the beasts and they will tell thee and the birds of the air shall declare unto thee." It was customary in his branch of the family to send the funeral flowers from friends as gifts to the local hospital. Only one wreath was placed in the hearse when it left Home Farm on its journey past the Museum and on to Willesden. The seeds of these plants had been brought home by Charles thirty years ago on his collecting expedition in Africa, when he had bagged the wild asses for Tring in 1900, and had bloomed miraculously at Ashton ever since. Walter loved them, and in nature their attendant flocks of bright pink flamingoes. They were the blue water lilies from Lake Victoria.

# CHAPTER 35

# Candle ends

WALTER ROTHSCHILD beat the system — in fact he beat two systems — in order to become the best known zoologist of his day and the greatest donor of all time to the British Museum of Natural History.[1] He was particularly inept at traditional book learning, but he broke loose from the toils of the conventional Victorian education. To the annoyance and irritation of the pedagogues and run-of-the-mill zoologists, he pursued a course of wilful but creative self-instruction, with his sights set on clear-cut objectives — ignoring the compulsory subjects at the University, ignoring the King's English, classics, mathematics, examinations, and rules, side-stepping his one-time idol and mentor, Alfred Newton, and believing with blind enthusiasm and zestful confidence that one had to ask of the birds and they would tell you. Moreover, with the birds and the butterflies, the sky was the limit. The second system he beat eventually, but for which, nevertheless, he had to pay a dreadfully heavy price — namely 18 years of the most creative period of his life, spent at his desk at New Court — was the combined pressures and expectations of his family and the Jewish community. This was a form of conditioning akin to brainwashing — which automatically prepared him for the heavy hereditary responsibilities which lay ahead. For Natty's position among the Jews, and the hard earned right he had won to enter every man's front door and the fearful vulnerability of his far-flung people, placed upon him exceptional demands and threatened, by implication, to do the same to his eldest son. There was also Walter's very real devotion to his parents — an exaggerated fear and respect for his father, and a profound love for his matriarchal mother. Fortunately for him he possessed a streak of rollicking, blinkered irresponsibility — totally lacking in his brother — which made Walter's withdrawal from the Bank and its associated commitments possible: for the sake of his father, Charles had relegated natural history and the promotion of conservation,[2] which were his principal interests in life, to the status of a hobby rather than a dedicated full-time occupation, but this Walter refused to do.

In addition to these major problems Walter had to contend with the relatively minor matter of the silver spoon. Finding this object so conveniently placed between his infant lips, he bit well and truly into the ladle with his milk teeth, and put it to good use — experiencing no neurotic sense of guilt or responsibility and taking full advantage of the freedom wealth can bestow. It must be clear to the objective onlooker that the man's smudge of mad genius could have gained full expression only with the financial support supplied by his father and grandfather, backed up by his mother's watchful and protective — at times over-protective — influence. But Walter always remained an isolated and enigmatical figure, even within his family, and also in the contemporary zoological scene. Furthermore a concatenation of circumstances — of which his own extreme secretiveness, coupled with a compulsory streak of deliberate carelessness, played a not inconsiderable part — resulting, as we have seen, in the loss of almost all his private and personal papers — added greatly to the impression of chaotic mystery. He must always remain a huge, benign figure, looming briefly and silently and unpredictably out of the mists, only to vanish with our questions unanswered, and leaving behind a sense of momentary bewilderment.

Not infrequently, when a man dies there seems to be a wordless, mindless conspiracy to obliterate his memory, and, what is more, in a curious way, the person himself appears to be a party to the act, for he behaves as if he were immortal and fails to make adequate arrangements — if any — for the preservation of his interests and life work, — and in the case of zoologists, for the future of their collections and the data relating to them. Mysteriously, the more active a man's life has been, and the more contacts he has enjoyed, the more people he has known and influenced, the more bewildering and dreadful is the sudden disintegration of the web of his activities and the more effectively death seems to fracture, fragment and pulverise his memory. Occasionally the dedicated *acharné* collector, feeling that his *objets d'art* or vintage cars or first editions are not only a perspicacious assemblage of interesting, rare and beautiful materials, but an expression of his own creativity — a projection or extension of himself — takes infinite trouble over the disposition of his treasures. He employs the best available lawyers to guide him and lays down elaborate rules and provides minute instructions for the care and protection of his legacy. Yet even so, matters seem to go awry, the legal experts differ in their interpretation of the relevant phrases, and for one reason or another his wishes are unintentionally misinterpreted or deliberately flouted. There is a mysteriously destructive element at work, which — so it seems — is determined to defeat the man's arrogant desire to cheat death by the survival of his spirit and his day dreams in the guise of his collections.[3] The bacteria take charge.

This familiar pattern is marvellously well illustrated by the departure of the Giant Land Tortoises, skins, bones, carapaces and eggs, from My Museum. Walter believed these animals were scientifically extremely valuable and unique,[4] and as far as their future was concerned he could not have been accused of carelessness. They are mentioned specifically in clauses 5 and 6 of his Will,[5] and their photographs in clause 9. Walter considered the 144 giants such an important and interesting part of My Museum, and so fragile,[6] that it never entered his mind that they could receive anything other then VIP treatment on the premises at Tring. And this is perhaps a classical example of the senseless, unnecessary, motiveless, stupidly destructive sequelae of death. Where are the 144 giant tortoises now — apart from the few types which are stored at South Kensington? Where is Rotumah — the centenarian who travelled round the world via the Marquesas Islands and the lunatic asylum in Sydney to Tring, and who died in the grand manner from unrequited sexual desire? Not as Walter took for granted, for ever safely ensconced in their place of honour in My Museum. One fervently hopes that one day the giants will find their ponderous way back to the exhibition galleries at Tring — where they rightfully belong — the dust of the Acton store-room swept for ever from their carapaces. The exhibit might be enlivened by a film of Aldabra or the Galapagos fauna, a good portrait of Harris or Webster and, maybe, a backcloth of the blown-up photograph of Walter, resplendent in his top hat and morning coat, astride Rotumah. At least there is an *envoi*, for the 100 year old giant "Presented to Lord Rothschild by the Ex-Queen of the Sandwich Isles" has returned — somewhat the worse for wear — after twenty years of miserable exile on the dusty, ill-ventilated, open shelves of Acton to the room in which Walter worked so enthusiastically for half a century. Historically, the tortoises belong to Tring. Evelyn Hutchinson, one of the most enlightened zoologists of our time — who appreciated every detail of its Victorian perfection, down to the red-ringed brass fire buttons and green and white lavatory tiles — once remarked: "Tring is the perfect museum of a museum".[7]

The Bill which enabled the British Museum to accept the gift of Tring, received the Royal Assent on July 29th, 1938. The Archbishop of Canterbury, who was Chairman of the Trustees, had laid "a matter of national importance" before the House of Lords. "....No scientific collection comparable to this, certainly no collection greater than this, has probably ever been given to a nation.... The scientific resources of the nation will be infinitely increased...."[1] He hoped the Trustees would be worthy of the gift since the Museum Bill marked an event of real importance in the scientific history of the country. Lord Crawford added that our entomological collections now became the most important in the world....

Walter prepared a brief memorandum for the Trustees on the gift of the Tring Museum.[8] The document was signed on the 16th July 1937; the first paragraph is set out below:

*My long connection with Zoology has convinced me that the methods of research in systematics — the ultimate object of which is the elucidation of relationship and evolution — require amplification. One point strikes me as particularly important: the systematist is constantly hampered by the fact that the material at hand is not sufficiently large, the conclusions drawn remaining provisional, requiring corroboration by the study of further material. It would be a great advance, and give the systematics of a particular group or order the necessary sound basis, if of some species at least the material of specimens from the entire area occupied by the species were available in such large numbers that it fully represented the range of variation in each district. Unfortunately, want of space restricts Museum collections to a comparatively small number of specimens. If more are sent, they have to be discarded as duplicates although the statistical method based on large numbers should be applied in systematics. I am particularly thinking of Lepidoptera, because my Museum contains a very large collection of these insects which lend themselves so admirably to the study of variation and descent and other problems. I consider it most regrettable that in the past large collections of Lepidoptera have been dispersed instead of being preserved for the purposes here indicated. The breaking up of the Joicey collection, for instance, could have been avoided, no doubt, if sufficient accommodation had been available and my own collection of Lepidoptera might finally suffer the same sad fate and science be the loser. I wish to prevent this eventuality and at the same time to create a centre for research with sufficient room for housing additional collections.*

Professor E.W. McBride in a letter written to *The Times*[9] in 1938, emphasised the significance of Walter's large series of specimens: "I should like to bear testimony to the unique character of this museum and to its enormous importance for the science of Zoology". One wishes he had voiced his opinion while the donor could have beamed his delight.

In the last decade over a million members of the public have visited the galleries at Tring. This part of the Museum, unlike the research departments, was conceived and executed by Walter single-handed, and he went out with furious energy to obtain especially large and fine examples of the species concerned. It is a remarkable fact that he virtually created the public galleries at Tring in a period of about ten years. "No one in their senses could have attempted such a *tour de force*," a visitor once remarked, and then added, "But have you noticed the way in which Lord Rothschild looks at an animal, alive or stuffed? I've never seen a man look at anything with such intense, all embracing understanding."

Walter never became a partner at New Court and his career as a banker was perhaps interesting but not distinguished. It is a trifle odd, however, that all traces of his activities have been expunged from the archives and that even the tin boxes containing his private papers have vanished mysteriously from the strong room. It is less surprising that his portrait — despite his good looks

— does not hang alongside those of Alphonse and Nathan Mayer in some discreet, thickly-carpeted corridor at the Bank. It is certain that long after those respected city bankers have been forgotten — did they breed champion pigeons at a place called Waterloo? — the distinguished Rothschilds associated with My Museum — the giraffe, the bird of paradise, the scarlet climbing lily, the blue/mauve orchid, the silk moth with the mother-of-pearl "windows" in its wings and the blind white intestinal worm — will be remembered. How one regrets that Millais — who had painted such a delightful portrait of Walter as a lively, imperious little boy — was no longer available, for surely he would have been inspired to immortalise the occasion when two brown bears bumbled across the august courtyard at New Court and sat patiently at the front door — now and again rattling their chains — waiting for Mr. Walter.

# Bibliographical Notes

**CHAPTER 2.      TRING IS FAIRYLAND (pages 3-8)**

1. Clutterbuck's Hertfordshire (Vol. 1, 1815) tells that the Manor is mentioned in the Doomsday Survey (possessed by Earl Eustace, Earl of Boulogne, whose daughter Matilda married Stephen, King of England). Matilda granted to the Abbot and Monks of St. Saviour of Faversham in Kent, the Manor of Trenges. In the reign of King James I it was granted by that King to his eldest son Henry, Prince of Wales, for the term of 99 years, and after the death of the Prince, to his second son Charles, Prince of Wales, for 99 years. Charles I (after his succession) settled it upon his Queen, Henrietta Maria, in dower for 99 years, upon whose death it came into the possession of Henry Guy, Groom of the Bedchamber and Clerk of the Treasury during the reign of King Charles II (Akerman 1851). It was he who paid Nell Gwynne her stipend and it was to *Elinors* (a house within the park) that she probably returned when the plague was raging in London. When Baron Lionel bought Tring Manor it included a number of farms and a total area of about 4000 acres. In addition there was stabling for 16 horses, coach houses, a Brew House, a Venison House, the *Green Man* inn, a silk mill, various shops and houses, and the 200 acre deer park which has recently been designated an "area of outstanding natural beauty." When Natty moved to Tring he organised some careful research into the legend that Charles II had really given Tring to Nell Gwynne, and that Henry Guy's ownership was a blind. He told Mr. Clements, the watchmaker in Tring, who was deeply interested in the subject, that in his view there was nothing in the tale and that he had "drawn a blank." Constance Battersea nevertheless refused to abandon this romantic story and clung to it firmly in her Reminiscences (Battersea 1922). As a girl she wrote an anonymous novelette based on the legend. We are told elsewhere that Henry Guy filched money from the Treasury in order to lay out the Park.

2. Tring Park changed hands once or twice before it was rented from Mr. Kay. Charles Orlando Gore sold it to Sir Drummond Smith and it was advertised for sale on August 15, 1820 by Mr. Christie, by order of the Executors of Sir Drummond Smith. *New Times*, 27 June 1820.

3. Walter's father was elected in 1865 as a Liberal for Aylesbury, Bucks, and held the seat until he was raised to the peerage in 1880. He joined the Liberal Unionists after he had taken his seat in the House of Lords.

4. LL. Emma notes this fact in a letter to Constance de Rothschild, a copy of which was entered in her notebook.

5. A portion of this estate was sold to Walter's uncle Leopold, on which he built his house Ascott. It now belongs, with a piece of the Park, to the National Trust. Mentmore was sold in 1977 by the Exors of the 6th Earl of Rosebery.

6. The home of Sir Anthony and Lady (Louise) de Rothschild.

7. RAL, RFamC/3. Natty to his parents. Several letters, 1859-1868.

8. RAL, RFamC/14/58.

9. Charles Gore, owner of Tring Park about 1670.

10. RAL, RFamC/14/58.

11. Harcourt Journals, 14 January 1881. MS Harcourt Dep. 349.

12. Newton Diaries, 30 January 1920.

13. Inherited by Emma from Baron Carl's collection (Rothschild, Carl, 1895). Sold by the 3rd Baron Rothschild at the 148 Piccadilly sale, 1937. It was described in Sotheby's catalogue (Plate 23) as the Lencker Tazza.

14. Painted by Sir Joshua Reynolds.

15. Haldane papers, Haldane MSS 5917. Natty to Haldane, 27 July (no year).

16. The excessive "tidyness" at Tring gave a sensation of suspended animation and a certain chilling lack of activity in the rooms. The fact that no books were ever left lying around and were only seen neatly arranged and sumptuously bound on the shelves was probably responsible for one grandchild's illusion that as a boy he was "not encouraged to read" (Rothschild, V. 1977, p. 18). When, aged 14, my grandmother advised me to become a subscriber to the Times Lending Library the Manager explained that he waived Lady Rothschild's subscription in deference to the unique number of volumes she had borrowed during the last 50 years. Moreover he only sent her brand new copies. Up to the day of her death Emma kept a diary with notes on the books she read. One of the last "Dreams" by Freud received a brief but favourable comment: "Interesting".

17. Meinertzhagen Diaries, Vol. 36, 11 February 1934. (The family does not wish the location of these diaries to be recorded.)

## CHAPTER 3    EMMA—WALTER'S MOTHER (pages 9-18)

This, and the following chapters, are based mainly on conversations with Emma and on letters to her mother-in-law Charlotte (Baroness Lionel) (RAL, RFamC/14/1-58), and to me (1917-1935). These have not been itemised, except in a few instances.

1. Walter had seven nieces and nephews, John, William and Peggy, children of his sister Evelina, and Miriam, "Liberty," "Nica" and Victor, children of his brother Charles, who periodically visited or lived at Tring.

2. This was recorded in the obituary notice in *The Times* newspaper (11 January 1935) which was written jointly by the author and Sir Anthony Lambert under the initials AEL.

3. RAL, RFamC/14/4.

4. LL. Emma to Baroness Lionel, 23 February (no year).

5. LL.

6. Rosebery Diaries, Dalmeny, 24 March 1882.

7. RAL, RFamC/3/40,44. Natty to his parents, 14-16 April 1867.

8. RAL, RFamC/19/17. Alfred to his parents, 15-16 April 1867.

9. RAL, RFamC/5/94. Leo to his parents, 17 April 1867.

10. *Connoisseur* 1920, Vol. 3, No. 10, p. 72.

11. Carnarvon 1976, p. 6.

12. RAL, RFamC/3/88.

13. LL. Charles to Emma from Vancouver, 25 April 1902.

14. RAL, RFamC/5. Leo to his parents. (This letter has been mislaid.)

15. RAL, RFamC/3/29. Natty to his parents, 16 January 1863.

16. Sybil (Rosebery's second daughter) claimed she was endowed with occult healing powers, and by a very strange coincidence one of the lives she claimed to have saved was the grandmother of the future Hannah Rothschild, granddaughter of the third Lord Rothschild.

17. Churchill 1937, pp. 23-24.

18. *Ibid.*

19. Emma considered it most unjust that her father cut Margaretha, his sixth daughter, out of his Will because she had married a Catholic, the Duc de Gramont, and had adopted his religion. The Duc took a very gloomy view of this measure himself and threatened legal action! Emma, however, respected her sister—for her change of religion was based on belief, not on expediency. She and Thérèse together subsequently allocated to Margaretha her share of the inheritance.

20. Harcourt Journals, 17 September 1893.

21. LL, letters from Emma to the author, 16 May 1922, 26 June 1922 and subsequently. She recorded a number of visitors who called or lunched at 148 Piccadilly—the Duchess of St. Albans, Lord Frederick Hamilton, etc. Her sister Thérèse was a regular guest both at Tring and in London.

22. RAL, RFamAD/2. Will of Mayer Amschel Rothschild, 17 September 1812.

23. Davis 1983.

24. Battersea 1922, p. 114.

25. In Mathilde's scrap album (in the possession of the late Minka Strauss) Chopin has written out for her one of his mazurkas. This composition had actually been published elsewhere, but he altered the first few lines in her honour.

26. de Rothschild, Baronne Willy (no date).

27. Newton Diaries, 3 July 1940.


## CHAPTER 4   EMMA—THE IRON APRONSTRINGS (pages 19-24)

1. Mr. Bob Grace, since the end of World War II, has given about ten lectures (unpublished) every winter, to various voluntary organisations, one talk titled "The Rothschild Estates" and the other "Old Tring." He has been able to check the statements in the following chapters which are based on my own recollections, and supply many interesting details.

2. LL.

3. Emma of course took a great interest in many Jewish charities and communal projects outside Tring. Natty was for many years President and a great benefactor of the Jewish Free School (which changed its status from charity to school in 1814) following Sir Anthony, who had presided for 30 years. In 1884 he arranged a fund raising dinner to which he invited Randolph Churchill (Jew's Free School Report 1884, 1885). In proposing the toast of the evening, the speaker paid tribute to the three Rothschilds serving on the ladies' committee: "I cannot sit down without bearing my testimony to the excellence of the ladies of the Rothschild family and the noble work they have done in this great city in connection not only with the people of their own race and faith but in all good works. They have been an example to the ladies of this country." Emma continued to take a personal interest in the Free School into ripe old age. In the twenties she was sending a general knowledge paper set at the school for her grandchild to answer, and she recorded that when she and Walter visited the school and a boys' club in Whitechapel in 1922, they were recognised, and their car quickly surrounded by a cheering crowd.

4. The allotment gardens were subsequently given to the town of Tring as a unit.

5. Bob Grace once asked a very old man in Tring if by chance he remembered anything about the first Lord Rothschild. "Ah," said the old man, "He was the fellow who gave you the sack if you went home without a rabbit in your pocket and a bundle of wood." This epitaph would not have displeased Natty—65 years after his death.

6. Battersea 1922, p. 38.

7. Bert Hinton had ten brothers and sisters, all of whom lived to be over 80. In retirement he collected old corks and fashioned them into marvellous decorations for pictures and mirror frames.

8. Morris 1853.

9. Rosebery Family Papers, Dalmeny. Natty to Rosebery, 12 November 1886 and other letters.

10. Magnus 1964, p. 27.

11. *Ibid.* p. 10; Longford 1966 (Pan Books edn.) pp. 214, 345.

12. Jordan 1937.

**CHAPTER 5    WALTER'S FATHER—*THE EMINENCE BLANCHE* (pages 25-40)**

1. RAL, RFamC/21. Journal written by Charlotte de Rothschild concerning Nathaniel Mayer (Natty) as a child.

2. RAL, RFamC/21. Charlotte to Louise de Rothschild, 1867.

3. Zangwill, *Jewish Chronicle*, 9 April 1915.

4. *East Anglia Daily News*, April 1915.

5. Egremont 1980, p. 205.

6. *Ibid.* p. 204.

7. The speech to which Max Egremont referred was delivered by Balfour in the House of Commons (10 July 1905): "Mr. A.J. Balfour said that this was, perhaps the most interesting, if not the most important, Amendment which was likely to be moved in the Bill, and he rose early in the debate in order to indicate what were the views of the Government in regard to it. The medieval treatment of the Jews was a permanent stain on European annals; and he agreed that if they could do anything to wipe it out, if they could even do anything to diminish its effects in the present time, it would be their bounden duty to do it. The right Hon. Baronet had condemned the anti-semitic spirit which disgraced a great deal of modern politics in other countries of Europe, and declared that the Jews of this country were a valuable element in the community. He was not prepared to deny either of these propositions. But he undoubtedly thought that a state of things could easily be imagined in which it would not be to the advantage of the civilisation of the country that there should be an immense body of persons who, however patriotic, able, and industrious, however much they threw themselves into the national life, still, by their own action, remained a people apart, and not merely held a religion differing from the vast majority of their fellow-countrymen, but only inter-married among themselves. He quite agreed that this country had not nearly reached the point when such a state of things became a serious national danger, but they must bear in mind that some of the undoubted evils which had fallen upon portions of the country from an alien immigration which was largely Jewish, gave those of them whom, like the right Hon. Baronet and himself, condemned nothing more strongly than the manifestation of the anti-semitic spirit, some reason to fear that this country might be, at however great a distance, in danger of following the evil example set by some other countries; and, human nature being what it was, it was almost impossible to guard against so great an evil unless they took reasonable precautions to prevent what was called "the right of asylum" from being abused." (Hansard: Parliamentary Debates, 4th series, 1905, Vol. 149, 10 July-21 July).

8. The brand of anti-semitism which characterised the Foreign Office in the nineteen thirties and forties was less overt—since open anti-semitism smacked of Nazism and was tacitly omitted from public utterances—but more vicious. Thus, for example, the Foreign Office instructed their Ambassador in Berlin in 1939 (Gilbert 1978, p. 223) to urge the German

Government to "check unauthorised emigration of Jews from the German Reich"—thus attempting to stop the refugees fleeing Hitler's persecution and, incidentally, embarrassing H.M.G. by attempting to land in Palestine—their National Home. In 1938, when Jews from Germany and Hungary, some of whom had relatives in England, were trying to flee from central Europe, Roger Makins (now Lord Sherfield) minuted a Foreign Office memorandum dealing with this refugee problem thus: "The pitiful condition to which German Jews will be reduced will not make them desirable immigrants" (Gilbert 1978, p. 192)—thus assisting in the passing of a death sentence on those striving to find a refuge in the U.K., Palestine or places outside Europe such as Western Australia. Among them chanced to be several close relatives of the Rothschild family who eventually died in Auschwitz and Theresienstadt death camps. By a melancholy coincidence a Jewish Commando—a friend of the family—fighting in the British Army, who entered Buchenwald with the victorious troops at the end of the war, heard an eye witness account of how Rozsika Rothschild's eldest sister was beaten to death by S.S. guards wielding meat hooks, on the railway line outside the camp.

9. Holmes 1979, p. 230. These two fanatical anti-semites were both Catholics. They were shrugged off by most people as Popish eccentrics. Chesterton once declared that as far as he was concerned the Jewish problem could be solved instantly if the whole community converted to Catholicism.

10. On the continent of Europe this was, socially, even more marked than in England. For instance in Austria it was unthinkable that an aristocrat or even an officer in the Army could address a Jew or a Jewess as "du" (thou): it was always "sie" (you) even among so-called good friends. Apart from anti-semitic tradition the professions followed by Jews—business in all fields—were regarded with even more contempt by the upper classes than they were in England. Anyone who worked for a living was looked down upon!

11. Foster 1981, p. 30.

12. RAL, RFamC/3/60 & 63.

13. RAL, RFamC/3/129.

14. There were one or two notable exceptions such as Dr. Robson Roose who became a general favourite and was invited to dinner parties, etc.

15. RAL, RFamC/3.

16. LL, Charles Rothschild's autograph collection.

17. *Pall Mall Gazette*, 28 October 1913. Davis, 1983.

18. *Pall Mall Gazette*, 1 April 1915.

19. Longford 1966, p. 610.

20. *Jewish Chronicle*, 30 April 1915.

21. Longford 1966, p. 443.

22. Roth 1939, p. 122.

23. Randolph Churchill papers (Dufferin Viceregal papers, 130/3 No. 76, Randolph Churchill to Dufferin 16 October 1885).

24. Hurst 1967, p. 99.

25. Rosebery sent his private correspondence to Natty at Tring locked in scarlet leather dispatch boxes marked "From the First Lord of the Treasury." All the letters were destroyed—only one of the boxes remains.

26. Rosebery papers, National Library of Scotland, MSS 10032. Hamilton to Rosebery, 13 January 1887.

27. Rosebery papers, National Library of Scotland, MSS 10032, Memorandum 8 September 1888.

28. Harcourt Journals 1894, 17 April. Loulou Harcourt also remarked: "He [Labouchère] talked a great deal about the Rosebery intrigue over the premiership; he knew that Massingham of the *Daily Chronicle* had been dining at Berkeley Square just before the crisis and had been squared... and believes the story that Massingham received indirectly from Rosebery a very large money bribe which is believed to have been passed to him through the Rothschilds."

29. *Morning Post*, 19 December 1883.

30. Jennings 1889.

31. Monypenny & Buckle 1929, Vol. 2, p. 816.

32. Gladstone papers, 44505, Folios 284-285.

33. Gladstone papers, 44505, Folio 70.

34. Gladstone Papers, 44490, Folios 164-165 & 197.

35. Gladstone papers, 44512, Folios 145-146.

36. Gladstone papers, 44523, Folios 295-298. Natty was always greatly interested in trees, and he had assembled a fine worldwide collection of conifers at Tring; this had attracted no attention until the Natural History Museum moved there during World War II, fifty years later. Emma was also a lover of trees and asked Gladstone for chips of the birch in question.

37. Gladstone papers, 44504, Folio 201.

38. Gladstone papers, Add. MS 44446, Folio 306.

39. LL, Inscribed volume.

40. RAL, RFamC/14/12.

41. LL, Copy in Emma's notebook in her own handwriting.

42. Gladstone papers, Add. MS 44511, Folios 215-217.

43. LL, Charles Rothschild's autograph collection.

44. Monypenny & Buckle 1929, Vol. 2, p. 733.

45. Haldane papers, FO 800110.

46. Balfour papers, British Library, Add. 49852, Folio 228-231.

47. Balfour papers, British Library. Add. 49858, Folio 165. Natty, as we have said, was really a dyed-in-the-wool free trader, and even when he disagreed violently with some of Lloyd George's policies, he was an anti-Tariff Reformer among the Unionists. Lloyd George's policies were too far to the left for Natty, for he did not believe in state ownership, but in private enterprise. But in many ways he endorsed Lloyd George's aims: It was his methods for achieving them that he thought mistaken. Thus in his speech in the Lords (he had been invited to speak by Lord Lansdowne (10 November 1909) since there was no one "with an authority equal to yours") Natty concentrated on the consequences of the measures suggested, but the press made sure that it was reported as a speech "against the budget."

48. LL, Charles Rothschild's autograph collection.

49. *Ibid.* "I am very sorry that any answer of mine should have given just cause of umbrage to the De Beers Company. I at once referred your letter and its enclosure to the War Office, and have received from Major Hanbury Williams the enclosed reply which shows how the mistake arose. I suppose I had better wait to make any corrections in the House of Commons until the War Office have got authentic particulars from South Africa. I could then make it on *their* authority instead of having to do it on that of the De Beers Company which is at present all I have got to go upon."

50. Royal Commission on Aliens Immigration Report, Vol. 1.

51. LL, Charles Rothschild's autograph collection.

52. Bleichröeder Archives, Rothschild to Gerson von Bleichröeder, 11 August 1890.

53. LL, Charles Rothschild's autograph collection.

54. FO Cypher to Spring Rice, 6 November 1905, No. 432 R., and LL, Charles Rothschild's autograph collection, Balfour to Natty, 6 November 1905: "I enclose you (1) a copy of the telegram which has just been sent off to St. Petersburg and which will, I hope, have the effect of stimulating the central government to put an end to these atrocious attacks on the Jews. (2) A brief note as to the precedents for giving consular help in the distribution of funds. If you will put yourself in communication with the Foreign Office, there will be no difficulty in obtaining consular assistance for sending help to the victims of the outrage."

55. *Standard*, 1 April 1915.

56. RA, W53/98. It was Leopold, spokesman for the three brothers, who brought the dire plight of the Jews in Russia to the King's attention, and it was Edward VII himself who suggested a memorandum which he could take with him on his visit to the Czar.

57. LL, Charles Rothschild's autograph collection, letter from Lord Knollys to Natty, 3 June 1908.

58. Magnus 1964, pp. 406-407.

59. RA, W53/105, letter from Charles Hardinge to Natty, 13 June 1908.

60. LL, Charles Rothschild's autograph collection, letter to Natty from Cardinal Merry del Val, 18 October 1913, reproduced in Roth 1935, illustrations at his pp. 36-37.

61. *Pall Mall Gazette*, 28 October 1913.

62. LL, Charles Rothschild's autograph collection, various letters; and Roth 1935.

63. Jahoda 1960.

64. Cook & Vincent 1974, p. 498 and note p. 483.

65. Ponsonby 1942, p. 278.

66. LL, Charles Rothschild's autograph collection.

67. Egremont 1980, p. 242.

68. LL, Charles Rothschild's autograph collection.

69. Dugdale 1936, p. 35.

70. Berlin 1980.

71. Ayer 1905. The year of the Ottoman 3% loan (£8,212,340).

72. A strike of dock and waterside labourers in 1889 for a rise in the time rates of wages from 5d to 6d an hour, the abolition of contract and piecework and the remedy of other grievances. The strike was carried on by the newly formed Dock Labourer's Union and was memorable as a demonstration that unskilled labourers could be successfully organised. (Encyclopaedia Britannica, 14th Edn. 1929).

73. Zangwill, *Jewish Chronicle*, 9 April 1915.

74. RAL, NMR/101/20. 6 (?) 7 September 1889 and 12 September 1889, Gustav and Alphonse to their cousins in London.

75. Haldane 1929, p. 163.

**CHAPTER 6     WALTER'S FATHER — NOBODY WANTED TO TOUCH LLOYD GEORGE (pages 41-51)**

1. Wolf, *Daily Chronicle*, 1 April 1915. Natty was elected to the House of Commons at the age of 25 as a Liberal — a good Victorian Liberal — and remained a Member for 20 years until elevated to the peerage (McCalmont 1971). He joined the Liberal Unionists when Home Rule split the party. Later Unionists and Conservatives united.

2. RAL, RFamC/3/13.
3. RAL, RFamC/3/90.
4. RAL, RFamC/3/16.
5. RAL, RFamC/3/135.
6. Speech at Holborn Restaurant, 24 June 1909. See *Daily Chronicle*, 25 June 1909.
7. Roth 1939, p. 129.
8. RAL, RFamC/3/57.
9. LL, Charles Rothschild's autograph collection. Letter from Winston Churchill to Natty, 3 December 1913.
10. LL, Charles Rothschild's autograph collection. Letter from Asquith to Natty, 4 November 1912.
11. LL, Charles Rothschild's autograph collection. Letter from Lloyd George to Natty, 22 March 1909.
12. *Times*, 18 November 1913; *Bucks Herald*, 22 November 1913, p. 7.
13. LL, Charles Rothschild's autograph collection. Letters from Lloyd George to Natty, 18 & 19 November 1913.
14. White 1980.
15. Lloyd George 1933-1936. Lloyd George 1938, Vol. 1, p. 70: "Among those whose advice I sought was Lord Rothschild... The nation was in peril, I invited him to the Treasury to talk. He came promptly. We shook hands. I said: "Lord Rothschild we have had some political unpleasantness..." He interrupted me: "Mr. Lloyd George this is no time to recall these things. What can I do to help?" I told him. He undertook to do it at once. It was done."
16. Lloyd George 1960.
17. Gardiner 1923, p. 290.
18. Lloyd George 1960.
19. *Reynolds Weekly Newspaper*, 4 April 1915.
20. *Daily Chronicle*, 8 April 1915.
21. LL, Natty to Miriam Rothschild, various letters 1911-1915.
22. *Country Life*, 3 February 1913.
23. LL, Natty to his wife, envelope dated by Emma January 24, 1912.
24. RAL, RFamC/3/43.
25. RAL, RFamC/14/35.
26. *Jewish World*, 7 April 1915.
27. *Daily Graphic*, 3 April 1915.
28. This assessment is based on the accounts of funerals and their accompanying photographs which have appeared in the press. In Paris in 1905 (*Figaro*, May 1905; *L'Illustration*, 3 June 1905), the funeral of Baron Alphonse de Rothschild also stopped the Paris traffic and 10,000 mourners filed past the coffin in what the *Figaro* described as *"Une immense manifestation de regrets."* Alphonse and Gustave were also known as friends of the ordinary man.

## CHAPTER 7   BECAUSE I AM SO HAPPY, MAMA (pages 52-56)

1. Mrs. Joan Cambridge, who analysed the handwritings of Walter, his parents and grand-parents. Unfortunately no letters from Walter, or examples of his handwriting as a child have been preserved, therefore Mrs. Cambridge emphasises that this is a tentative sugges-

tion. Both she and another expert in the graphological field agreed on the lack of motor control.

2. RAL, RFamC/14. Various letters from Emma to her mother-in-law, Baroness Lionel, 1868-1878.
3. Günthersburg, Baron Carl's Frankfurt house.
4. He complained to Sokolow in 1917 of being unwell (CZA Z4/117), but the indisposition did not seem to have interfered with his working day.
5. Ruddy Shelduck (*Casarca ferruginea*). A species of semi-marine duck.
6. Bonhote, J.L. & Rothschild N.C., 1895-1897.
7. TMC, various letters from Walter to his Museum curators, Hartert and Jordan.
8. Emma first refers to the arrival of his holiday tutor when Walter was 16; the precise date is uncertain. Subsequently Mr. Althaus lived with the family at Tring for several years.

**CHAPTER 8    I HAVE NEARLY DRIVEN ALTHAUS OUT OF HIS SENSES**
**(pages 57-69)**

1. RAL, RFamC/5/185.
2. Alfred Minall may have been helping Walter only "part time" at this period. Owing to the destruction of the Tring Estate papers (see Chap. 32) we do not know the date on which the "museum" became a full time job for Minall. Up to 1908 all the Museum accounts were kept at the Estate Office.
3. Günther, A.C.L.G. 1910; Günther, A.E. 1975.
4. GC, Rothschild to Günther, 28 January 1885. The following pages are based on the Günther correspondence (a total of about 320 letters) from Walter to A.C.L.G. Günther between 1882-90.
5. Günther, A.E. 1975. Rothschild to Robert Günther, 2 February 1914.
6. GC, Rothschild to A.C.L.G. Günther, 25 December 1884 and subsequent letters.
7. Rothschild, James 1884, Vol. 1. Biographical note by Emile Picot, pp. i-xix.
8. Anon (Constance Battersea) 1887.
9. RAL, RFamC/21, Journal written by Charlotte de Rothschild concerning Nathaniel Mayer (Natty) as a child.
10. M(iriam) R(othschild) 1955.
11. Rothschild 1932.
12. RAL, RFamC/5, Leopold to his parents.
13. GC, Rothschild to A.C.L.G. Günther, 28 July 1885 and several further letters 1882-87.
14. Jordan, in his brief biography (Jordan 1938B) made a mistake when he said that Walter never learned to make a skin; he did, but in later years delegated this task to skilled taxidermists.
15. During his life Walter amassed a fine collection of guns, including several walking stick guns which he used when shooting small birds on walks in the woods both in England and in foreign countries. He also possessed a rhinoceros rifle, several Mauser rifles and pistols, and a pair of three-barrelled guns. When shooting on the Tring Reservoir in later years he used a pair of Atkinsons 12 bore shotguns, with his initials inlaid in gold.
16. Alfred Newton correspondence, Lilford to Newton, 25 October 1887.
17. RAL, RFamC/3/55.
18. Rosebery family papers, Natty to Rosebery, 16 December 1886.

### CHAPTER 9   WHO IS THE PINK AND GOLD BOY IN THE CORNER? (pages 70-79)

1. GC, Rothschild to Günther, 19 December 1887.
2. LL, Walter to his mother, 25 November 1887.
3. GC, Rothschild to Günther, 19 December 1887.
4. Corti 1928A p. 450. (There is a *lapsus calami* in line 24, Beaconsfield for Rothschild.)
5. Systematist = one who studies the classification of animals and plants.
6. Günther, A.C.L.G. 1910.
7. *Ibid.*
8. GC, Rothschild to Günther, 29 June 1889 and subsequent letters. Note: only 12 letters from Günther to Walter are preserved in the Lane Library; they are altogether lacking in the Günther correspondence at the British Museum (see also Günther 1975).
9. Wollaston 1921.
10. Rothschild 1893.
11. Walter published papers in due course on the birds from the Papuan Islands, Goodenough Island, Rook Island, Admiralty Islands, Dampier Island, Vulcan Island, Rossel Island, Louisiade Group, Sandwich Islands, Natuna Islands, Galapagos Islands, Laysan and neighbouring islands, Narborough Island, Macquarie Islands, Sula Islands, Kermadec Islands, Antipodes Islands, Talant Islands, Bounty Islands, Aru Islands, Kulambangra and Florida Islands, Timorlaut Islands, Isabel Island, Coiba Island, Treasury Islands, to mention only a few.
12. Rothschild 1891.
13. Alfred Newton correspondence, Lilford to Newton, 25 October 1887.
14. Rug no. BM(NH) No. 1939, 2929a.
15. Cambridge University Archives, Graces of the Senate and Examination Lists 1887-88; 1889-90.
16. Walter's father also did not sit his finals, probably on the advice of his tutors, but he and Lord Grey received an Honorary Degree at the same ceremony in Cambridge in 1911.
17. RAL, RFamC/3. Various letters from Natty to his parents from Trinity College, 1859-62.
18. Alfred Newton correspondence and LL. Letters Rothschild to and from Newton, in particular 14-21 December 1891. These are the original letters from Alfred Newton to Rothschild but the Professor made longhand rough copies and these are lodged in the Department of Zoology, Cambridge.

### CHAPTER 10   I AM NOT A LIEUTENANT, I AM A CAPTAIN (pages 80-85)

1. From 1903 onwards he was a military member of the Bucks Territorial Forces Association.
2. Lord Balcarres' personal diaries, 29 January 1908.
3. TMC, Rothschild to Hartert, various letters.
4. RAL, RFamC/3/28, Natty to his mother and father, no date.
5. Balfour Papers, British Library, Addn. 49854, Folios 142-144.
6. Balfour Papers, British Library, Addn. 49854, Folios 150-151.
7. LL, Charles Rothschild's autograph collection.
8. LL, Froude, no date (1883?) Privately printed.

9. LL, Charles Rothschild's autograph collection.

10. Pakenham 1979, p. 606.

11. Balfour papers, British Library. Add. 49859, Folios 229-231.

12. Swann 1930.

13. Extracted from Embarkations, South African Campaign 1899-1902, also departure of troops from South Africa 1899-1903.

14. Richards 1910, Vol. 1, p. 68.

15. *Bucks Herald*, 22 June 1901.

16. Although Walter disappears from the Army lists in 1892, in October 1915 he was appointed Sub Commandant of the 2nd (Mid Bucks Battalion) of the Bucks Volunteer Defence Corps.

## CHAPTER 11    PLEASE ASK WALTER (pages 86-93)

This chapter is based mainly on letters in the Lane Library from Charles to various members of his family, Hugh Birrell and Walter's curators. The letters are not listed separately.

1. Natty's aspirations for his second son were sensed by his numerous friends, illustrated in a letter from Lord Esher which ran as follows:
   "I am sure that it will be a source of much pleasure and gratification to you to hear that very strong expressions were used in the presence of the King and Queen yesterday as to the ability, tact and capacity of Charlie in dealing with the work of the Red Cross committee. It was very pleasant to hear, and I am personally more pleased than I can say to be the means of repeating it to you". LL, Letter from Lord Esher to Natty, 21 November 1905.

2. LL, letters from George Carter to Natty, 8 & 17 July 1891, and letters to Emma, 9 July and 27 October 1891.

3. *Jewish Chronicle*, 23 May 1890.

4. LL, E.E. Bowen, letter to Lord and Lady Rothschild, 2 July 1895.

5. LL, letter George Trevelyan to Miriam Rothschild, 11 November 1950.

6. Nuttall Diaries, 16 January 1909.

7. Günther, A.C.L.G. 1910.

8. Rozsika enjoyed Walter's confidence and would never have revealed the identity of the blackmailing couple. She and I had discovered their identity by accident after Walter had successfully guarded his secret for nearly half a century. It is curious that during Walter's life there were persistent rumours that he was the victim of blackmail but these were assumed to be fairy tales — like so many of the absurd stories circulated about the family, from the days of the Battle of Waterloo onwards.

9. No copy of this letter is extant.

10. Rothschild 1900.

11. The British Museum were puzzled by the fact that Lord Rothschild's Museum at Tring contained 65 specimens of mounted cassowaries. Perhaps there were emotional elements in the size of this collection as well as a strictly scientific one. The cost of mounting them came to over £2,000, which was about ten times Dr. Jordan's annual salary.

12. Charles was acutely aware of this problem. A month after he arrived in Switzerland he experienced the first brief remission in his long and agonising illness and he wrote to his wife that he not only felt better, but the *joie de vivre* was coming back The following day he felt well enough to travel to Geneva for the day, with the faithful Jordan, to inspect a large collection of insects which could be of interest to his brother. He bought the butterflies for Walter.

329

**CHAPTER 12     ROZSIKA — WALTER'S SISTER-IN-LAW (pages 94-99)**

The contents of this chapter are based chiefly on family letters and conversations between Rozsika, her sisters and the author.

1. LL, Charles Rothschild to Hugh Birrell, 23 January, 28 February and 22 March 1908.

2. Battersea 1922, p. 40.

3. Rozsika Rothschild's home in Hungary (now Roumania), confiscated during World War II by the authorities.

4. It was customary in Austria-Hungary, as in England, to add a place name to one's title thus: Lord Montagu of Beaulieu. Wertheimer of Wertheimstein eventually dropped the Wertheimer, and they used only Wertheimstein as their surname. The family came from Worms in the eighteenth century. The male branch has now been extinguished.

5. A good eye, a steady hand and muscle control and quick reflexes are qualities demanded by both surgeons and a certain type of athlete. An excellent example is found in the Colledge family. Lionel Colledge was a renowned ENT surgeon (*Who Was Who* 1941-1950), and his daughter Cecilia a world champion skater (Arlott 1975). His son Maule was a first class lawn tennis player.

6. She initiated three such loans, one in 1924, the year after Charles's death, for £7,902,700 (Kingdom of Hungary 7½% sterling bonds) "for the purposes of carrying into effect the financial reconstruction of Hungary under a scheme which was drawn up under the auspices of the League of Nations"; a second (£1,250,000 7½%) to be applied by the counties for road construction, and the remainder for hospitals, electricity works, railways, water works and housing; for an extension of these schemes a third loan (£1,000,000 at 6%) was issued jointly by Rothschilds, Barings and Schroders in 1927. Rozsika's life long friend Count Teleski, then Hungarian Chancellor of the Exchequer, stayed frequently at Ashton and helped her with the negotiations with her cousins. Her son was a minor at this time.

7. LL, letters from Emma to Rozsika, and Charles to Rozsika, 1907.

8. The only friends in the party were the agent Richardson Carr, his daughter Kitty and Karl Jordan. Charles wrote to Rozsika that Jordan was delighted with the invitation and added, "He is a dear old man." At the time Jordan was 45 years old! At the turn of the century you were considered old at forty.

**CHAPTER 13     MY MUSEUM (pages 100-109)**

This and the following chapter are based on personal recollections and the Tring Museum correspondence. These letters are not referred to separately, and only a few additional references are supplied. Details of the Felder and Sir Walter Buller transactions, and the cost of the building of the Museum are in the Lane Library.

1. Whitehead & Keates 1981, p. 26.

2. There were lesser circumstances which influenced the Trustees. Thus the Keeper of Entomology told me that he strongly opposed the suggestion of moving the Lepidoptera from South Kensington to Tring and supported the reverse operation, when the two collections were about to be amalgamated. It was a sad lack of judgement on his part, for, in the comparatively cramped, dimly lit conditions of the British Museum, one loses just those indefinable qualities to which Richard Meinertzhagen drew attention and which fostered enthusiasm and ideas at Tring. The Keeper, however, sounded so eminently sensible and objective that he managed to persuade those who should have known better. Who can blame him? He liked a quiet life and his wife had decided she wanted to remain in Wimbledon. So that was that.

3. Only a few years later the Trustees decided to move the Department of Ornithology to Tring (Minutes of the Trustees of the British Museum (Natural History), No. 38, 7 May 1970, para. 28) and in order to find the necessary building space pulled down the entomological wing added to the museum in 1910 by Charles! It is difficult to imagine a more comic/tragic situation. The correspondence and deliberations about this proposed gift can be found in the British Museum (Natural History) Archives (Gift of Tring to the British Museum) and in the Balcarres diaries (Vol. 52, p. 69 (25.3.1939); Vol. 53, pp. 23-27 (24.6.1939, 22.7.1939)).

4. Whitaker's Almanack 1890.

5. Hartert 1898.

6. Sawyer 1971.

7 *Novitates Zoologicae*: 40 volumes were published with Walter Rothschild, Ernst Hartert and Karl Jordan as editors. The great bulk of the articles concerned the Collections and ran to some 23,000 pages, 600 plates and numerous text figures.

8. TMC. Oscar T. Baron, a railway engineer, collected 1019 humming birds in Mexico, California and Ecuador, which he mounted himself after studying them in nature. Walter bought his private collection in 1893 for £615. This was a very large sum at the time, but the collection was a revelation to ornithologists; until then no-one had seen these birds mounted in natural positions. Walter brought off a great *coup* by the purchase of this collection. Later he financed a special expedition for Baron, who then discovered many new species in the Andes of Peru. The reference in the British Museum's guide to the "Zoological Museum, Tring" (1979, Pub. no. 816 ISBN 0565 008161) is somewhat misleading for it states that the superb collection of humming birds in the octagonal display case was "presented" to the Museum in 1893. As we have seen, Baron's humming birds were bought by Walter Rothschild for his museum and this exhibit formed part of the gift/bequest in 1936/37 to the British Museum. The bulk of Baron's humming birds were sold with the rest of the birds to the American Museum of Natural History.

9. *Harcourt Journals*, Friday, 3 August 1894.

10. Gladstone papers, Vol. 432. Add. MS 44517, Folio 244.

11. If Minall's diaries had not been destroyed in the tidying up bonfire one would know this, since he kept a list of all visitors.


**CHAPTER 14     THE RUMP OF OUR ONLY OLD MALE IS RATHER DICKY
(pages 110-119)**

Various publications concerning the French Rothschilds' collections (including catalogues produced by the collectors themselves) are listed in the Catalogue of the Bibliotheque Centrale des Musées Nationaux. A brief article on members of the Rothschild family as collectors compiled by Nelly Munthe (daughter of Baron Elie de Rothschild) has appeared in Le Petit Larousse de la Peinture, Vol. 2. (See also bibliographical notes for Chapter 13 regarding the Tring Museum correspondence.) The French House has presented its Archives to the Archives Nationales and these are located at Fontainebleau. They are classified up to 1868 only. The rest have not yet been sorted. These papers were moved to the premises before the outbreak of World War II, but it would appear that some documents, photographs and newspaper clippings were mislaid or destroyed during the rebuilding of the Bank at Rue Lafitte. The late Bertrand Gille has published a history of the Archives in which he refers to other sources of Rothschild papers. He told the author that he believed that Count Corti did not exhaust the material in the Viennese Archives and that although the Archives from the Viennese House and the Frankfurt House were destroyed during the war, some papers are still preserved in East Germany. M. Gille had been refused access to the Archives of the English House.

1. Henri's collection of têtes-de-mort was bequeathed to the Musée des Arts Decoratifs by his widow Mathilde (livres d'inventoire 25618-25798) and he donated his collection of 5000 autographs (mounted in 70 bound volumes) to the Bibliotheque Nationale (1933). See also Catalogue de l'exposition "Trois cents autographes de la donation Henri de Rothschild" 1933. Paris, Ed. des Bibliotheques Nationales de France.

2. Baroness Salomon's (Adèle — Emma's sister's) collections have been scattered. A number of individual objects were bequeathed to the Musée de Cluny (20501-20603 etc.) and were exhibited on various occasions including an exhibition "Chefs-d'oeuvre de l'art juif" on the 5th June 1981.

3. James's library was bequeathed to the Bibliothèque Nationale (see Rothschild [N.] J.E. 1884-1920).

4. Ferdinand bequeathed a collection of Renaissance objects known as the Waddesdon Bequest (see Read, London, British Museum 1902), to the British Museum. The contents of Waddesdon Manor consist mainly of the collection of Ferdinand, but also of *objets d'art* collected by his sister Alice, and Baron Edmond (see James A. de Rothschild Collection at Waddesdon Manor, 1967-77).

   Béatrix de Rothschild's collection forms the Fondation Ephrussi de Rothschild at St. Jean, Cap Ferrat. (See Fondation Ephrussi de Rothschild 1969.)

   A number of Walter's relations collected works of art. Some of the more notable collections were those of Baron Carl of Frankfurt (Catalogue of Baron and Baroness Carl von Rothschild's art collection at Günthersburg and Untermainquai No. 15. Privately printed, Frankfurt 1895), Baron Nathaniel of Vienna (Notizen uber einige Meiner Kunstgegerstände, Privately printed, Wien 1903), Baron Gustave and Alphonse of Paris (see N[elly] M[unthe] 1979), Alfred of London (Davis: A description of the works of art forming the collection of Alfred de Rothschild, London 1884), etc., etc.

5. A vignette of Edmond as a collector is included in "La Collection d'Estampes Edmond de Rothschild au Musée du Louvre." (Coblentz, Ed. des Musée Nationaux Paris 1954), See also "Un Musée de la Gravure. La Collection Edmond de Rothschild" (Blum, l'Art et les Artistes, Paris 1936).

6. The two principal Rothschild collections of rings were bequeathed to the Musée des Arts Decoratifs (legs. 9782 & 10005) and the Musée de Cluny (legs. 15479-15600). Baronne Nathaniel (Charlotte) herself no mean artist, also collected pictures, a number of which she bequeathed to the Louvre.

7. Gift to the Musée des Arts Decoratifs (1908) (livres d'inventoire 14955-15096).

8. Miriam's nephew Edmond sold those manuscripts which had survived World War II (for the major part of her collection disappeared during the German occupation) and these included the corrected proofs of *Les Fleurs du Mal* by Beaudelaire and similar rare items. No catalogue was issued for the sale. She also collected first editions in their original bindings from Villon to Rimbeau, Montaigne to Proust, and also many letters and musical scores from writers and musicians: "Dans la domaine étroite et raffiné de la bibliophilie on peu dire qu'elle fut un example hors du commun."

9. Alphonse de Rothschild's collection of postage stamps, which was one of the finest in existence, was sold piecemeal by the owner when he left Austria as a refugee from Nazi oppression in 1940 (see Mueller 1942). Baron Henri wrote a book about postage stamps "Les timbres poste et leurs amis" (Paris, 1938).

10. The collections of living and pressed Iris were bequeathed and donated to the Royal Botanic Gardens, Kew in 1923 and 1967.

11. The collection of ectoparasites was given to the British Museum of Natural History in 1913 (see Hopkins & Rothschild 1953, pp. 357-8).

12. Catalogue of objects of gold and silver etc., in the collection of the Baroness James (Jones 1912), Mentmore Catalogue, privately printed, 2 Vols. 1884 (Hannah Rosebery), Catalogue of 18th century printed books and manuscripts, privately printed (Rothschild, V., 1954).

13. Rothschild [N.] J.E. 1884, Vol. 1, Introduction by Emile Picot. Picot was a great admirer of the Baron James and noted that the success of his library was due to his taste, not his fortune. James compiled most of his own catalogue, but the task was finished by Emile Picot.

14. LL, Sage & Co. estimates 28 and 29 September 1910.

15. Manson Bahr 1951 (short history of the B.O.C.).

16. Pick = to select: a term used professionally among dealers at the time.

17. Haldane Papers, MS 6016, Folio 105. Rothschild to Miss Haldane, 5 December 1896. Miss Haldane, like Emma, was interested in Hegel and other German philosophers.

18. British Museum (Natural History) Archives. Register of Directors Incoming letters books, Rothschild to Ray Lankester, 20 December 1906.

19. Quite recently I was walking in an hotel garden in Beverley Hills when I was stopped by a man who asked me if it was true that I had some connection with the English Rothschilds. "I am Wood's (the valet) nephew," he explained. "I owe a lot to the second Lord Rothschild. When I was a boy he interested me in Red Indians and I emigrated to the States and I've made a life study of their artefacts. He was so very kind to me and started me off collecting."

20. Roth 1939, p. 132.

21. Meinertzhagen Diaries, August 1937.

22. TMC, Rothschild to Hartert, 30 July 1908.

23. Balcarres Diaries, 22 March 1924.

24. And remunerative! Over 7 million copies — not counting the de luxe edition — have been sold to date of the book version of this film.

25. British Museum (Natural History). Tring Museum Ledgers. Nos. 7, 8 & 10.

26. Between 1888-1915 an annual expenditure on housekeeping of £2000 p.a. was considered quite modest for a country house like Taplow (Mosley 1976, p. 76).

27. Smit 1980.

28. He thinks it unlikely that, since the old entomological wing built for Walter by Emma and Charles had been demolished, and a modern ornithological block erected in its place, Walter's footsteps will be heard in those new surroundings. The garden where he swung in his hammock has also been built over.

### CHAPTER 15    THE FELLOW IS ALWAYS RIGHT (pages 120-127)

1. Meinertzhagen Diaries, Vol. 41, 8.1937.

2. Hinton 1937-38, p. 336.

3. Jordan 1938A.

4. Riley 1960.

5. Hopkins & Rothschild 1953; foreword by Miriam Rothschild.

6. A revision of the Lepidopterous Family Sphingidae (Rothschild & Jordan 1903); A Revision of the American Papilios (Rothschild & Jordan 1906); Revision of the Papilios of the Eastern Hemisphere, exclusive of Africa (Rothschild 1895); A monograph of Charaxes and allied prionopterous genera (Rothschild & Jordan 1898, 1899, 1900 & 1903); Notes on

Saturnidae; with a Preliminary Revision of the Family down to the genus Automeris, and descriptions of some new species (Rothschild 1895); The Avifauna of Laysan and the neighbouring islands; with a complete history to date of the Birds of the Hawaiian possessions (Rothschild 1893 & 1900); Extinct birds (Rothschild 1907); Ornithological explorations in Algeria (Rothschild & Hartert 1911, 1914, 1923); A review of the Ornithology of the Galapagos Islands, with notes on the Webster-Harris Expedition (Rothschild & Hartert 1899).

7.  Hinton 1937-38, p. 337.
8.  Karlsson 1978.
9.  MMPI = Minnesota Multiphase Personality Inventory.
10. Hartert was a Dane by birth, but his father decided to join the Prussian Army and was naturalised. Jordan was German by birth.
11. Wigglesworth 1965 (Presidential address).
12. Jordan 1938B.
13. Even Jordan got confused by the Nathans and Nathaniels of the Rothschild family. But it was not only Walter who failed to show his contributions to his co-editor! Had I seen *In memory of Lord R* either in typescript or in proof, it would have been corrected and amended. But I did not. Karl Jordan himself could be rather autocratic — in fact it would never have occurred to him to consult anyone but Rozsika, who gave him some facts and the dates — that was all.
14. When the number of visitors to the Natural History Museum in London was falling off in 1974, those at the Tring Museum continued to rise steadily. (Report on the British Museum (Natural History) 1972-74; Appendix IX).
15. Wolf, 1915.
16. The pot calling the kettle black, as far as the microtome was concerned! Much to my regret Jordan never cut a single serial section during the 70 years he worked at Tring and his classification of fleas as well as moths and butterflies was based on hard portions of the animals concerned — external rather than internal features.
17. Audubon 1969.
18. Roth 1939.
19. LL.
20. Pye-Smith 1965.

## CHAPTER 16    WALTER'S CURATORS: Ernst Hartert & The Birds (pages 128-142)

This chapter is based essentially on my recollections and conversations with the persons concerned, but Hartert's expeditions are described in some detail in *Novitates Zoologicae*. The Günther correspondence and Tring Museum correspondence have provided much valuable information (these letters are only occasionally referred to separately), and so have Professor Ernst Mayr and Dr. David Snow.

1.  Rothschild 1916.
2.  Stresemann 1975.
3.  Hartert 1903-1922.
4.  Rothschild 1934.
5.  Royal Society Year Book 1912. -
6.  Rothschild 1905.
7.  Meinertzhagen diaries, Vol. 36, 17.11.1933.

8. Rothschild 1934, p. 357.
9. GC, Rothschild to A.C.L.G. Günther 1885 (no precise date).
10. Alfred Newton correspondence, Lord Lilford to Newton, 25 October 1887.
11. Rollin 1959.
12. Snow 1958.
13. Lack 1947.
14. Mayr 1975.
15. Mayr 1942.
16. Barclay-Smith in litt. Notes for the history of the British Ornithological Club.
17. TMC, Rothschild to Hartert, 4 April 1898.
18. The parents of these cranes had been bought many years previously from Lord Newton. Dr. Murphy records their wild shrieks when he visited Tring in 1932.

**CHAPTER 17    WALTER'S CURATORS: Karl Jordan & The Butterflies (pages 143-153)**

1. Jordan 1929.
2. Hopkins and Rothschild 1953, p. 2.
3. M[iriam] R[othschild] 1955. This biography (upon which p. 145-148 are based) together with an outline and appreciation of Karl Jordan's work, and his bibliography (up to his 94th year) will be found in the *Trans. R. Ent. Soc.* Vol. 107, 1955, published in his honour.
4. Riley 1955.
5. Rothschild 1905.
6. Zimmermann 1955.
7. Mayr 1955, 1982.
8. Jordan, 1896, 1905.
9. Sympatric species = species having the same or overlapping areas of geographical distribution; opposite of allopatric species = having separate or mutually exclusive areas of distribution. Originally spelled synpatric by Karl Jordan. *Novit Zool.* 23, 1916, p. 150.
10. Jordan 1916.
11. Jordan 1911. Mimetic polymorphism = distinct genetic varieties of the same species capable of existing together because they superficially resemble (= mimic) other species with inherent protective qualities such as poisonous blood.
12. TMC, Rothschild to Hartert, 23 July 1904.
13. LL, Manuscript draft of "Beginners Guide to International Finance."
14. Hopkins 1955.
15. Riley 1960.
16. Riley 1937.
17. In fact this was eventually a Revision to which Jordan should have been accorded joint authorship, since he learned at amazing speed once Walter had aroused his interest, and he contributed enormously to the general section and provided the illustrations. This was fully explained in the text.
18. Rothschild & Jordan 1903.
19. For example, *Entomological Record*, Vol. 15, 1903, pp. 309-312; Vol. 16, 1904, pp. 5-10; 44-47; 75-78.
20. Occasionally an unknown scribe, probably one of the Goodsons, seems to have written out a few paragraphs of some of Walter's fair copies.

## CHAPTER 18     WALTER'S COLLECTORS (pages 154-169)

The Tring Museum correspondence and the account in *Novitates Zoologicae* of Rothschild and Hartert's expedition to Algeria in 1908, 1909 and 1911 have been used extensively in writing this chapter.

1. Schama 1978.
2. Draeger, Paris (no date).
3. TMC, letters to Hartert, Jordan and Rothschild. These letters (about 40,000) were annotated on my behalf by Nicholas Wollaston, son of A.F.R. Wollaston. Another 40,000 (approximately) from later years have not been examined in detail. Among them are about 500 letters from A.S. Meek which have been consulted by J.S. Heron (1975) and others for their papers concerning this intrepid collector.
4. If the flea collectors were included another 100 names could be added. But this enterprise, although shared by Jordan, was mainly carried out by Charles himself at home. The latter also owned a small house in Akeman Street, Tring, where his assistant F.J. Cox mounted flea material and set butterflies.
5. Rothschild 1893.
6. Rothschild & Hartert 1899.
7. LL, Charles Rothschild to Birrell, 19 April 1902.
8. TMC, Rothschild to Hartert, 10 June 1902.
9. LL, various contracts with collectors.
10. TMC and LL.
11. Stresemann first introduced Mayr to Tring.
12. Whitaker 1897 and Wilsher 1970.
13. Charles collected intensively in the U.K., Hungary and Switzerland, and for brief periods in France, Portugal, Italy, Austria, Egypt, Japan, Ceylon and North America.
14. Rothschild & Hartert 1911.
15. Alfred Newton correspondence, Department of Zoology, Cambridge.

## CHAPTER 19     WALTER'S COLLECTORS: A.F.R. Wollaston & N.C. Rothschild
##               (pages 170-176)

This chapter is based on conversations with Wollaston and my father, Wollaston's diaries and letters published posthumously by his wife, the Rothschild/Birrell correspondence and a number of letters written by Charles to Emma, stored in the Lane Library.

1. The caterpillar of *Arctia caja*—a very hairy caterpillar—is known to entomologists colloquially as the Woolly Bear.
2. Rothschild, N.C. 1903. The flea was named in honour of the Egyptian Pharoah Cheops (Hellenized Version, Khufu) who was the 2nd King of Dynasty IV ca. 2600 BC.
3. Traub & Starcke 1980 (Miriam Rothschild, pp. 1-9).
4. Hirst 1925, p. 16; Hirst 1953.
5. Amschel Mayer (1773-1855), Walter's ancestor, had one great passion—the cultivation of plants in his Frankfurt garden (N[elly] M[unthe] 1979).
6. Dykes 1913.
7. Druce 1924.
8. This rose is now considered a variety, endemic in the Midlands, *Rosa obtusifolia* Derv. var. *Rothschildii* (Druce) Wooley-Dod.

9. Andrassy Kuzmann: killed in the 1914-18 war.

10. Charles Rothschild left his collection of Hungarian butterflies to his wife in memory of the many happy holidays they enjoyed together in Hungary. She gave it during her life time to the Tring Museum, and his collection of living Iris was bequeathed to the Royal Botanic Gardens at Kew. The collection of pressed Iris was given to the Natural History Museum (British Museum) but destroyed in the 2nd World War by bombing. A duplicate collection which belonged to Rozsika was passed on by me to the Royal Botanic Gardens. It contained 194 specimens (43 species and varieties). There was a card index of over 100 "Iris" correspondents. Charles's live orchid collection also went to Kew Gardens and the paintings made by Lady Davy were left to the National Collection.

11. Von Degen 1923.

## CHAPTER 20    WALTER'S COLLECTORS: William Doherty (pages 177-180)

This chapter is based on Hartert 1901 and correspondence and papers in the Tring Museum. Unfortunately Doherty's photographs were destroyed in the bonfire.

## CHAPTER 21    WALTER'S COLLECTORS — THE KING OF BULGARIA IS COMING ON FRIDAY.... (pages 181-184)

This chapter is based on my recollections and conversations with the persons concerned.

1. The specimen upon which the description of a new species is based is called the type. Types are the most valuable entities in a Museum collection and every serious worker on a particular bird or group of birds has to examine the types and base other descriptions upon a comparison with them.

## CHAPTER 22    HURRAH! WE ARE OFF.... (pages 185-196)

This chapter and the next are based mainly on:

a)  The bowdlerised version of both Harris's and Drowne's diaries which Walter published in *Novitates Zoologicae*, 1899, vol. 6, pp. 85-142 (unfortunately the originals were destroyed, see Chapter 35);

b)  Beck's notes (corrected by Walter) which he extracted from his own diaries (*Novitates Zoologicae* 1902, Vol. 9, pp. 373-380);

c)  The Harris/Webster/Rothschild correspondence (in the Lane Library).

1.  The collection made in the Galapagos Islands by Professor Bauer in 1890-92 had been purchased by Walter and was housed at Tring. In *Review of the Ornithology of the Galapagos Islands* he says: "The study of their collections made the desire for more material more ardent." Presumably Bauer had collected South Albemarle, a fact known to Harris.

## CHAPTER 23    THE GIANTS (pages 197-205)

In addition to the correspondence listed in the bibliographical notes for Chapter 22, the Tring Museum correspondence and Rothschild/Günther correspondence in the British Museum (Natural History) have been used extensively.

1. Harris's most serious and valid complaint was that he knew that the material stored in spirit urgently needed sorting, arranging and repacking. Webster insisted that customs regulations made this impossible, and to Harris's consternation it was sent right through to Tring, and as a result these preserved specimens did not arrive in good condition. This was deeply disappointing from the scientific angle, and very expensive for him, since he was paid for individual specimens, not for time spent collecting.

2. We have no evidence on this point since all but twelve of Günther's letters to Walter were accidentally lost long before the bonfire (see Günther 1975).

3. Stoddart 1971.

## CHAPTER 24    THE ROTHSCHILDS AND ANIMALS (pages 206-214)

Note: The detailed account of the Tring Agricultural Scheme (milk records, tuberculin tests, feeding methods, etc.) was lost when the Tring Park Estate papers were destroyed during World War II. The farming enterprise was a commercial success, for the first Lord Rothschild worked hard to prove his theory that high quality animals cost no more to keep than stock of poor quality.

1. *Times*, 3 March 1939 (High Court of Justice).

2. LL, letter from Charles to Rozsika, 6 March 1906.

3. LL, letter from Charles to Emma, 4 March 1901.

4. Zangwill, *Jewish Chronicle*, 9 April 1915.

5. Thornton 1880, p. 78.

6. Cowles 1973, p. 155.

7. Pascal 1933.

8. The family as a whole were interested in providing medical assistance for those in need, but the Baroness James was particularly dedicated to this cause.

9. Baron Edouard's wayward younger son Maurice, who was better known in the family for unconventional behaviour and his fabulous collection of works of art and some successful racehorses, was also interested in animals from other viewpoints. He organised and took part in a zoological expedition to Africa and together with Henri Neuville wrote scientific papers on giant warthogs, Okapis, etc. (Rothschild, Maurice de & Neuville 1906, 1909, 1909-11) He also collected fleas for Charles.

10. Mohamed Shah H.H. Aga Khan, an exceptionally successful racehorse breeder and owner.

11. Many stately homes had, in those days, farms and gardens which were expensive luxuries. One of the outstanding features of the Tring Park farm and garden was that they were run as economically successful and highly profitable enterprises. This was one of the reasons why visitors from abroad were guided to Tring, apart from the high quality of the farm stock that was bred on the premises.

12. Chivers 1976.

13. RAL, B17 Rh 1895/27.

14. LL, Charles Rothschild's autograph collection, letter from Baron Konkigo Takahashi to Lord Rothschild (Natty) 6 July 1906.

15. *Country Life*, 1900, Vol. 8, pp. 86-88, 325-327, 749-751.

16. The Field, Tring Park Agricultural Scheme, 1897-1908. Tring pioneered one of the earliest and most detailed milk recording projects.

17. Tubbs 1946.

18. Many garden varieties of flowers were named after members of the family in France and England. There were Baron and Baroness Rothschilds among the roses and orchids and

rhododendrons (Phillips & Barber), and Cymbidiums Rozsika, Nathaniel, Miriam, etc., etc. The latest addition to the ranks comes from France, a rose, Baroness Edmond de Rothschild, the result of a cross between Baccara x Crimson King x Peace.

19. Thomas & Hinton 1920.
20. *National-Zeitung*, Basel, Nr. 160, 14 April 1963, p. 19. "Rothschild mit blauen Auge."
21. These names have been checked in various scientific catalogues and journals such as Peters' check list of birds of the world, Vols. 1-15, 1931-62. They have not been collected together in a single publication. In addition to specific names there are seven genera named after Walter or his brother Charles. (See Appendix 1)
22. Grand Larousse Encyclopédique, Vol. 9, p. 387, 1964.
23. Günther 1975, Chapter 31, p. 417.
24. Between August 1896 and February 1897 10,456 game birds were shot at Tring— 1828 of these were wild duck.
25. Prof. F.A. Lindemann FRS (later Lord Cherwell).
26. Rothschild, V. 1973, 1981.
27. R[othschild], M[iriam] 1955.
28. Beston, 1974 edn., pp. 19-20.

## CHAPTER 25     THE GREAT ROW (pages 215-223)

1. Karlsson 1978.
2. Murphy 1932.
3. *Harcourt Journals*, 7 August 1892.
4. Farrar 1884.
5. Several secretaries were employed to investigate all those cases which appeared to be genuine and in need of assistance. It was said that Natty and Emma spent over £100,000 per annum on personal care arising from letters received from strangers.
6. LL, Rothschild to Fredensen, 20 November 1905, 27 November 1905, 7 December 1905, 20 December 1905 (copies).
7. Egremont 1980, p. 123.
8. RAL, RFamC/14/26, Emma to Baroness Lionel.
9. *Times* Newspaper, 11 January 1930.
10. *Bucks Herald*, January 1906 (various numbers).
11. Rothschild & Jordan 1906.
12. TMC, Charles Rothschild to Hartert, Christmas Day 1907.
13. TMC, Rothschild to Hartert, 15 June 1907.
14. LL, Fredensen to Rothschild, various letters.
15. Ballad of Reading Gaol: "Yet each man kills the thing he loves, By each let this be heard. Some do it with a bitter look, Some with a flattering word. The coward does it with a kiss, The brave man with a sword!"

## CHAPTER 26     CATHERINE WHEELS (pages 224-232)

1. Jordan 1938B.

2. Bob Grace (unpublished lecture).
3. Bob Grace (unpublished lecture).
4. *Vanity Fair*, caricature by Spy.
5. Baron Lionel, Mayer Amschel, Natty (1st Lord Rothschild), Ferdinand, Walter (2nd Lord Rothschild), Lionel (son of Leopold) and James (son of Edmond). All were Liberals or Liberal Unionists, except Lionel who was elected as a Conservative.
6. Poliakoff 1968.
7. *Anti Suffrage Review* 1908-12.
8. LL, Charles Rothschild to Birrell, 25 January 1901.
9. Hansard, Parliamentary Debates, 4th series, Vol. 82, 26 April 1900-11 May 1900. Sea Fisheries Bill, 7 May 1900, Column 954.
10. Hansard, Parliamentary Debates, 4th series, Vol. 93, 29 April 1901-13 May 1901, column 131.
11. Bucks Herald, 24 March 1906.
12. Gladstone Papers, Vol. 80, Add. MS 46064, Folio 27. Rothschild to Gladstone, 17 March 1906.
13. Gladstone Papers, Vol. 80, Add. MS 46064, Folio 170. Rothschild to Gladstone, 18 March 1906.
14. There was a serious difference of opinion between the English and French houses on the question of relations with Russia. Alphonse and Gustav believed that a loan to Russia would increase the Rothschild influence, and that they would then be able to do more to help their persecuted co-religionists. Natty absolutely refused to have any business dealings with a country perpetrating such outrages against the Jews and drew attention to the situation by forgoing the estimated £2,000,000 profit on the deal. Natty furthermore was a shrewd financier and it is probable that quite apart from his horror and indignation he had serious doubts about the future of the Russian Empire, especially as a sound investment for his firm.
15. Balfour papers, British Library, 49858, Folios 193-196 Balfour to Rothschild, 18 March 1906.

"My dear Walter Rothschild,

I have read with great interest your speech of the 15th of March, as it raises a question about which there seems to be some misconception in the public mind.

You express a regret, which I also feel, that the only observations which I was able to make from the Front Opposition Bench were critical rather than constructive in their character. The fact is so; but responsibility for it does not rest with me, nor am I to be blamed because the debate concluded without the Opposition leaders having an opportunity of discussing, or even of dividing on, the economic policy which they recommend to the country.

Those who suppose that our primary duty in connection with the fiscal resolution moved last Monday on behalf of the Government was to expound our own views, mistake the whole situation. Our primary duty was to find out what the Government meant; to elicit the true import of the solemn repudiation of a "protective system" which they asked the House to record upon its journals. When this was discovered it would have been both proper and expedient to debate and vote upon an alternative policy, if this seemed right in our eyes.

As you are aware, we could neither discover the policy of the Government, nor were we permitted to vote upon our own. What happened was this. As soon as the seconder had resumed his seat, I rose to ask certain questions of the Government as to what they meant by a "Protective System"—stating at the same time that I should deal with my own views on fiscal policy when we came to the Wyndham's Amendment, which was moved on behalf of the late Administration. My appeal was vain. It was evident that the Government had not the slightest idea as to what they meant by a "Protective System." They could not answer the simplest question upon it. Their resolution had apparently been drawn up for the sole purpose of embarrassing the Opposition: and its authors forgot that whether or not it fulfilled their amiable intention, they would have to stand cross-examination as to the precise meaning of the pledge by which they purposed to bind the House of Commons.

What happened is, I believe, unexampled in parliamentary history. To the questions put by the Leader of the Opposition, no answer was given—indeed, no answer was attempted. I have since been informed that they did not deserve an answer because they were 'mere dialectics'. 'Dialectics' in the modern jargon of political controversy is a pompous and inaccurate description of arguments to which a reply cannot easily be found. But when a Government are foolish enough to start an unnecessary controversy, they should at least be wise enough to consider how the controversy is to be carried on. Their answer to obvious questions need not necessarily be good answers. With a majority of three hundred and fifty no high standard of logic is required. But decencies should be preserved. There should at least be a show of replying to the Leader of the Opposition, and if his reasoning be feeble (as no doubt it may be) what an opportunity for a debater like the Prime Minister."

16. GC, Rothschild to Günther.
17. GC, Rothschild to Günther (undated, received 25 February 1892).
18. Rothschild 1893.
19. Bowdler Sharpe 1907.
20. British Museum (Natural History), Report.
21. In recent years there has been a disastrous tendency to appoint older and older Trustees. In Walter's day nine members of the elected board had been appointed when they were under fifty. In my year the average age was 65, and no-one under fifty had been considered.
22. BM(NH)A, Directors' Incoming and Outgoing Letter books 1895-1922. Lankester to Rothschild 10 January (175) and 19 January (185) 1903.
23. BM(NH)A. Telegram, Rothschild to Lankester, 17 January 1903 (191).
24. Balcarres Diaries, 29 January & 9 February 1908.
25. *Daily News*, 24 November 1908.
26. Roth 1939, p. 249.
27. *The Times*, 26 November 1908.
28. Walter, who lived with his mother, had no commitments apart from My Museum and his modest little family. Emma, on the other hand, had a multitude of responsibilities which included orphanages, hospitals, old folks' homes and ex-employees of her father in Frankfurt. She was thoughtful and careful with money, frugal and self-disciplined, and the very antithesis of self-indulgent, yet the fact remained with her multitude of good works and so forth she was a far bigger spender than Walter.
29. The first non-family director of N.M. Rothschild & Sons when it became a Limited Company.
30. Morton 1962, p. 224.
31. Cowles 1973, p. 199.
32. Cowles 1973, p. 279.


CHAPTER 27    IF HIS MAJESTY'S GOVERNMENT WILL SEND ME A MESSAGE....
(pages 233-248)

1. Schama 1978 Chap. 6, p. 190.
2. Roth 1939, Chap. 14, p. 280.

3. Meinertzhagen Diaries, Vol. 41, 8.1937.
4. Druck 1928, pp. 189-193. Edmond affirmed his profound belief in the "rebirth of Eretz Israel" in the speech he delivered in 1925 (during his fifth and final visit to Israel) from the pulpit of the Great Synagogue in Tel-Aviv. See Appendix 3.
5. WA, Rothschild to Weizmann, 10 April 1917.
6. Stein 1961, p. 470.
7. Rothschild to Balfour, FO, 371/3083, File 143082, Folio 45.
8. Stein 1961, p. 548.
9. Weizmann 1949, p. 262.
10. CZA, Z4/117; Weizmann 1977, Vol. 8, p. 2.
11. Samuel 1945, p. 146.
12. Dugdale 1936, p. 213.
13. Asquith 1928, Vol. 2, p. 65.
14. Winston Churchill in Friedman 1973, pp. 6-7.
15. Moses Gaster Papers, Memorandum of a Conference held on 7 February 1917, sent by Sokolow to Gaster, 21 February 1917. (Walter Rothschild attended this conference.) Also WA.
16. Haldane 1929, Chap. 5, p. 162.
17. LL, Tring Park Visitors Book; Charles Rothschild to Hugh Birrell, various letters; Gardiner 1923, p. 546; Randolph Churchill papers, various letters.
18. Milner, Letters and Papers, MS Dept. 28, Folio 88/89; MS Dept. 200, Folio 111/113.
19. LL, Charles Rothschild's autograph collection.
20. Vereté 1970.
21. Stein 1961, p. 183.
22. Zangwill, 1915.
23. Weizmann 1974, Vol. 6, letter 229.
24. Théâtre Pigalle, Draeger Imp., Paris.
25. Gilbert 1978, p. 42.
26. WA, Charles Rothschild to Weizmann, 9 June 1915.
27. Weizmann 1975, Vol. 7, letter 45.
28. Cohen 1935, p. 250.
29. RAL, RFamC/14.
30. LL, Charles Rothschild's autograph collection.
31. Egremont 1980, p. 123.
32. CZA, Z4/117, Rothschild to Sokolow, 13 October 1917.
33. WA, Rothschild to Weizmann, 22 August 1917.
34. Schama 1978, p. 191.
35. Cabinet Office Memorandum on the Balfour Declaration 1923. Recorded in the National Register of Archives (Scotland) Survey No. 12 (Earl of Balfour), p. 4, folder 5.
36. Weizmann 1949, p. 205. He was vaguely aware that something was amiss for he remarked that Charles Rothschild was "inclined to melancholy."
37. CZA, Z4/117.
38. In Hungary the number of Jews admitted to the University was limited to the proportion of members of the various ethnic and national groups in the total population.
39. Stein 1961, p. 186.

40. WA, Rozsika Rothschild to Weizmann, 13 December 1915.
41. WA, Rozsika Rothschild to Weizmann, 11 November 1917.
42. WA, Rothschild to Weizmann, 15 May 1917.
43. WA, Rozsika Rothschild to Weizmann, 23 December 1917.
44. Weizmann 1949, p. 205.
45. Weizmann 1977, Vol. 10, letter 131.

## CHAPTER 28    DEAR LORD ROTHSCHILD (pages 249-269)

1. McCalmont 1971.
2. WA, Charles Rothschild to Weizmann, 9 June 1915 [& Weizmann 1975, Vol. VII, footnote p. 208].
3. Weizmann 1975, Vol. VII, p. 315, letter 291.
4. Meinertzhagen 1959, pp. 8 & 9.
5. Schama 1978.
6. Druck 1928.
7. Friedman 1973.
8. Friedman 1973, p. 130.
9. Stein 1961, p. 298.
10. WA, Rozsika Rothschild to Weizmann, 26 January 1917.
11. *Ibid.*
12. WA, Rothschild to Weizmann, 8 February 1917. Actually written on the 6th February, the day before the Gaster Conference.
13. WA, Memorandum of conference of 7 February 1917. Also CZA Z4/90/1.
14. Friedman 1973, pp. 142-144, 156. Bertie 1924, Vol. 2.
15. Some of these concepts were eventually embodied in a modified form in "The Statement of the Zionist Organisation regarding Palestine: Proposals to be presented to the Peace Conference" of which Walter was the first signatory, and which was handed in, in Paris, on the 4th February 1919 (WA).
16. See above, 13.
17. Friedman 1973, p. 131, states "Lord Rothschild's pointed question as to whether any pledges had already been given to the French concerning Palestine" is a *lapsus calami* for Mr. James de Rothschild (see p. 252) who pressed Sykes on this point.
18. Weizmann 1975, Vol. VII, letter 333.
19. WA, Rozsika Rothschild to Weizmann, 16 April 1917.
20. Friedman 1973, p. 130.
21. See in particular Stein 1961 and Friedman 1973.
22. Wolf, 1 April 1915; Zangwill, 9 April 1915.
23. Meinertzhagen 1959, p. 9.
24. Wolf, *Edinburgh Review*, April 1917.
25. A representative organisation of British Jewry, dating from 1760 (see *Encyclopaedia Judaica* 4 (13) p. 1150). The Conjoint Foreign Committee was formed by the Board (together with the Anglo-Jewish Association) in 1878.
26. The Anglo-Jewish Association. Established in 1871 with the object of protecting Jewish rights in backward countries by diplomatic measures.

27. *Times*, 24 May 1917.
28. *Times*, 28 May 1917, written 25 May.
29. *Jewish Chronicle*, 20 July 1917.
30. WA, Rothschild to Weizmann, 12 June 1917.
31. *Jewish Chronicle*, 22 June 1917. Minutes of meeting of London Committee of the Deputies (of the Board of Deputies of British Jews) 17 June 1917.
32. WA, Rothschild to Weizmann, 17 June 1917.
33. WA, Rothschild to Weizmann, various letters and telegrams, June-November 1917.
34. CZA, Z4/117 (1026?), 14 September 1918.
35. Rothschild to Balfour, June 1917.
36. To meet the Morgenthau Mission from the U.S.A., at Balfour's request. Weizmann 1975, Vol. VII, footnote p. 449.
37. Zoological term for a name given to an animal without a description of the species concerned, *i.e.* a naked name.
38. FO 371/3083, File 143082, Rothschild to Balfour, 18 July 1917 (new draft Declaration enclosed). Stein suggests that this letter (reproduced on p. 235) was for some reason delayed for a couple of days, but this does not seem to have been the case. Walter answered letters immediately or not at all, and always sent Balfour's letters by messenger.
39. Stein 1961, p. 472.
40. Lloyd George 1933-36, Vol. III, p. 117.
41. FO 371/3083, File 143082, Balfour to Rothschild, 19 July 1917.
42. CZA, Z4/117, Rothschild to Sokolow, 13 October 1917.
43. WA, Rothschild to Weizmann, 18 September 1917.
44. WA, Rothschild to Weizmann, 21 September 1917.
45. FO 371/3083, File 143082, Folio 98. Rothschild to Balfour, 22 September 1917. On September 21st the *Jewish Chronicle* had also drawn attention to German propaganda along these lines. Sir Ronald Graham subsequently also emphasised this point.
46. FO 371/3083, Folio 89 and following.
47. In the meantime Hankey had circulated copies of the Milner-Amery draft declaration (afterwards modified) to "representative Jewish leaders" to obtain their views in writing. In addition to Lord Rothschild, copies were sent to Stuart Samuel, Leonard Cohen, Claude Montefiore, Philip Magnus, the Chief Rabbi of England, Nahum Sokolow and Chaim Weizmann.
48. CZA, Z4/117 (126), Sokolow to Rothschild, 4 July 1917.
49. CZA, Z4/117, Sokolow to Rothschild, 4 March 1918.
50. WA, Rothschild to Weizmann and Sokolow, 1 November 1917.
51. WA, Rothschild to Weizmann, 11 November 1917.
52. *Jewish Chronicle*, 7 December 1917.
53. Swann 1930, pp. 65-71.
54. WA, Rozsika Rothschild to Weizmann, 17 November? (Thursday a.m.).
55. WA, Z4/117, Rothschild to Weizmann, 17 November 1917.

## CHAPTER 29     WHAT HAS BECOME OF TWO OSTRICHES? (pages 270-282)

1. *Jewish Chronicle*, 7 December 1917, p. 14.

2. Meinertzhagen 1959, pp. 190, 205, 354.
3. *Ibid*. p. 354.
4. WA, Rothschild to Weizmann, 1 November 1917.
5. WA, Rothschild to Weizmann, 6 January 1918.
6. WA, Rothschild to Weizmann, 1 March 1918.
7. WA, Rothschild to Weizmann, 6 December 1918.
8. Now part of the British Museum (Natural History) collection.
9. Stein 1961, p. 638.
10. Walter Rothschild's invitation to Weizmann to attend the dinner at the Ritz on 21st December is in the Weizmann Archives, but there is a singular dearth of correspondence about the event.
11. WA, Rothschild to Weizmann, 7 November 1918.
12. WA, Rothschild to Weizmann, 10 November 1918.
13. Balfour Papers. Whittinghame (folders Nos. 599, unsorted correspondence). Balfour to Rothschild, 26 September 1919.
14. Balfour Papers, Whittingehame (folders Nos. 5 & 9, unsorted correspondence), Balfour to Rothschild, 29 September 1919.
15. Balfour papers, Whittingehame (folders Nos. 5 & 9, unsorted correspondence), Rothschild to Balfour, 3 October 1919.
16. Meinertzhagen 1959, pp. 57-60.
17. *Ibid*. pp. 3-4.
18. Weizmann 1949, pp. 228-229.
19. Meinertzhagen 1959, p. 254.
20. *Ibid*. pp. 37-42.
21. Lawrence 1935, p. 384.
22. Balfour Papers, Whittingehame (folders 5 & 9, unsorted correspondence), Rothschild to Balfour, 30 December 1919. An autograph hunter had cut out Walter Rothschild's signature.
23. Lothian Papers, GD40/17/206, items 34 & 35. An enclosure with Walter Rothschild's letter of 30 December 1919, a memorandum for the Prime Minister from Weizmann on the subject of the Litany Waters in Palestine.
24. Balfour Papers, Whittingehame (folder No. 5, 1919-1922, unsorted correspondence), Balfour to Rothschild, 2 January 1920.
25. Lothian Papers, GD40/17/206, item 33, 2 January 1920.
26. *Ibid*. item 38, 6 January 1920.
27. Balfour's letters are printed in full since the owner of the copyright granted permission to publish only if no quoting "out of context" occurred.
28. Lothian Papers, GD40/17/206, item 46.
29. Friedman 1973.
30. WA, Z4/16037, Rothschild to Weizmann, 20 May 1920.
31. WA, Z4/16037, Rothschild to Weizmann, 3 June 1920.
32. CZA, Z4/117 6994.
33. CZA, Z4/117 5402.
34. WA, Z4/16037, Rothschild to Weizmann, 20 May 1920.
35. WA, Copy of letter to Prime Minister (British Delegation, San Remo Conference) 21 April 1920.

36. Meinertzhagen 1959.
37. WA, Rothschild to Weizmann, 8 February 1917.
38. For a history of the Economic Council see Weizmann 1977, Vol. X, pp. 101-137.
39. Meinertzhagen 1959.
40. WA, Z4/16055, items 262 and 263. Telegram with revision.
41. See also WA, Z4/16037, Rothschild to Weizmann, 27 June 1920 and 29 June 1920.
42. WA, Statement of the Zionist Organisation regarding Palestine (Paris, 3 February 1919) pp. 10-11.
43. Weizmann 1949, p. 329.

### CHAPTER 30     A FIGURE IN THE BACKGROUND (pages 282-286)

1. Cabinet Office memorandum on the Balfour Declaration, N.R.A.(S.) Survey No. 12, folder 5.
2. Vereté 1970.
3. FO, 371/3083, File 143082, Balfour to Rothschild, 7 November 1917. "Many thanks for your two letters of November 5th. I quite agree to the publication of my letter taking place in *"The Times"* on Friday the 9th. It seems to me that the decision of H.M. Government should be made known as soon as possible. I much appreciate the proposal that a deputation should come here to express the thanks of the Zionist Federation for the decision taken by H.M. Government, but is not this really somewhat superfluous? While I do not wish to discourage the idea if you support it, I am afraid that next week is going to be a particularly busy one as far as I am concerned."
4. FO, 371/3083, File 143082, Balfour to Rothschild, 9 November 1917. "As you press for a deputation I will certainly try to receive it on some date such as you suggest, but I cannot yet fix a definite day. If I may I will let you know nearer the time. I do not think you need have any fear that H.M. Government will rescind the policy which has been deliberately adopted after full consideration and discussion by the Cabinet."
5. Balfour Papers, Whittingehame (folders Nos. 5 & 9, unsorted correspondence), Balfour to Eddy Marsh, 19 February 1923.
6. WA, Weizmann to Rothschild.
7. Rothschild 1923 A,B & C. Rothschild & Hartert 1923.
8. *Nature*, Vol. 112, p. 697, 10 November 1923.

### CHAPTER 31     YOU WILL BE PAINTED BLUE AND YELLOW AND EXPOSED IN THE HIGH STREET (pages 287-294)

This chapter is based on personal recollections and letters written to Charles Rothschild to his wife during the years 1916-1923, and to his friend Hugh Birrell during the years 1899-1908.

1. Rothschild, Miriam 1979, Charles had an amazing fund of funny stories and aphorisms and he thoroughly enjoyed the roars of laughter he could evoke in audiences of every description.
2. A lumper = systematist who amalgamates the description of several so-called distinct species, thus making one where there were previously several.
3. Sawyer 1971.
4. A selection of Walter's local interests around Tring is given below: He was Member of Parliament for Aylesbury from 1899-1910; J.P. for Bucks; member of the Tring U.D.C.

from the day of its inception until he resigned during World War I (see p. 244); President of and frequent lecturer at the Victoria Club, Aylesbury (a recreational club for working men, founded by his father and given by him to the town in 1887 — Walter made up the deficit on the club accounts at the end of every year); President of the Tring men's Conservative Association; Tring & District Conservative Association, Bowling Club, Cricket Club, Hockey team, etc.

5. A selection of Walter's community interests is given below: He represented the Manchester Great Synagogue on the Jewish Board of Deputies from 1890-1934; was Member of Council of the United Synagogue and Board of Management; member of the London Jewish Board of Guardians; member of the Deputies of British Jews (for 46 years) representing Manchester Old Hebrew Association (Vice President of the Deputies 1917-34); Chairman of the Judean's Regiment Committee; President of the Union of Jewish Literary Societies, the Jewish Peace Society, the Hayes (Approved) School for Jewish boys, London Jewish Hospital, the Maccabean's Society and the Ort-Oze; Trustee and Chairman of the Executive of the Jewish Health Organisation of Great Britain, President of the first *Jewish Chronicle* Music Festival. He was a great supporter of the North East London Jewish Institute and prime mover in the erection of the old South Hackney Synagogue.

6. Hartert 1933.

7. Jordan 1938B.

## CHAPTER 32    THE PRIMROSE WAY—THE TRUTH ABOUT THE ROTHSCHILDS (pages 295-301)

1. Rothschild, V. 1982.

2. Butler [1865.] 1912.

3. Walter's father was described in a newspaper article (*Sunday Herald*, 4 April 1915) as having "altered much in appearance in recent years. His removal of his beard in particular made an enormous difference." Neither Walter nor Natty ever shaved after their teens, but the Baron Henri was clean shaven from middle age onwards.

4. Over and over again the statement appeared in the press that Walter's brother Charles or Walter himself had paid £1,000 and sometimes £10,000 for a flea off a grizzly or a polar bear. As late as September 1939 this fabrication appeared in Ripley's column "Believe it or not" in a Washington newspaper.

5. Lucien Wolf, who was given access to the Rothschild (New Court) Archives by Natty, debunked the blackmailer Dairnvaell's story (articles in the *Westminster Gazette*, 26 June 1909, p. 13; *Daily Telegraph*, 17 January 1913; and later in his "Essays in Jewish History" 1934 (edn. Cecil Roth)). In "Histoire édifiante et curieuse de Rothschild Ier Roi des Juifs" (Paris 1820-50), Dairnvaell had asserted that after obtaining prior knowledge of the Allied victory at Waterloo, Nathan Mayer amassed a huge fortune by deliberately spreading a rumour of defeat, thus enabling him to buy stock at a low price on a depressed market and sell it later at an inflated price when the truth became known. The writers of popular books (f.e. Morton's "The Rothschilds" (1962) prefer Dairnvaell's invention since the true story is prosaic and Nathan Mayer behaved with exemplary, if dreary probity. He did indeed receive news of the victory at Waterloo before the official dispatches arrived but (according to Wolf, but see also Rothschild, V., 1982) immediately sent on the messenger to the Prime Minister with information which was published in the *Courier* for June 21 (1851) prior to the official bulletin. The news that Nathan was selling Stock was reported in the *Courier* for the *previous day*. What is more curious is that serious reviewers of Morton's book, such as Geoffrey Nicholson (*Guardian*, 14 September 1962) seem to have no knowledge of the true

story and added censorious comments of their own: "It was one or other Rothschild, who, getting first wind of the victory at Waterloo, enriched himself at the expense of the nation's Consols." Thus long cherished apocryphal tales are clung to tenaciously whether true or false.

6. Corti 1928B, Vol. 2, p. 450.
7. Rothschild, Arthur de 1873.
8. Rothschild, Henri de 1938.
9. Cowles 1973, p. 193.
10. Roth 1939, p. 119.
11. Blair, BBC film, 6 January 1981.
12. An anecdote often re-told by Emma.
13. Derby Diaries, Vol. 3, 1 December 1889. Natty had great difficulty in finding a suitable author to write the Life of Disraeli. He first approached Lord Rosebery but by September 13, 1904 he was writing to Archie: "Since all men of light and learning had refused the work and we could find no statesman who would undertake... a work of three years and would consent to any form of surveillance, we thought it right to accept the proposal of *the Times* to entrust the task to a leading journalist."
14. Wolf, *Daily Chronicle*, 1 April 1915.
15. Rosebery Papers, Dalmeny. Natty to Rosebery, 27 February 1897.
16. Rosebery Papers, Dalmeny. Natty to Rosebery, 23 November 1886.
17. LL. Copy of letter from Walter to Marie Fredenson, 25 June 1934.
18. These form the basis of chapters 22 & 23.
19. The author's home: built by Charles Rothschild in 1900.
20. This account was given to me by both the late Leonard Rance and Frank Smith, verbally and in letters.
21. An estimate made by Nicholas Wollaston, who annotated those written between 1898 and 1908.
22. Günther, A.E. 1975.
23. WA, Walter to Weizmann, 7 (8?) November 1917.
24. WA, Walter to Weizmann, 11 November 1917.
25. Berenson 1964.

### CHAPTER 33    THE BIRDS CROSS THE ATLANTIC (pages 302-304)

1. Archives of the American Natural History Museum. Correspondence Rothschild to Dr. L.C. Sanford, 8 November 1931.
2. Jordan 1938B.
3. Mrs. H.P. Whitney and her family bought the Rothschild collection of birds for the American Museum in memory of her husband — also a noted bird collector.
4. LL, correspondence Ernst Mayr to Miriam Rothschild, 19 September 1979 and Hartert 1930, Mayr 1930.
5. Snow 1973.
6. Archives of the American Natural History Museum. Correspondence Dr. R.C. Murphy to Dr. L.C. Sanford, 1932 (no exact date).
7. Murphy 1932.

8. Stresemann 1975.
9. *Museums Journal*, Vol. 32, No. 1, pp. 3-4, April 1932. Ed. comment "Noble Collectors": *Morning Post*, 16 March 1932, "Lord Rothschild's Birds."
10. Lizzie Tenderson, although in the throes of one of her psychotic episodes, realized what the sale of the birds meant to Walter, for although she immediately demanded a substantial slice of the sum he had received from the American Museum, she, pathetically, began saving her money "To help you buy back the birds."

### CHAPTER 34    HOME FARM (pages 305-312)

1. LL, Baron Carl's Testament.
2. *The Times*, 12 November 1936, p. 17.
3. Cain 1981.
4. LL, Copy of Walter's first Testament.
5. LL.
6. Roeder was a specialist in insect physiology and its significance in the interaction between insects and nocturnal predators. He had made a particular study of the ears of moths.
7. BM (NH)A, Directors reports to Trustees. Comm. 24 July 1937, 23 October 1937.
8. BM (NH)A.
9. Rothschild, Miriam 1967.

### CHAPTER 35    CANDLE ENDS (pages 313-317)

1. Hansard, 5 July 1938 (Museum Bill).
2. Rothschild, Miriam 1979.
3. An extraordinary example of this was the fact that Eagle Clarke's widow sent all his amazingly beautiful distribution maps and lighthouse returns of migrating birds to salvage—to be pulped for the war effort—in 1940 (Rothschild, Miriam 1962).
4. The combined Rothschild and British Museum material constitutes the most comprehensive collection of recent giant tortoises in the world. Apart from its importance as a basis for conventional systematic investigation, it is now realized that it has proved extremely valuable for studies of adaptation to insular environments. It is indeed a unique collection.
5. LL, copy of Walter Rothschild's Will, 19 November 1923 and codicils.
6. Dr. J.G. Sheals (Keeper of Zoology at the British Museum (Natural History)) agrees and thinks that mounted specimens particularly are surprisingly delicate and very easily damaged by handling. Furthermore fluctuations in humidity and temperature, he believes, lead to cracking and cause the horny shields on the shell to spring away from the underlying bone.
7. Minutes of the Trustees of the British Museum (Natural History), No. 38, 7 May 1970, para. 28. In this document the quotation is wrongly attributed to me—not to Evelyn Hutchinson who made the comment.
8. LL, copy of Walter Rothschild's memorandum to the Trustees of the British Museum (Natural History). Signed 16 July 1937.
9. *The Times*, 26 April 1938.

# Bibliography

## Books

We have included a few published books in the Bibliography which are not actually quoted in the text, but which have proved useful in compiling the volume.

The National Register of Archives, London, has proved invaluable for locating collections of papers, and various documents in the Public Records Office and Scottish Record Office have also been consulted. For printed books published by members of the Rothschild family, or books and catalogues about their collections, I have consulted the British Museum Catalogue of Printed Books, the Catalogue General de la Bibliothèque Nationale (up to 1935, and various supplements), and the Catalogue of the Bibliothèque Centrale des Musées Nationaux.

ANON. (Battersea, Constance (Lady)). *A Propos d'un Portrait.* Traduit de l'Anglais par L.C. de Mornex. Paris, 1877.

ARLOTT, JOHN (Ed.). *The Oxford Companion to Sports and Games.* Oxford University Press: London, 1975.

ASQUITH, HERBERT HENRY (Earl of Oxford and Asquith). *Memories and Reflections,* 1852-1927. Edited by Alexander Mackintosh. 2 vol. Cassell & Co.: London, 1928.

AUDUBON, JOHN JAMES. *Audubon, by Himself;* A Profile of John James Audubon from Writings Selected, Arranged, and Edited, by Alice Ford. Natural History Press: Garden City, New York, 1969.

AUDUBON, JOHN JAMES. *The Birds of America;* from original drawings by John James Audubon. 4 vols. The Author: London, 1827-38.

AYER, JULES. *Century of Finance,* 1804-1904. *The London House of Rothschild.* Neely: London, 1905.

BATTERSEA, CONSTANCE (Lady). *Reminiscences.* Macmillan: London, 1922.

BERENSEN, BERNHARD. *Sunset and Twilight.* From the diaries of 1947-58. Edited and with an epilogue by Nicky Mariano. Introduced by Iris Origo. Hamish Hamilton: London, 1964.

BERLIN, ISAIAH. *Personal Impressions.* Edited by Henry Hardy; with an introduction by Noel Annan. Hogarth: London, 1980.

BERTIE, FRANCIS LEVESON (Viscount Bertie of Thame). *The Diary of Lord Bertie of Thame, 1914-1918.* Edited by Lady Algernon Gordon Lennox. 2 vol. Hodder & Stoughton: London, 1924.

BESTON, HENRY. *The Outermost House; A Year of Life on the Great Beach of Cape Cod.* Ballantine Books: New York, 1974 edn.

BLAKE, ROBERT NORMAN WILLIAM. *Disraeli*. Eyre & Spottiswoode: London, 1966.

BUCKLE, GEORGE EARLE—in succession to William Flavelle Monypenny. *The Life of Benjamin Disraeli, Earl of Beaconsfield*. Vol. V & VI. John Murray: London, 1920.

BUTLER, SAMUEL. *The Note-Books of Samuel Butler, Author of "Erewhon."* Selections arranged and edited by Henry Festing Jones. A.C. Fifield: London, 1912.

CARNARVON, HENRY GEORGE ALFRED MARIUS VICTOR FRANCIS HERBERT (6th Earl). *No Regrets*. Weidenfeld & Nicolson: London, 1976.

CHIVERS, KEITH. *The Shire Horse: A History of the Breed, the Society and the Men*. J.A. Allen: London, 1976.

CHURCHILL, WINSTON LEONARD SPENCER. *Great Contemporaries*. Thornton Butterworth: London, 1937.

CLUTTERBUCK, ROBERT. *The History and Antiquities of the County of Hertford*; compiled from the best printed authorities and original records, preserved in public repositories and private collections; embellished with views of the most curious monuments of antiquity, and illustrated with a map of the county. Vol. I. Nichols, Son & Bentley: London, 1815.

COHEN, LUCY. *Lady de Rothschild and her Daughters, 1821-1931*. John Murray: London, 1935.

COOKE, ALISTAIR BASIL and VINCENT, JOHN. *The Governing Passion: Cabinet Government and Party Politics in Britain, 1885-86*. Harvester Press: Brighton, 1974.

CORTI, EGON CAESAR (Count). *The Rise of the House of Rothschild*. Translated from the German by Brian and Beatrix Lunn. Victor Gollancz: London, 1928 A.

CORTI, EGON CAESAR (Count). *The Reign of the House of Rothschild*. Translated from the German by Brian & Beatrix Lunn. Victor Gollancz: London, 1928 B.

COWLES, VIRGINIA. *The Rothschilds. A Family of Fortune*. Weidenfeld & Nicolson: London, 1973.

DAVIS, R.W. *The English Rothschilds 1799-1915*. Collins: London, 1983.

DRUCK, DAVID. *Baron Edmond Rothschild. The story of a practical idealist*. Translated by Leo M. Glassman. With an introduction by Nathan Straus. Hebrew Monotype Press: New York, 1928.

DUGDALE, BLANCHE ELIZABETH CAMPBELL. *Arthur James Balfour, First Earl of Balfour*. 2 vols. Hutchinson & Co.: London, 1936.

DYKES, WILLIAM RICKATSON. *The Genus Iris*. University Press: Cambridge, 1913.

EGREMONT, MAX. *Balfour. A Life of Arthur James Balfour*. Collins: London, 1980.

*Encyclopaedia. Grand Larousse Encyclopédique*. Vol. 9, 1964. Augé, Gillon, Hollier-Larousse, Moreau et Cie: Paris.

*Encyclopaedia Britannica*. 14th edition, edited by J.L. Garvin. Encyclopaedia Britannica Co.: London, New York, 1929.

*Encyclopaedia Judaica*. Keter Publishing House Ltd.: Jerusalem, 1971.

FARRAR, J.M.. *Mary Anderson. The Story of Her Life and Professional Career*. D. Bogue: London, 1884.

FOSTER, R.F.. *Lord Randolph Churchill*. Clarendon Press: Oxford, 1981.

FOWLER, JOHN KERSLEY (of Aylesbury). ('Rusticus') *Records of Old Times, Historical, Social, Political, Sporting and Agricultural*. Chatto & Windus: London, 1898.

FRIEDMAN, ISAIAH. *The Question of Palestine, 1914-18: British-Jewish-Arab Relations*. Routledge & Kegan Paul: London, 1973.

GARDINER, ALFRED GEORGE. *The Life of Sir William Harcourt*. 2 vols. Constable & Co.: London, 1923.

GILBERT, MARTIN. *Exile and Return: The Emergence of Jewish Statehood.* Weidenfeld & Nicolson: London, 1978.

GILLE, BERTRAND. *Histoire de la Maison Rothschild.* Librairie Droz: Genève, Vol. 1, Des origines à 1848, 1965; Vol. 2, 1848-1870, 1967.

GOULD, JOHN. *The Birds of Australia.* 7 vols. Printed by R. & J.E. Taylor; Published by the author: London, 1840-48.

GOULD, FRANCIS CARRUTHERS. *Froissart's Modern Chronicles* (F.C.G.'s Froissart's Modern Chronicles, 1902; 1903-6.) Told and pictured by F.C. Gould. 3 vols. T. Fisher Unwin: London, 1902-08.

GRIGG, JOHN. *Lloyd George: The People's Champion, 1902-1911.* Eyre Methuen: London, 1978.

GUNTHER, ALBERT EVERARD. *A Century of Zoology at the British Museum.* Through the Lives of Two Keepers 1815-1914. Dawsons: London, 1975.

HALDANE, RICHARD BURDON. *An Autobiography.* Hodder & Stoughton: London, 1929.

HARRIS, MOSES. *The Aurelian, A Natural History of English Moths and Butterflies, Together With the Figures of Their Transformations and of the Plants on Which They Feed.* Drawn, engraved, and coloured from natural objects. New edition, with their systematic names, synonyms, and additional observations upon the habits of the species figured by John O. Westwood. Henry G. Bohn: London, 1840.

HARTERT, ERNST JOHANN OTTO. *Die Vögel der paläarktischen Fauna.* Systematische Übersicht der in Europa, Nord-Asien und der Mittel-meerregion vorkommenden Vögel. 3 Bd. Berlin, 1903-22.

HIRST, LEONARD FABIAN. *The Conquest of Plague.* A Study of the Evolution of Epidemiology. Clarendon Press: Oxford, 1953.

HOLMES, COLIN. *Anti-Semitism in British Society, 1876-1939.* Edward Arnold: London, 1979.

HOPKINS, GEORGE HENRY EVANS and ROTHSCHILD, MIRIAM. *An Illustrated Catalogue of the Rothschild Collection of Fleas (Siphonaptera) in the British Museum (Natural History).* With keys and short descriptions for the identification of families, genera, species, and subspecies. Vol. I, Tungidae and Pulicidae. London, 1953.

HURST, MICHAEL CHARLES. *Joseph Chamberlain and Liberal Reunion.* The Round Table Conference of 1887. Routledge & Kegan Paul: London, 1967.

JAHODA, MARIE. *Race Relations and Mental Health.* UNESCO: Paris, 1960.

JENNINGS, LOUIS JOHN (Ed.). *Speeches of the Right Honourable Lord Randolph Churchill, M.P., 1880-1888;* collected, with notes and introduction, by L.J. Jennings. Longmans & Co.: London, 1889.

KARLSSON, JON L. *Inheritance of Creative Intelligence.* Nelson-Hall: Chicago, C. 1978.

LACK, DAVID LAMBERT. *Darwin's Finches.* University Press: Cambridge, 1947.

LAWRENCE, THOMAS EDWARD. *Seven Pillars of Wisdom; A Triumph.* Doubleday, Doran & Company Inc.: Garden City, New York, 1935.

LESLIE-MELVILLE, BETTY and JOCK. *Raising Daisy Rothschild.* Simon & Schuster: New York, 1977.

LLOYD GEORGE, DAVID (Earl Lloyd George of Dwyfor). *War Memoirs of David Lloyd George.* 6 vols. Ivor Nicholson & Watson: London, 1933-36. 2 vols. Odham Press Ltd.: London, 1938.

LLOYD GEORGE, RICHARD (2nd Earl). *Lloyd George.* Frederick Miller: London, 1960.

LONGFORD, ELIZABETH PACKENHAM (Countess of). *Victoria R.I.* Pan Books: London, 1966.

McCALMONT, FREDERICK HAYNES. *McCalmont's Parliamentary Poll Book: British Election Results, 1832-1918.* 8th edition with introduction and additional material by J. Vincent and M. Stenton. Harvester Press: Brighton, 1971.

MAGNUS, PHILIP MONTEFIORE (Bart.). *King Edward the Seventh.* John Murray: London 1964.

MAYR, ERNST. *Systematics and the Origin of Species from the Viewpoint of a Zoologist.* Columbia University Press: New York, 1942.

MAYR, ERNST. *The Growth of Biological Thought. Diversity, Evolution, and Inheritance.* Harvard University Press: London,1982.

MEINERTZHAGEN, RICHARD. *Middle East Diary, 1917-1956.* Cresset Press: London, 1959.

MONYPENNY, WILLIAM FLAVELLE and BUCKLE, GEORGE EARLE. *The Life of Benjamin Disraeli, Earl of Beaconsfield.* Vol. 2, 1860-1881. New and revised edition. John Murray: London, 1929.

MORRIS, FRANCIS ORPEN. *A History of British Butterflies.* Groombridge & Sons: London, 1853.

MORTON, FREDERIC. *The Rothschilds; a family portrait.* Atheneum: New York, 1962.

MOSLEY, NICHOLAS. *Julian Grenfell. His life and the times of his death 1888-1915.* Weidenfeld & Nicolson: London, 1976.

PAKENHAM, THOMAS (Hon.). *The Boer War.* First American edition Random House: New York, 1979.

PASCAL, ANDRE. *(Henri de Rothschild) Croisière autour de mes souvenirs*; préface de Mme Collette. Émile-Paul Frères: Paris, 1933.

PETERS, JAMES LEE. *Check-List of Birds of the World.* (Continuation by various authors) Edited by Ernst Mayr and James C. Greenway. Harvard University Press: Cambridge, Mass., 1931 onwards.

PEYREFITTE, ROGER. *Les Juifs.* Flammarion: Paris, 1965.

PHILLIPS, CECIL ERNEST LUCAS and BARBER, PETER NORMAN. *The Rothschild Rhododendrons: A record of the gardens at Exbury.* Cassell: London, 1967.

POLIAKOV, LÉON. *Histoire de l'antisemitisme.* Calmann-Levy: Paris, 1968.

PONSONBY, ARTHUR AUGUSTUS WILLIAM HARRY (Baron). *Henry Ponsonby, Queen Victoria's Private Secretary. His life from his letters.* Macmillan & Co.: London, 1942.

READ, CHARLES HERCULES. *The Waddesdon Bequest.* The collection of jewels, plate and other works of art, bequeathed to the British Museum by Baron Ferdinand Rothschild. London, 1899.

READ, CHARLES HERCULES. *The Waddesdon Bequest.* Catalogue of the Works of Art bequeathed to the British Museum by Baron Ferdinand Rothschild. London, 1902.

RICHARDS, WALTER. *His Majesty's Territorial Army.* A descriptive account of the Yeomanry, Artillery, Engineers, and Infantry, with the Army Service and Medical Corps, comprising "the King's Imperial Army of the Second Line". 4 vol. Virtue & Co.: London, 1910.

ROTH CECIL. *The Magnificent Rothschilds.* Robert Hale: London, 1939.

ROTH, CECIL (ed.). *The Ritual Murder Libel and the Jew.* The report by Cardinal Lorenzo Ganganelli (Pope Clement XIV). Woburn Press: London, 1935.

ROTHSCHILD, ARTHUR DE (Baron). *Histoire de la poste aux lettres depuis ses origines les plus anciennes jusqu'à nos jours.* 1st edition, Librairie Nouvelle: Paris, 1873; 2nd edition, Librairie Hachette & Cie: Paris, 1873.

ROTHSCHILD, HENRI DE. *Exposé des Travaux Scientifiques.* Doin: Paris, 1906.

— *Les Timbres poste et leurs amis.* Champion, Paris, 1938.

— *Cent ans de Bibliophilie 1839-1939. La bibliothèque James de Rothschild.* Paris, 1939.

— *Un bibliophile d'autrefois, le Baron James-Edouard de Rothschild.* Droz: Paris, 1934.

ROTHSCHILD, (NATHAN) JAMES EDWARD DE (Baron). *Catalogue des livres composant la bibliothèque de feu M. le Baron J. de Rothschild.* (Chiefly compiled by him and completed by E. Picot.) 5 Vols. Paris, 1884-1920. (Privately printed)

ROTHSCHILD, VICTOR (Lord). *Rothschild Family Tree, 1450-1973.* Curwen Press: London, 1973; Revised edition 1981. (Privately printed)

ROTHSCHILD, VICTOR (Lord). *'You have it, Madam'*. The purchase, in 1875, of Suez Canal shares by Disraeli and Baron Lionel de Rothschild. London, 1980. (Privately printed)

ROTHSCHILD, VICTOR (Lord). *The Shadow of a Great Man*. New Court: London, 1982. (Privately printed)

ROTHSCHILD, WALTER (Hon.). *The Avifauna of Laysan and the Neighbouring Islands*; with a complete history to date of the Birds of the Hawaiian Possessions. R.H. Porter: London, 1893.

ROTHSCHILD, WALTER (Hon.). *Extinct Birds*. An attempt to unite in one volume a short account of those birds which have become extinct in historical times — that is, within the last six or seven hundred years. To which are added a few which still exist, but are on the verge of extinction. Hutchinson & Co.: London, 1907.

ROTHSCHILD, WILLY DE (Baronne). *30 Melodies de la Baronne Willy de Rothschild,* Durand Schoenewerk & Co.: Paris (no date).

SAMUEL, HERBERT LOUIS (Viscount). *Memoirs*. Cresset Press: London, 1945.

SCHAMA, SIMON. *Two Rothschilds and the Land of Israel*. Collins: London, 1978.

SHAKESPEARE, WILLIAM. *Macbeth,* 1606.

SNOW, DAVID WILLIAM. *A Study of Blackbirds*. George Allen & Unwin: London, 1958.

STEIN, LEONARD. *The Balfour Declaration*. Vallentine Mitchell: London, 1961.

STRESEMANN, ERWIN. *Ornithology from Aristotle to the Present*. Translated by Hans J. and Cathleen Epstein; edited by G. William Cottrell; with a foreword and an epilogue on American ornithology by Ernst Mayr. Harvard University Press: Cambridge, Mass., 1975.

SWANN, JOHN CHRISTOPHER. *The Citizen Soldiers of Buckinghamshire, 1795-1926*. Compiled with the assistance of many officers of the corps concerned. Hazell, Watson & Viney: London & Aylesbury, 1930.

THORNTON, JOHN. *The English Herd Book of Jersey Cattle*. English Jersey Cattle Society: London, 1880.

TRAUB, R. and STARKE, H. (Eds.). *Fleas*. Proceedings of the international conference on fleas, Ashton Wold, Peterborough, UK. Balkema: Rotterdam, 1980.

TUBBS, LAURENCE GORDON. *The Book of the Jersey*. Printed by W.E. Baxter: Lewes. 1946.

WEIZMANN, CHAIM. *Letters and Papers of Chaim Weizmann*. Edited by M.W. Weisgal. Vols. VI-X. Israel Universities Press: Jerusalem, 1974-1977.

WEIZMANN, CHAIM. *Trial and Error*. The autobiography of Chaim Weizmann. Hamish Hamilton: London, 1949.

WHITAKER'S ALMANACK. *1890 & 1897*. Whitaker: London.

WHITE, JERRY. *Rothschild Buildings: Life in an East End tenement block, 1887-1920*. Routledge & Kegan Paul: London, Boston, 1980.

WHITEHEAD, PETER JAMES PALMER and KEATES, COLIN. *The British Museum (Natural History)*. P. Wilson: London, 1981.

WHO WAS WHO. Vol. IV, 1941-50. Adam & Charles Black: London, 1967.

WILSHER, PETER. *The Pound in Your Pocket, 1870-1970*. Cassell: London, 1970.

WOLF, LUCIEN. *Essays in Jewish History... With a memoir*. Edited by Cecil Roth. Jewish Historical Society of England: London, 1934.

WOLLASTON, ALEXANDER FREDERICK RICHMOND. *Life of Alfred Newton, Professor of Comparative Anatomy, Cambridge University, 1866-1907*. John Murray: London, 1921.

WOLLASTON, ALEXANDER FREDERICK RICHMOND. *From Ruwenzori to the Congo: a naturalist's journey across Africa*. John Murray: London, 1908.

WOLLASTON, MARY (Ed.). *Letters and Diaries of A.F.R. Wollaston*. Selected and edited by Mary Wollaston; with a preface by Sir Henry Newbolt. University Press: Cambridge, 1933.

# Bibliography

## Newspapers, Journals, Catalogues and Reports

AKERMAN, JOHN YONGE. Moneys received and paid for secret services of Charles II and James II from 30 March 1679 to 25 December 1688. (Edited from a MS in the possession of William Selby Lowndes Esq) *Camden Society Papers*, Vol. 52, 1851.

[Aliens Commission]. *Report of the Royal Commission on Alien Immigration:* with Minutes of evidence, appendices, index and analysis of evidence. Vol. 1, 10 Aug. 1930, Cd. 1741; Vols. 2-4, 1903, Cd. 1742, Cd. 1741-1, Cd. 1743. In: *Reports from Commissioners (Royal Commission).* Session 17 Feb. 1903 - 14 Aug. 1903. Vol. 9.

*Anti Suffrage Review*, London. No. 1, Dec. 1908, p. 8; No. 20, July 1910, p. 7; No. 21, Aug. 1910, p. 2; No. 39, Feb. 1912, p. 34.

BLAIR, LES. (Producer) *Beyond the pale.* BBC1 Film, 6 January 1981. ("Play for today" series. Editor: J. Gregory; Co-writers: Les Blair & Jon Amiel.)

BLUM, ANDRÉ. Un musée de la Gravure. La collection Edmond de Rothschild. *L'Art et les Artistes, Paris, October 1936.* (Revue Mensuelle d'Art ancien et moderne.)

BONHOTE, J.L. & ROTHSCHILD, N.C. *Harrow Butterflies and Moths.* Harrow School Scientific Society Memoirs. Vol. 1, 1895; Vol. 2, 1897.

BOWDLER SHARPE, R. The Hon. Walter Rothschild, MP, Zoologist, Philanthropist, Sportsman. *North London Illustrated,* Vol. 1, No. 5, 23 December 1907.

British Museum (Natural History). *Zoological Museum, Tring.* British Museum (Natural History), London, 1979, Publ. No. 816.

*Buckinghamshire Herald*, Aylesbury. 22 June 1901; January 1906 (various numbers); 24 March 1906; 22 November 1913.

Cabinet Paper. *Palestine and the Balfour Declaration.* Memorandum by the Secretary of State for the Colonies, January 1923, with Extract of Minute by Mr. Ormsby Gore, 24 December 1922, and memorandum: Palestine. History of the Negotiations leading up to the Balfour Declaration of 2 November 1917.

CAIN, A.J. Zoological ideas 1831-1981. *Nature*, 293: pp. 15-16, 1981.

COBLENTZ, SUZANNE. *La Collection d'estampes Edmond de Rothschild au Musée du Louvre.* Préf. de GEORGES A. SALLES. Editions des Musées Nationaux, Paris, 1954.

*Connoisseur* London. Notable collections: the collection of Mr. Alfred de Rothschild in Seamore Place, by Mrs. Steuart Erskine. *Connoisseur*, Vol. III, No. 10: pp. 71-79, June 1902.

*Country Life*, London. Vol. 8, 1900; Vol. 33, 3 February 1913.

*Courier*, London. 21 June 1915.

*Daily Chronicle*, London. 25 June 1909; 1 April, 1915; 8 April 1918.

*Daily Graphic*, London. 3 April 1915.

*Daily Mail*, London. 3 November 1919.

*Daily Mirror*, London. 3 April 1915.

*Daily News*, London. 24 November 1908.

DAIRNVAELL, GEORGES. *Histoire édifiante et curieuse de Rothschild Ier, Roi des Juifs par Satan* [*G. Dairnvaell*]. Paris, 1846.

DAVIS, CHARLES. *A description of the works of art forming the collection of Alfred de Rothschild.* 2 Vols. London, 1884.

DEGEN, A. VON. In memory of Nathaniel Charles Rothschild. Translation of memoir published in *Pester Lloyd, Budapest*, November 1923.

DRUCE, G. CLARIDGE. Obituary. The Hon. N. Charles Rothschild. *J. Northants nat. Hist. Soc.* Vol. 22: pp. 135-141, 1924.

*East Anglian Daily News*, April 1915.

*English Illustrated Magazine*, London. The Hon. Walter Rothschild at Tring Park. In: Morning Calls, pp. 210-216, 1896.

*Entomological Record*, Vols. 15 & 16, 1903-4.

*Field*, London. 1897-1908.

*Figaro*, Paris. May 1905.

*Fondation Ephrussi de Rothschild, Villa-Musée, Ile-de-France.* Genève, Paris, Munich, 1969.

FROUDE, TOM (Trooper). *The charge of Kassassin.* Windsor, 1883(?) (Privately printed.)

FULLER, E. *She Championed Children. The Story of Eglantyne Jebb.* Save the Children Fund, London, 1967 (2nd edn.).

GÜNTHER, ALBERT C.L.G. *The Autobiography of Dr. Albert C.L.G. Günther FRS,* 1910. Typed from MS (untraced) by Robert T. Günther, 1914. Annotated by A.E. Günther, 1964-70. Presented to the British Museum, April 1970.

Hansard Record of Parliamentary Debates: House of Commons: 4th series, Vol. 82, 26 April - 11 May 1900. Vol. 93, 29 April - 13 May 1901; Vol. 149, 10-21 July 1905. House of Lords: 5th series, Vol. 110, 5 July 1938.

HARTERT, E. *Guide to the Hon. Walter Rothschild's Zoological Museum at Tring.* Tring, 1898.

— William Doherty. Obituary (with extracts from diary). *Novit. Zool.*, Vol. 8: pp. 494-506, 1901.

— Journey to Algeria and Marocco in 1929. *Novit. Zool.*, Vol. 38: pp. 331-335, 1933.

HERON, J.S. A.S. Meek's journeys to the Aroa River in 1903, and 1904-05. *Emu*, Vol. 75: pp. 232-233, 1975.

HINTON, M.A.C. Obituary notice, Lionel Walter Rothschild. *Proc. Linn. Soc. Lond.* 150th Session, pp. 334-337, 1937-38.

HIRST, L. FABIAN. Plague fleas with special reference to the Milroy lectures 1924. *J. Hygiene, Cambridge*, Vol. 24, No. 1: pp. 1-16, July 1925.

HOPKINS, G.H.E. Bibliography of scientific publications by H.E.K. Jordan. *Trans. R. ent. Soc. Lond.* Vol. 107: pp. 77-94, 1955.

*Jewish Chronicle*, London. 23 May 1890; 30 April 1915; 22 June 1917; 20 July 1917; 7 December 1917.

*Jewish World*, London. 7 April 1913.

Jews' Free School. *Report for 1884*. London, 1885.

JONES, E. ALFRED. *Catalogue of the objects in gold and silver and the Limoges enamels in the collection of the Baroness James de Rothschild.* London, 1912.

JORDAN, K. On mechanical selection and other problems. *Novit. Zool.* Vol. 3: pp. 426-525, 1896.

— Der Gegensatz zwischen geographischer und nicht-geographischer Variation. *Z. wiss Zool.* Vol. 83: pp. 151-210, 1905.

— The systematics of some Lepidoptera which resemble each other, and their bearing on general questions of evolution. *Int. Congr. Ent. I.* (Brussels 1910), pp. 385-404, 1911.

— Notes on Arctiidae. *Novit. Zool.* 23: pp. 124-150, 1916.

— On some problems of distribution, variability and variation in North American Siphonaptera. *Int. Congr. Ent. IV* (Ithaca, N.Y. 1928), pp. 489-499, 1929.

— Obituary of L.W. Rothschild. *Nature*, Vol. 140: p. 574, 1937.

— Lord Rothschild 1868-1937. *Obit. Not. Fellows R. Soc. Lond.* Vol. 2: pp. 385-386, 1938 (A).

— In memory of Lord Rothschild PhD, FRS, JP. *Novit. Zool.* Vol. 41: pp. 1-41, 1938 (B).

— Reminiscences of an Entomologist. *Int. Congr. Ent. X* (Montreal 1956) Vol. 1: pp. 59-60. 1958.

*L'illustration*, Paris. 3 June 1905.

MANSON BAHR, P.H. A short history of the Club. *Bull. Br. Orn. Club*, Vol. 71: pp. 2-4, 1951.

MAYR, ERNST. My Dutch New Guinea Expedition 1928. *Novit. Zool.* Vol. 36: pp. 20-26, 1930.

— Karl Jordan's contribution to current concepts in systematics and evolution. *Trans. R. ent. Soc. Lond.* Vol. 107: pp. 45-66, 1955.

— Materials for a history of American Ornithology. In Stresemann, E. *Ornithology from Aristotle to the present*, Cambridge, Mass. 1975.

*Morning Leader*. 23 November 1908.

*Morning Post*. 9 December 1883; 16 March 1932.

MUELLER, EDWIN. Baron Alphonse de Rothschild. A great collector is dead. *Stamps,* a weekly magazine of Philately. New York, No. 523, pp. 405 & 426, 19 September 1942.

M[UNTHE], N[ELLY]. *Les Rothschilds, famille de collectionneurs originaire de Francfort.* Vol. 2, pp. 1616-1618. Petit Larousse de la Peinture, Paris 1979.

MURPHY, ROBERT CUSHMAN. Moving a Museum: The Story of the Rothschild Collection of Birds, presented to the American Museum of Natural History in Memory of Harry Payne Whitney by his wife and children. *Natural History*, Vol. 32, No. 6: pp. 509-523, 1932. (Printing error in journal gives pagination as 497-511)

*Museums Journal*, Vol. 32, No. 1: pp. 3-4, April 1932.

*National-Zeitung*, Basel. No. 160, p. 19, 14 April 1963.

*Nature*, London. Obituary notice. The Hon. N.C. Rothschild, by E.E.A. Vol. 112: p. 697, 10 November 1923.

*New Times*, 27 June 1820.

NICHOLSON, GEOFFREY. "Books of the day" *Guardian*, No. 36141, p. 4, 14 September 1962.

*Novitates Zoologicae*. A journal of zoology. Vols. 1-40. Edited by Walter Rothschild, Ernst Hartert and K. Jordan. 1894-1938, Vol. 41. Edited by Miriam Rothschild & K. Jordan, 1938-1939. Tring, Herts.

[Old Age Pensions] *Report of the Committee on Old Age Pensions, 7 June, 1898,* presented to both Houses of Parliament. C-8911, 1898.

*Pall Mall Gazette*, London. 28 October 1913; 1 April 1915.

PARKER, SHANE. A.S. Meek's three expeditions to the Solomon Islands. *Bull. Brit. Orn. Club.* Vol. 87, No. 8: pp. 129-135, 1967.

PYE-SMITH, G.H.R. Memories of Pastor Jourdain. *Oologists' Record*, Vol. 39, No. 2: pp. 2-4 May 1965.

[Review] The Revision of the Sphingides — Nomenclature, Classification. *Entomologist's Record*, Vol. 15: pp. 309-312, 1903; Revision of the Sphingides — Nomenclature, classification, geographical distribution, Vol. 16: pp. 5-10, 44-47, 75-78, 1904.

*Reynolds Weekly Newspaper*, London. 4 April 1915.

RILEY, N.D. Lord Rothschild FRS. *Entomologist*, Vol. 70: pp. 217-220, October 1937.

— Karl Jordan and the International Congresses of Entomology. *Trans. R. ent. Soc. Lond.* Vol. 107: pp. 15-24, 1955.

— Heinrich Ernst Karl Jordan 1861-1959. *Biographical Memoirs of Fellows of the Royal Society*, Vol. 6: pp. 107-133, 1960.

ROLLIN, N. White plumage in blackbirds. *Bull. Brit. Orn. Club.* Vol. 79: pp. 92-96, 1959.

ROSEBERY, HANNAH. *Catalogue of Mentmore and its contents.* 2 Vols. Edinburgh, 1884 (Privately printed).

ROTHSCHILD, CARL VON (Baron & Baroness). *Catalogue of Baron and Baroness Carl von Rothschild's art collection at Günthersburg and Untermainquai No. 15.* 5 parts. Frankfurt, 1895 (Privately printed).

ROTHSCHILD, HENRI DE. *Trois cent autographes de la donation Henri de Rothschild.* Catalogue de l'exposition. Editions des Bibliothèques Nationales de France, Paris, 1933.

[BIBLIOTHÈQUE NATIONALE] Rothschild, Henri de. *Lettres, autographes et manuscripts de la collection Henri de Rothschild. T.I. Moyen Age — XVIe siecle.* Textes publiés et annotés par Roger Gaucheron. Paris, 1924.

ROTHSCHILD, JAMES DE. *Prèmiere réponse officielle de M. le Baron James Rothschild au pamphlet intitulé: Histoire édifiante et curieuse de Rothschild Ier, Roi des Juifs (par Satan (G. Dairnvaell)).* Bruxelles & Paris, 1846 (2nd edn.).

ROTHSCHILD, MAURICE DE & NEUVILLE, HENRI. *L'Hylochoerus meinertzhageni* O. Ths. *Bulletin de la Société philomatique de Paris*, Ser. 9, Vol. 8: pp. 141-164, 1906.

— Remarques sur l'okapi. *C. r. hebd. Séanc. Acad. Sci., Paris*, Vol. 149: pp. 693-695, October 1909.

— Recherches sur l'okapi et les girafes de l'Est africain. *Annls. Sci. nat. (Zool.)* 9th Series, Vol. 10: pp. 1-93, 1909; Vol. 13: pp. 1-185, 1911.

R[OTHSCHILD], M[IRIAM]. Karl Jordan — a biography. *Trans. R. ent. Soc. Lond.* Vol. 107: pp. 1-9, 1955.

ROTHSCHILD, MIRIAM. Several large white moths commonly called buterflies seen for the first time this year. *Entomologist's mon. Mag.* Vol. 98: pp. 167-170, 1962.

— The courage of an eccentric. *Jerusalem Post*, Special Supplement, p. 12, 2 November 1967.

— *Nathaniel Charles Rothschild 1877-1923.* Cambridge, 1979 (Privately printed).

ROTHSCHILD, NATHANIEL VON. *Notizen uber einige Meiner Kunstgegenstände.* Wein, 1903 (Privately printed).

ROTHSCHILD, N.C. New species of Siphonaptera from Egypt and the Soudan. *Entomologist's mon. Mag.* Vol. 14: pp. 83-87, 1903.

ROTHSCHILD, SALOMON (Baroness). *Chefs-d'oeuvre de l'art Juif.* Catalogue of the exhibition at the Musée de Cluny, 5 June 1981.

ROTHSCHILD, VICTOR (Lord). *The Rothschild Library. A catalogue of the collection of eighteenth century printed books and manuscripts formed by Lord Rothschild.* 2 Vols. Cambridge, 1954 (Privately printed).

ROTHSCHILD, WALTER. Description of a new Pigeon of the genus *Carpophaga. Proc. Zool. Soc., Lond.* pp. 312-313, 1891.

— Descriptions of three new birds from the Sandwich Islands. *Ibis*, Vol. 5, 6th Series: pp. 112-114, 1893.

— A revision of the Papilios of the Eastern Hemisphere, exclusive of Africa. *Novit. Zool.*, Vol. 2: pp. 167-473, 1895.

— Notes on Saturnidae; with a preliminary revision of the Family down to the genus *Automeris*, and descriptions of some new species. *Novit. Zool.*, Vol. 2: pp. 35-51, 1895.

— A monograph of the genus *Casuarius:* with a dissertation on the morphology and phylogeny of the Palaeognathae (Ratitae & Crypturi) and Neognathae (Carinatae) by W.P. Pycraft. *Trans. Zool. Soc., Lond.* Vol. 15, Part 6: pp. 109-148, 1900.

— On the advantage of trinomials. *Ibis*, Vol. 5, 8th Series: pp. 135-137, 1905. (A letter).

— Henry Eeles Dresser. *British Birds*, Vol. 9: pp. 194-196, 1916.

— The President's Address: Algeria and its fauna. *Proc. ent. Soc. Lond.* 1922: pp. 131-162.

— The President's Address: On some aspects of variation in Lepidoptera. *Proc. ent. Soc. Lond.* 1923: pp. 122-134 1923A.

— On a second collection sent by Mr. George Forrest from N.W. Yunnan. *Novit. Zool.*, Vol. 30: pp. 33-58, 1923 B.

— Remarks on a Mountain Gorilla from near Lake Kivu. *Proc. Zool. Soc., Lond.* pp. 176-177, 1923 C.

— (Presidential address, Zoology Section) The pioneer work of the Systematist. *Brit. Assoc. Adv. Sci. York Meeting,* pp. 89-102, 1932.

— Ernst Johann Otto Hartert (1859-1933): An appreciation. *Ibis*, Vol. 4 (13th series): pp. 350-377, 1934.

— List of scientific publications. *Novit. Zool.*, Vol. 41: pp. 17-41, 1938.

ROTHSCHILD, WALTER & HARTERT, ERNST. A review of the ornithology of the Galapagos Islands with notes on the Webster-Harris expedition. *Novit. Zool.*, Vol. 6, No. 2: pp. 85-205, 1899.

— Further notes on the fauna of the Galapagos Islands: "Introductory Remarks" and "Field Notes on the Tortoises of the Galapagos Islands" by R.H. Beck. *Novit. Zool.*, Vol. 9: pp. 373-380, 1902.

— Ornithological explorations in Algeria. *Novit. Zool.*, Vol. 18: pp. 456-550, 1911.

— A zoological tour in West Algeria. *Novit. Zool.*, Vol. 21: pp. 180-204, 1914.

— An ornithological autumn journey to Algeria. *Novit. Zool.*, Vol. 30: pp. 79-88, 1923.

ROTHSCHILD, WALTER & JORDAN, K. A monograph of *Charaxes* and allied prionopterous genera. *Novit. Zool.*, Vol. 5: pp. 545-601, 1898; Vol. 6: pp. 220-286, 1899; Vol. 7: pp. 281-524, 1900; Vol. 10: pp. 326-342, 1903.

— A revision of the lepidopterous Family Sphingidae. *Novit. Zool.*, Vol. 9 (Suppl.): pp. cxxxv, 972, 1903.

— A revision of the American Papilios. *Novit. Zool.*, Vol. 13: pp. 411-752, 1906.

Royal Society. *Year Book for 1912.*

SAWYER, F.C. A short history of the Libraries and list of manuscripts and original drawings in the British Museum (Natural History). *Bull. Br. Mus. nat. Hist. (hist. ser.)* Vol. 4: pp. 77-204, 1971.

SCHESTAGE, FRANZ. *Katalog der Kuntsammlung des Freiherrn Anselm von Rothschild in Wien.* Wien, 1866.

SELME, PIERRE. L'oeuvre scientifique d'Armand de Gramont. *Revue des Questions Scienti-fiques*, Vol. 124, pp. 30-35, 20 January 1963.

SMIT, F.G.A.M. The Ghost of Tring Museum. *Flea News*, No. 20, May, 1980.

SNOW, D.W. Obituary: Robert Cushman Murphy and his 'Journal of the Tring trip'. *Ibis*, Vol. 115: pp. 607-611, 1973.

Sotheby & Co. *Catalogue of the celebrated collection of German, Dutch and other continental silver and silver-gilt of the 15th, 16th, 17th and 18th centuries. Mainly inherited from the late Baron Carl von Rothschild of Frankfurt; also collected by the Baron Lionel de Rothschild and by the first Lord and Lady Rothschild. Also massive English silver and silver-gilt of the 18th and early 19th centuries. Removed from 148 Piccadilly, W.1. Sold by order of Victor Rothschild.* London, 1937.

*Standard.* 1 April 1915.

STODDART, D.R. Settlement, development and conservation of Aldabra. *Phil. Trans. Roy. Soc. Lond.* B., Vol. 260: pp. 611-628, 1971.

*Sunday Herald:* 4 April 1915.

SWATHLING (Lord). *Jewish Chronicle,* 30 April 1915.

*Théâtre Pigalle.* Draeger Imp. Paris.

THOMAS, OLDFIELD & HINTON, M.A.C. Captain Angus Buchanan's air expedition. 1. On a series of small mammals from Kano. *Novit. Zool.,* Vol. 27: pp. 315-320, 1920.

*Times Newspaper,* The, London. 26 November 1908; 18 November 1913; 24 May 1917; 28 May 1917; 11 January 1930; 11 January 1935; 12 November 1936; 26 April 1938; 6 July 1938; 3 March 1939.

Tring Park Estate: Bill of Sale. *Particulars and conditions of sale.* Prepared by Chinnock, Galsworthy and Chinnock, Land Agents and Surveyors. London, Tuesday, 7th May 1872.

*Vanity Fair.* September 1900.

VERETÉ, MAYIR. The Balfour Declaration and its Makers. *Middle Eastern Studies,* Vol. 6: pp. 48-76, January 1970.

WOLF, LUCIEN. *Westminster Gazette,* p. 13, 26 June 1909.

— *Daily Telegraph,* 17 January 1913.

— Lord Rothschild: an appreciation. *Daily Chronicle,* 1 April 1915.

— *Edinburgh Review,* April 1917.

— The Rothschilds. *Jewish Chronicle Supplement,* No. 88, p. i. April 1928.

— The political problems of the Rothschilds. *Jewish Chronicle Supplement,* No. 103, pp. i-ii, July 1929.

ZANGWILL, I. *Jewish Chronicle,* 9 April 1915.

ZIMMERMAN, E.C. Karl Jordan's contribution to our knowledge of the Anthribid beetles. *Trans. R. ent. Soc. Lond.,* Vol. 107: pp. 67-68, 1955.

# Appendix 1

SYNOPSIS OF SPECIES AND SUBSPECIES OF PLANTS AND ANIMALS
NAMED IN HONOUR OF WALTER ROTHSCHILD*

| Organism | No. of species and subspecies | Genera |
|---|---|---|
| Plants | 14 | — |
| Birds (Synonymy ignored) | 58 | 2 |
| Mammals | 18 | — |
| Fish | 3 | — |
| Reptiles & Amphibia | 2 | 1 |
| Insecta (includes 78 Lepidoptera) | 153 | 1 |
| Arachnida | 3 | — |
| Millipedes | 1 | — |
| Acarina | 3 | — |
| Nematoda | 1 | — |

* Some of the dedications are to "Baron Rothschild" and it is not always certain which member of the family is indicated. Furthermore synonyms confuse the picture. So these numbers are approximate. Charles Rothschild had 3 genera and 7 species of fleas, 1 tick, 1 Nycteribid fly and 1 spider as well as several plants named in his honour. Miriam Rothschild had 1 larval trematode worm, 1 genera and 2 species of fleas and 1 Nycteribid fly named in her honour.

Note: Details of these species are filed in the Lane Library.

# Walter Rothschild's immediate family

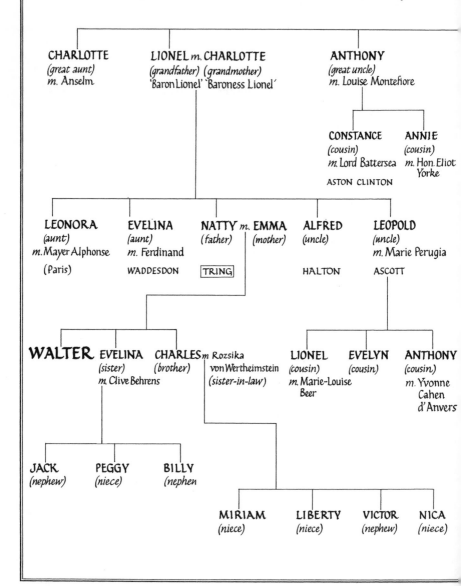

**CHARLOTTE**
*(great aunt)*
*m.* Anselm

**LIONEL** *m.* **CHARLOTTE**
*(grandfather)* *(grandmother)*
'Baron Lionel' 'Baroness Lionel'

**ANTHONY**
*(great uncle)*
*m.* Louise Montefiore

**CONSTANCE**
*(cousin)*
*m.* Lord Battersea
ASTON CLINTON

**ANNIE**
*(cousin)*
*m.* Hon. Eliot
Yorke

**LEONORA**
*(aunt)*
*m.* Mayer Alphonse
(Paris)

**EVELINA**
*(aunt)*
*m.* Ferdinand
WADDESDON

**NATTY** *m.* **EMMA**
*(father)* *(mother)*
TRING

**ALFRED**
*(uncle)*
HALTON

**LEOPOLD**
*(uncle)*
*m.* Marie Perugia
ASCOTT

**WALTER**

**EVELINA**
*(sister)*
*m.* Clive Behrens

**CHARLES** *m.* Rozsika
*(brother)* von Wertheimstein
*(sister-in-law)*

**LIONEL**
*(cousin)*
*m.* Marie-Louise
Beer

**EVELYN**
*(cousin)*

**ANTHONY**
*(cousin)*
*m.* Yvonne
Cahen
d'Anvers

**JACK**
*(nephew)*

**PEGGY**
*(niece)*

**BILLY**
*(nephew)*

**MIRIAM**
*(niece)*

**LIBERTY**
*(niece)*

**VICTOR**
*(nephew)*

**NICA**
*(niece)*

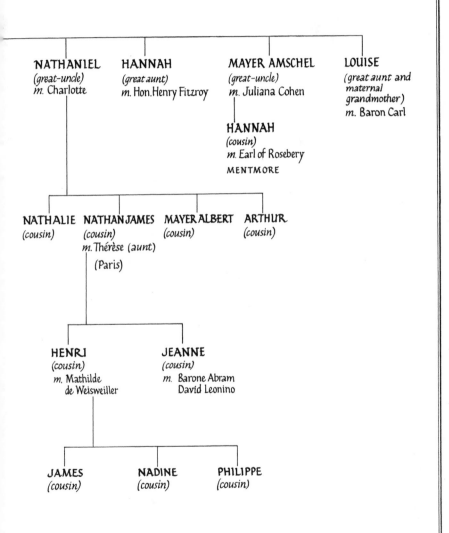

**NATHANIEL**
*(great-uncle)*
*m.* Charlotte

**HANNAH**
*(great aunt)*
*m.* Hon. Henry Fitzroy

**MAYER AMSCHEL**
*(great-uncle)*
*m.* Juliana Cohen

**LOUISE**
*(great aunt and maternal grandmother)*
*m.* Baron Carl

**HANNAH**
*(cousin)*
*m.* Earl of Rosebery

MENTMORE

**NATHALIE**
*(cousin)*

**NATHAN JAMES**
*(cousin)*
*m.* Thérèse *(aunt)*

(Paris)

**MAYER ALBERT**
*(cousin)*

**ARTHUR**
*(cousin)*

**HENRI**
*(cousin)*
*m.* Mathilde
de Weisweiller

**JEANNE**
*(cousin)*
*m.* Barone Abram
David Leonino

**JAMES**
*(cousin)*

**NADINE**
*(cousin)*

**PHILIPPE**
*(cousin)*

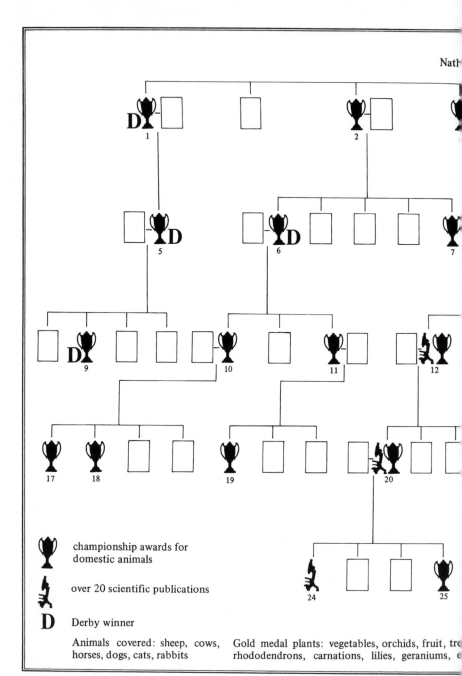

championship awards for domestic animals

over 20 scientific publications

**D** Derby winner

Animals covered: sheep, cows, horses, dogs, cats, rabbits

Gold medal plants: vegetables, orchids, fruit, tre rhododendrons, carnations, lilies, geraniums, e

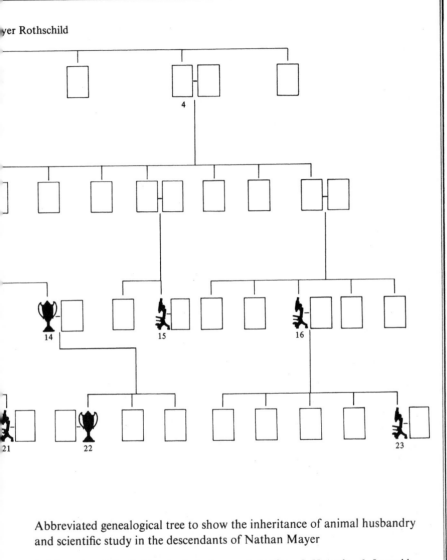

ьyer Rothschild

4

14  15  16

21  22  23

Abbreviated genealogical tree to show the inheritance of animal husbandry and scientific study in the descendants of Nathan Mayer

1. Mayer Amschel   2. Lionel   3. Anthony   4. Louise   5. Hannah   6. Leopold
7. Natty   8. Emma   9. Harry (Rosebery)   10. Lionel   11. Anthony   12. Charles
13. Walter   14. Evelina   15. Henri   16. Armand (de Gramont)   17. Rosemary (Seys)
18. Edmund   19. Renée   20. Miriam   21. Victor   22. William (Behrens)
23. Corise (de Gramont)   24. Charles (Lane)   25. Rozsika (Lane)

# Index

Rothschild entries in the index are in alphabetical order according to the personal name (in capitals) by which that member of the family was usually known: in subheadings, for convenience, members of the Rothschild family are referred to by this name only.

A letter "*n.*" after a page reference indicates a footnote on that page. The bibliographical notes have been indexed with the number in brackets after the page number. Both are referred to only when they introduce new material. References in italic type are plate numbers, not page numbers.

Aaronsohn, Aaron, 276
Abingdon (island), 200
Acton, Natural History Museum store, 315
Aga Khan, 210
Agnew, William, 7*n.*
Aharoni, J., 271, 272
Aitken, Bill, 279*n.*
Albemarle (island), 186, 193, 200, 201, 202, 366, 367
Albert Hall, Zionist meeting at, 281–2
albinos, 63, 138
Aldabra (island), 114, 197, 205
Alexander, D. L., 255–6, 257
Alexandra, Princess (later Queen Alexandra), 109
Algeria, collecting in, 93, 127, 157, 163–6, 233, 293
Allenby, General Edmund (later Field Marshal; 1st Viscount Allenby; captured Jerusalem), 245, 272, 280
Alps, *see* Switzerland
Althaus, Mr. (son of Frederick Theodore Althaus), 61–2
Althaus, Mrs. (mother of Frederick Theodore Althaus), 56*n.*, 61–2
Althaus, Frederick Theodore (tutor), 56, 61, 63, 65, 67, 68, 69; *13, 17*
American Museum of natural History, 139, 140, 302–3
Amery, Leopold, 270, 344(47)
Anderson, Mary, 23, 216
Anglo-Jewish Association, 256, 258
Ansus (island), 179
Anthribiidae, 149
anti-semitism, persecution of Jews, 26–30 *passim*, 32–7, 90, 231, 234, 236–7, 244

Arabs
   on Walter's Algerian expedition, 165–6
   and Zionists, 272–3, 274, 275
Archbishop, *see* Canterbury
Ascott, 319(5)
Ashton Wold, 104, 246, 298
Asquith, H. H. (1st Earl of Oxford and Asquith; 1858–1928; Prime Minister 1910–16), 31, 42*n.*, 237*n.*, 250
Aston Clinton, 4, 5, 14, 15
asymmetry, 143
Audubon, John James (1785–1851), 105, 125
Auk, Great, 105
Avebury, Lord, 123*n.*
*Avifauna of Laysan, The* (Rothschild), 73, 76–7, 78, 109, 152
Aylesbury, 5, 249, 293

Baker, Stuart, 154*n.*
Balcarres, Lord (later Earl of Crawford), 80, 231
Balfour, Arthur James (1st Earl of Balfour; 1848–1930; Prime Minister 1902–6,
   Foreign Secretary 1916–19)
   and anti-semitism, 26, 37, 38
   and Dreyfus case, 217*n.*, 240
   Emma on, 240*n.*
   Jews, sympathy for, 33, 38, 217*n.*, 240
   lost letters, 300
   Natty, relations with, 31, 31–2, 33, 37–8, 38, 81–2, 83–4, 225, 240, 242
   resigns party leadership (1911), 37–8
   Walter, relations with, 226–7
   and Zionism, Balfour Declaration (*qq. v.*), 38, 235–6, 237, 238, 240–2, 253, 261–2, 263–6, 272, 273–5, 277–8, 280–1, 281–2, 283–5, 286, 300
   mentioned, 227, 243*n.*, 246
Balfour Declaration, 52, 93, 234–71, 279, 280, 281, 283–6, 300–301; *50*
   Walter and campaign for, 52, 93, 234–69 *passim*, 280, 281, 283–5, 286, 300–301
   why addressed to Walter, 234–5, 235–6
   Balfour invites draft declaration, 235
   Rozsika and, 241–7 *passim*, 251
   Sokolow takes charge of drafting, 260*n.*, 261, 265*n.*
   Montagu's opposition, 262–3
   War Cabinet considers, 265–6
   Walter receives Balfour's communication, 266; *50*

Thanksgiving Meeting (December 2nd
 1917), 266–9
 opinions concerning, 270–1
 campaign to rescind, 279–81
 confirmed at San Remo conference
 (1920), 280
 ratified by League of nations (1922),
 283
 dearth of official records, 283, 284–5
Balzac, Heim de (zoologist), 157
Bang-Haas (dealer), 157
Bannerman, David (ornithologist), 141,
 303
Barber, William, 105
Barnes, George, 280
Barnes, Oscar T., 331(8)
Battersea, Lord, see Flower, Cyril
Bauer, Professor, 193, 197, 202n.
bears, 159, 317
 fleas from, 295
Beaverbrook, Lord, 279
Beck, R. H., 193, 199, 200, 201, 202,
 204, 299
beetles, Coleoptera, 107, 134, 145, 148,
 149, 158, 177
Behrens, Evelina (Walter's sister), see
 Rothschild, Evelina
Behrmann, Sam, quoted, 101
Belloc, Hilaire, 26
Beni Israel Trust, 17n.
Bentley Priory, 5
Bentwich, Herbert (historian), 252
Berenson, Bernard ("B.B."), 301
Bergmann, David, 247
Berlepsch, Count (ornithologist), 129, 156
Berlin, Isaiah, 38
Beston, Henry, 214
Biddy (Charles's collie), 206
birds
 albinos, 63, 138
 collectors, collecting trips, 73, 76, 156,
 157–8, 178, 233
 eggs, 2, 105, 114, 126, 137, 158, 311–12
 extent of collection, 71, 102, 137, 137–8
 extinct, 91, 152
 Ibis, Walter's relations with, 132,
 133–4, 228
 importance of collection, 91, 130,
 138–41
 Palaearctic, 130–1, 140
 of Paradise, 158, 302, 311–12
 sale of collection, 59, 92, 126, 135,
 139, 140, 215, 302–4
 theft from collection, 181–4

Tring fauna, 134
trinomial system, 128–9, 132–4
 mentioned, 62, 63, 67, 81, 142, 262
 see also cassowaries; humming birds;
 kiwis; ostriches; pheasants
Birrell, Hugh, 90, 91, 94, 101, 123n.,
 156, 171
Bismarck, Herbert, 29
Bismarck, Otto von, 47n., 217n.
Bittern, 57
Blackie (dog), 13
blacks, prejudice against, 27
Blake, William, 7–8
Bleichröder, Gerson von, 33
Blenheim, 14
Board of Deputies, 235, 255, 256,
 258–60, 273
Boden Kloss, C. (collector), 156
Boer War, 81–2, 83, 84–5
Bonn, University of, 65, 67–8
Bonny (Walter's St. Bernard), 206
botany, 173–4, 336
Boulanger, Professor G. A. (zoologist),
 203–4
Bowen, E., 89–90
Bradford, Lady, 32n.
Brandeis, Judge Louis Dembitz, 237n.,
 251, 264
Brett, Reginald Baliol (2nd Viscount
 Esher), 29 (see also Esher)
Brighton, Walter's visit to, 63
British Association, 308–9
British Museum, 306, 332(4)
British Museum (Natural History)
 and continuity, 117
 criticisms of, 101–2, 103n., 117–18
 King George V at, 47
 theft from, 183–4
 Tring Park refused by, 103
 Tring records destroyed by, 63, 185,
 298–300
 Walter's bird collection not bought
 by, 304
 Walter's collection given to, 52,
 310–11, 315–16
 Walter's relations with, 57, 63, 113,
 203–4, 228–9
 Walter a trustee of, 59, 117, 228, 293
 mentioned, 59, 65, 262, 349(4)
British Oological Association, 126
British Ornithological Club, see
 Ornithological Club
broom, "extinct", 95
Brown, Keith (entomologist), 150

Buckingham Palace, 108–9
Buckle, George Earle (Editor of *The Times* newspaper), 31
Bucks, Michael, 230
Bucks Imperial Yeomanry, 61, 80–1, 84–5; *32*
Bucks Volunteer Defence Corps, 329(16)
Buller, Sir William (ornithologist), 100
Bullfinch, Desert, 171
Bullock, Otis E., 188, 365
Bulwer-Lytton, Sir Edward (1st Baron Lytton), 41
Butler, Montagu, 36
Butler, Samuel (1835–1902), quoted, 295
butterflies and moths, Lepidoptera
British, 150–2
Charles and, 148, 174, 329(12), 337(10)
collecting trips, 148, 160–1, 163, 166, 178, 233
display cases, 112
Emma's interest in, 21
*Euchloë pechi*, 166
extent of collection, 67, 71, 124, 137, 148
Hawkmoths, 152–3
importance of collection, 139, 147, 149–50, 316
Macrolepidoptera and Micros, 148
*Papilios* (Swallowtails), 145, 148, 220
Walter and others at Ashton hunt, *46*
Walter's main entomological interest switched to, 148–9
mentioned, 105

Cain, Professor Arthur (zoologist), 308*n.*
Calcroft, Henry, 216
Cambridge, Duke of, 85
Cambridge, Joan (graphologist), 52
Cambridge University, 14, 26–7; *see also under* Rothschild, Nathan Mayer *and* Walter
Campbell-Bannerman, Sir Henry (Prime Minister 1905–08), 226
Canterbury, Archbishop of (Cosmo Gordon Lang), 59, 158, 304, 315
Carnarvon, Lord ("Porchy"; 6th Earl of Carnarvon), 12
Carr, Richardson (agent), 106, 107, 211, 211–12, 222, 244, 305, 330(8)
cassowaries, 93, 103, 104, 111, 140, 291, 302
Catholics, anti-Catholic feeling, 27
Cecil, Robert (1st Viscount Cecil of

Chelwood; Under Secretary for Foreign Affairs 1915–18) 237, 238, 245, 266–7, 267, 270, 273, 280, 281
Chamberlain, Joseph (Secretary of State for Colonies 1895), 238, 238*n.*
Chamberlain, Neville (Prime Minister 1937–39), at butterfly hunt, *46*
Champneys, 287
Chapman, Frank M. (ornithologist), 139, 303
Charles (island), 200, 201, 366
Charles II, King, 4, 319(1)
Chatham (island, Galapagos group), 200, 201, 366
Chatham Islands (east of new Zealand), 73, 76–7
Chesterton, G. K., 26
Chief Rabbi of England, 267*n.*, 281, 344(47)
Chiltem Hills, 3, 4, 7
Chopin, Frédéric, 17
Christopher (driver), 305
Churchill, Lord Randolph (Chancellor of the Exchequer 1886), 26, 29–30, 47*n.*, 321(3); *35*
Churchill, (Sir) Winston (Prime Minister 1940–45, 51–55), 15, 16, 42*n.*, 45, 64, 237, 238, 246, 301
Claire, Sister, 300, 301, 306–9, 311, 312
Clarke, Cyril (physician and geneticist), 151–2
Clarke, Eagle (zoologist), 349(3)
Cockayne, E. A. (pediatrician and entomologist), 151
Cocos Island, 193–4, 201, 202
Cohen, Leonard, 344(47)
Cohen, Lionel, 30
Coleoptera, *see* beetles
Colette, 210
Colledge family, 330(5)
Common, E. F. B. (entomologist), 149
congresses, international, 144
Conjoint Foreign Committee, 235, 255–60
conservation, 114, 139, 174, 197, 205, 313
Cormorant, flightless, 202
Cornell, J. M., 188, 190, 365
Corporaal (collector), 155
Corti, Count, 295, 296
Coryndon, R. S., 167
Costermongers and Street-sellers Union, 46*n.*
Coues, Elliot (ornithologist), 128

*Country Life*, 48*n*., 211
Cowen, Joseph, 252
Cowles, Virginia, 232, 295*n*.
Cox, F. J., 336(4)
cranes, 142
Crawford, Lord, 117, 231, 315 (*see also* Balcarres)
Creditanstalt Bank crisis, 18
Crewe, Peggy (Lady Crewe), 245, 250
Crewe-Milnes, Robert Offley Ashburton (Marquess of Crewe), 31, 36, 270, 273, 281
crocodile, 171
Cromer, Lord, 36
Cunningham, Dr. Cyril (ornithologist), 169, 181–4
Curzon, Lord (Secretary of State for Foreign Affairs 1919–24), 273, 275, 280

D'Abemon, Lord, at Tring shoot, *34*
*Daily Chronicle*, 46–7
Dairnvaell, Georges, 347(5)
Dalmeny, Harry (later Lord Rosebery), 64, 208, 248, 274
Davies, Ron, 139
De Beers Company, 32, 238*n*.
death camps, 323(8)
Degen, A. von (botanist), 174
Derby, Lady, 296*n*.
Dérolle (dealer), 157
Devonshire, Duke of, 284
Dickenson, V., portrait by, *17(a)*
Dingo, 62, 63, 102
"Diplomaticus", 46
Disraeli, Benjamin, 30, 31, 32*n*., 37, 41, 71, 296–7
dock strike (1889), 39, 301
Doggett (taxidermist), 111
dogs, 206–7, 307, 308; *13*
Doherty, William (collector), 154*n*., 156, 169, 177–80, 187
Dollman, G. (zoologist), 74
donkeys, wild, 170, 172–3
Dresser, Henry (ornithologist), 129
Dreyfus, Alfred, 217–18, 240
Drowne, Frederick Peabody, 193, 198–9, 200–201, 366
Druce, Claridge (botanist), 174
Dugdale, Blanche, 236*n*., 285
Duncan (island), 197, 198–9, 200, 202*n*.
Dunn, Lady Mary, 320(16)
Dykes, William Rickatson (botanist), 173–4

East Africa, collecting in, 178, 179, 312; *see also* Sudan
*East Anglian News*, 25*n*.
East End of London, 37, 45–6, 242
Eastern Archipelago, collecting in, 178
Edward VII, King (formerly Prince of Wales), 23, 26–7, 34–6, 122–3, 210–11, 216
Edwards, George, 216
egg collection, *see* birds
Egremont, Max (2nd Baron Egremont), 26
Egypt, 30, 81, 82
collecting expedition, 91
Eichorn, A. F. (collector), 155
Einstein, Albert with Walter at public dinner, *51*
Elinors, 4
Elton, Charles (ecologist), 139
Entomological Club, 109
Erlanger, Carlo von, 132
Esher, Lord, 329(1)
Everett, Alfred (collector), 154*n*., 187
Ewart, Colonel, 82
*Extinct Birds* (Rothschild), 91, 152

farm stock, 208, 210–12
Faroult, Victor (collector), 155, 157, 233, 298
fata morgana, 164
Feisal, Emir, 272–3, 274, 275
Felder affair, 107
Fellows, Major, 298
Ferdinand, King, of Bulgaria, 156, 183; 62
*Field, The*, 211
Fischer, Dr. Heinrich, 27
fish, collected by Walter at Brighton, 63
fleas
    Charles and, 110, 146, 160, 166, 171, 173, 233, 336(4), 338(9)
    off grizzly bears, 295
    Jordan and, 143, 146, 173, 233
    plague-carrying, 171, 173
    Tring collection, 147
Flower, Cyril (later Lord Battersea), 15, 215
Ford, E. B. (geneticist), 171
Forrest, George, 157
Foster, Sir John, 249
Fowler, Tom, 211
Fredensen, Mr. (father of Marie Fredensen), 217–18
Fredensen, Marie, 216–23, 297; *58*

Freud, Sigmund, 320(16)
Friedman, Isaiah, 279
Frohawk, F. W. (artist), 201, 228, 288
Froude, James Anthony, 31, 296*n.*
Froude, Trooper Tom, 82

Gainsborough, Thomas, 7 (*Morning Walk*)
Galapagos Islands, 138, 154*n.*, 185–6, 187–96, 197–205, 299, 365–9
   expedition to, 185–202
gannets, 63
Gardiner, Professor Stanley (zoologist), 67
Gaster, Moses, 237*n.*, 251, 252, 258, 267*n.*
Gayner, Francis, 170
George V, King, *47*
Geospizids, 138, 200
Giffard, Captain, 187
Gille, Bertrand, 331(note)
Giraffe, Rothschild's, 229
Gladstone, W. E., 29, 30, 30–1, 41, 109, 296
Gladstone, W. E. (son of preceding), 226
Glünder, Ellie, 55–6, 57, 61, 90; *27*
Godman, Dr., 163
Goebel, Karl, painting by, *2*
Goldsmith, Nellie, 98
Goodson, Arthur, 115, 116, 119, 302
Goodson, Fred, 115
Goodson, Leslie (son of Fred Goodson), 115
goosander, 63
Gore, Charles, 6
Gore, Charles Orlando, 319(2)
gorilla, 286; *31*
Gosford, Lady, 7*n.*
Grace, Bob, 321(1, 5)
Graham, Sir Ronald, 266*n.*, 345(45)
Gramont, Duc de, 321(19)
Grant, Sybil (née Primrose), 15, 50
Greuze, Jean-Baptiste, painting by, 11*n.*
Grey, Edward (1st Viscount Grey of Fallodon; 1862–1933; Foreign Secretary 1905–16), 238, 246, 255*n.*, 262*n.*, 328(16)
Grosstephen, father and son, 30, 49*n.*
Grosz, Emil, 95
Gunnersbury House, 9, 297
Günther, Dr. Albert C. L. G., 57
   death, 57
   Natty, relations with, 72, 91, 149, 204
   puzzled by Walter, 65, 131

relations with Walter: curators recommended, 57, 65; tortoises, 103, 113, 197, 198, 202, 203, 204; Tring visits, 57–8, 63–4, 213; Walter's early enthusiasm, 57, 60, 63, 67, 69, 71; Walter loses confidence in, 72–3; miscellaneous, 75, 107, 113, 117–18, 138, 149, 169, 228, 311
Günther, Albert Everard (grandson of Dr. Albert Günther), 204
Günther, R. T. (son of Dr. Albert Günther), 73, 203
Günthersburg, 53
Gutteridge, Mrs., 20, 21

Haldane, Miss Elizabeth (Lord Haldane's sister), 114
Haldane, Richard Burdon (1st Viscount Haldane; 1856–1928; Secretary of State for War 1905–12, Lord Chancellor 1912–15), 27–8, 32*n.*, 40, 163*n.*, 238, 245, 288
Hamilton, E., 29
Hankey, Lord, 344(47)
Harcourt, Loulou (Lewis), 7, 16, 45, 108, 216, 297*n.*, 324(28)
Hardinge, Sir Charles, 35–6
Harlech, Lord, *see* Ormsby Gore, William
Harmer, Sidney, 163, 167, 168, 169
Harmon, Dr. J., 163
Harris, Charles Miller, 139, 155–6, 169, 185–6, 187–96, 197, 198–202, 365
Harrow School, 89–91
Hartert, Mr. (son of Ernst Hartert), 135, 244
Hartert, Ernst (ornithologist), 42
   a bird specialist, 132, 134, 135, 141, 143–4
   capabilities, 121, 129–30, 131–2, 134, 135
   collecting experience, 130, 156
   collecting trips to Algeria, 93, 163–6, 222, 233, 293
   death, 135
   and Doherty, 177, 180
   honours received, 130*n.*, 132
   and international congresses, 144
   Jordan, relations with, 124, 130*n.*
   publications, 130–1, 134, 303
   responsibilities, 115–16
   retirement, 134–5, 302
   and sale of bird collection, 126, 303
   self-taught, 130, 177
   and trinomial system, 128–9, 132–3

Walter, introduced to, 57, 107, 129
Walter, relations with, 72, 108, 121,
    129–32, 221
with Walter and Jordan at Tring, 45
Walter's influence on, 119, 121
Walter's memory admired by, 116–17
mentioned, 81, 85, 86, 100, 103, 105,
    106, 111, 112, 125, 137, 155, 158, 159,
    161–3, 167, 169, 198, 203–4, 271, 291
Hartert, Mrs. Ernst, 129, 130n., 131, 135
Hastoe, 19
Hawaiian Islands (formerly Sandwich
    Islands), 73, 76–8
Hawkmoths, 152–3
Haystaff (carpenter), 21
Hebrew University, 285
Hellmayr, Carl, 132
Hennrick, Professor J., 186–7
Henry, Sir Charles, 279
Heron, Dr., 54
Heron, J. S., 336(3)
Herzl, Theodor (1860–1904; Zionist
    leader), 237n., 239
Hillingdon, Lord, 181
Hinton, Bert, 21
Hinton, Howard (entomologist), 177
Hinton, M. A. C., 120, 121
Hirsch, Baron, 250
Hirst, Fabian, 173
Home Farm, 58, 305, 306–7
Home Rule, 225
Honour, J., & Sons, 104
Hopkins, G. H. E. (entomologist), 143
horses, shire, 211
housing, Lloyd-George's proposals, 43–4
Huckvale, William (architect), 104
Hull, Galen D., 193, 198
humming birds, 108
Hungary, 18, 94–5, 174, 244, 245, 288
Husseini, Haj Amin al, 280
Hutchinson, Evelyn, 315
Huxley, Sir Julian, 205, 298

Ibis, 132, 133–4, 228
Ibises, 114
Indefatigable (island), 200
Indian Ocean, islands in, 114
Innocent IV, Pope, 36
insects, 107, 112, 222; see also beetles;
    butterflies; fleas
Iris, 110, 173–4, 337(10)
Isaacs, L. L., 30
island faunas, 73, 137, 139, 185
Israel, 17n.

James (island), volcano on, 199
James, Lord, of Hereford, 38
Japan, Charles's collecting trip to, 56(b)
Jebb, Eglantine, 245n.
Jeeves (groom at Tring), 103
Jersey Cattle Society, 211
Jersey Cow, The, 211
Jettmar (collector), 155
Jewish Board of Guardians, 293
Jewish Chronicle, 89, 344(45)
Jewish Free School, 45, 89, 321(3)
Jewish Peace Society, 293
Jews
    and animals, 213
    Rothschild concern for, 19, 22, 26–31,
        32–7, 313, 321(3) (see also under
        Zionism)
    status of, in England, improvement in,
        26–31, 37, 238
    see also anti-semitism; Zionism
Jobi Island, 179
Jopling, Louise, portrait by, 12
Jordan, Ada, 244
Jordan, Dr. Karl (entomologist), 32
    appearance, 59n.
    and bird collection, eggs, 303, 304,
        311–12
    capabilities, 125, 144, 144–5, 146,
        213–14
    Charles, relations with, 93, 146–7,
        160–1, 213–14, 271, 330(8)
    collecting experience, 159
    collecting trips, 160–1, 233
    contribution to science, 148
    drawing by, 24
    Emma's recollection of, 59
    entomological curator, 91, 107,
        115–16, 145, 148–9, 155
    an F.R.S., 132, 147–8
    Hartert, relations with, 124, 130n.
    Hartert succeeded by, 302
    house in Tring, 92, 118
    and international congresses, 144
    personality, 146
    publications, 143, 147, 152–3, 220
    recognition, belated, 132, 144–5, 147–8
    remoteness, in later life, 147
    salary, 105, 118
    and Tring Museum, transfer to British
        Museum, 309–10
    Walter, introduction to, 143
    Walter, relations with, 72, 120–1,
        126–7, 130, 131, 133, 144, 152–3,
        293

on Walter, 23, 87, 120, 122, 123, 124–5, 127, 133, 151, 206, 213, 224, 291, 327(14)
with Walter and Hartert at Tring, *45*
mentioned, 139, 299, 308
Jordan, Minna, 130*n*., 147
Jourdain, Pastor (ornithologist), 126

kangaroos, 102; *27*
Karlsson, Jon, 122, 215
Kay, Mr., 4
Keren Hayesod, 280*n*.
Kerr, Philip (later Lord Lothian), 278
Kettlewell, Bernard, 151
Keulemans, Mr. and Mrs. J. G. (artists), 73, 137
Kitchener, Lord, 81–2, 83
kiwis, 70, 206; *30*
Kleinschmidt, O. (zoologist), 128
Knollys, Lord, 35
Kühn, Heinrich, 187

Lack, David (ornithologist), 138–9
Lambert, Sir Anthony, 320(2)
Lampyridae, 149
Lankester, Ray (zoologist), 114, 228–9
Lansdowne, Lord, 36, 238, 324(47)
Lanyon, Wesley E., 141
Lawrence, Miss, 21
Lawrence, T. E., 273, 276–7
Laysan, Palmer's expedition to, 156; *see also Avifauna of Laysan*
Le Moult (dealer), 155, 157
League of British Jews, 279
League of Nations, 280, 283
Lencker Tazza, *see* Orpheus Cup
Leopold, Prince (son of Queen Victoria), 23
Lepidoptera, *see* butterflies and moths
Levick, Thomas, 200
Lilford, Lord, 66*n*., 73, 138
Lindemann, Professor F. A. (later Lord Cherwell), 213
Lipschitz, Elizabeth (Walter's landlord's daughter at Bonn), 68–9; *18*
Llandaff, Lord, 36
Lloyd George, David (later 1st Earl Lloyd-George of Dwyfor; 1863–1945; Chancellor of the Exchequer 1908–15, Minister of Munitions 1915–16, Prime Minister 1916–22)
an East End view of, 46
European geography, his ignorance of, 245

housing scheme, 43–4
Natty, relations with, 41–5, 225, 324(47)
and old age pensions, 42–3
and Zionism, Balfour Declaration, 237, 238, 246, 251, 252, 261*n*., 270, 275, 280, 281
Lloyd George, Richard, 45
Loreburn, Lord, 36
Louisa Cottages, Tring, 20
Lyttleton, Alfred, 36

McBride, Professor E. W. (zoologist), 316
MacDonald, Alexander, 202
Magdalene College, Cambridge, 57, 70
Magnus, Philip, 344(47)
Makins, Roger (later Lord Sherfield), 323(8)
Mallet, Louis, 238
Manning, Cardinal, 39
Manning, Dr., 202–3
Marsh, Edward, 284
Massingham, Henry William, 324(28)
Maxse, Leo, 38
Mayer, A. B., 132
Mayr, Ernst (zoologist), 139–40, 145, 158, 166, 302
Meath, Lord, 270
Meek, Albert S. (collector), 156, 157, 187, 336(3)
Meinertzhagen, Colonel Richard Balfour, criticism of, 271
ornithologist, 8, 118, 270
Walter, relations with, 8, 135, 271 *n*., 277, 280–1
on Walter, 105, 115*n*., 120, 234*n*., 250
and Weizmann, 276
and Zionism, 8, 234*n*., 270–1, 273, 275–6, 280–1
mentioned, 134, 255
Melon (collector), 161
Memory, *see* Charles Rothschild, James de Rothschild, Walter Rothschild
Mentmore, 4, 5, 10, 15, 64
Merry del Val, Cardinal Rafael, 36
Milner, Lord, 238, 240, 273, 344(47)
mimetic polymorphism, 145
Minall, Alfred (joiner and journeyman carver), 1, 2, 54, 54–5, 57, 60–1, 62, 71, 104, 111, 129, 298–9; *15, 33*
Minall, Mrs. Alfred, 19, 62, 64, 104–5, 137
Moa, Giant, 113; *55*

Mond, Alfred Moritz, 280*n.*, 281, 285
Monné (Walter's Pyrenaean Mountain
    Hound), 206, 232, 307, 308; *92*
Montagu, Edwin, 246, 249, 254, 262,
    263, 272
Montefiore, Claude, 255–6, 257, 258,
    344(47)
Montefiore, Sir Moses, 239*n.*
Morley, John (Viscount Morley), 31, 36
Morocco, Jews in, 33
Morton, Frederic, 232
Moss, Rev. Miles (collector), 157
Munroe, Dr., 141
Munthe, Nelly, 331(note)
Murphy, Dr. R. C., 303
M'zab (Arabs), 165

Napoleon, letter signed by, 306; *56*
National Trust, 319(5)
Natural History Museum, *see* British
    Museum (Natural History)
*Nature*, 24, 286
Nauheim, Joseph, 220, 230
naval armaments, 42
Nelson, George, 188, 190–1, 366
Nelson, Horatio (Lord Nelson), letter
    signed by, 306; *56*
Neuville, Henri, 338(9)
New Court, *see* Rothschild, N. M., &
    Sons
New Guinea, *see* Papua New Guinea
Newton, Alfred (1829–1907,
    ornithologist), 66*n.*, 73, 75, 75–6,
    76–9, 128, 138, 156, 167–9, 177
Newton, Thomas Wodehouse Legh (2nd
    Baron Newton), 7, 18
Nicholls, Captain, 187
Nicholson, Sir Arthur, 35–6
Nicholson, Geoffrey, 348(5)
Nicolson, Harold, 261
Nissen, Dr. Charles, 127, 162, 164
Nordau, Max, 281
Norfolk, Duke of, 36
North Africa, collecting in, 160, 222; *see
    also* Algeria *and under* Egypt
*North London Illustrated, The*, 228
*Novitates Zoologicae* (Tring Museum
    journal), 76–7, 108, 116, 132
Nuttall, Professor (zoologist), 91

Ockenden, George Richard, 154*n.*
Odessa, pogrom in, 275–6
Ogilvie, Robert, 202
Okapi, 76, 105

O'Keefe, Dr., 307–8
Olivier, Professor (entomologist), 149
opossum, 63
orchids, 212, 337(10)
Ormsby Gore, William George Arthur
    (4th Baron Harlech; 1885–1964;
    Assistant Secretary War Cabinet
    1914–19), 241*n.*, 270, 279, 280, 281,
    284
Ornithological Club, 112–13, 303,
    311–12
ornithology, *see* birds
Orpheus Cup (Lencker Tazza), 7
ostriches, 108, 271–2

Pacific Islands, South Sea islands, 73,
    75, 156, 157
Packenham, Thomas, 83*n.*
Palestine
    Anglo-French condominium plan,
        251–2
    boundaries, 275, 277–8
    British Mandate, 280–1
    colonisation of, 154
    Jewish national home/state in, *see*
        Zionism
*Pall Mall Gazette*, 28–9, 36
Palmer, Henry (collector), 73, 76–8, 139,
    155–6
Papilios, *see under* butterflies
Papua New Guinea, 158, 179, 302
pensions, old age, 42–3
Perugia, Marie, *see* Rothschild, Marie de
pheasants, 67, 67–8, 114, 133
    Formosan, 116–17
148 Piccadilly, 22, 207, 296, 320(13); *49*
Picot, Emile, 111
Pinto, Dorothy, *see* Rothschild, Dorothy de
Piper, Mr., 26
Poliakoff, Léon, 225*n.*
Ponsonby, Henry, 36–7
Poulton, Arthur (zoologist), 115, 119
Prag, J., 259
press, the, 229
Primrose, Neil, 266
Primrose, Peggy, 50
Primrose, Sybil (later Sybil Grant), 15,
    50
Proust, Marcel, 26
Pye-Smith, Major (ornithologist), 126
Pyrenaean Mountain Hound, Walter's,
    *see* Monné

Quaritch, Bernard, 125

Rothschild, CONSTANCE de ("Connie"; 1843–1931; born in England; daughter of Sir Anthony de Rothschild; married Cyril Flower, later Lord Battersea; Walter's cousin), 4–5, 14, 14–15, 58, 94, 240n., 319(l)

Rothschild, DOROTHY Matilde de (née Pinto; 1895– ; born in London; married James Armand de Rothschild; Walter's cousin by marriage), 38, 240, 242, 247, 267n.

Rothschild, EDMOND Adolphe Maurice Jules Jacques (1926– ; son of Maurice Edmond Charles de Rothschild), 332(8)

Rothschild, EDMOND James de ("Baron Edmond"; 1845–1934; born in Paris; son of James Mayer de Rothschild; married Adelheid von Rothschild; Walter's cousin; partner in Rothschild Frères, Paris; initiated Jewish re-colonisation in Palestine, founded twenty-seven agricultural settlements in Palestine), 39n., 110, 235, 239n., 250n., 301, 332(4), 372–4

Rothschild, ELIZABETH Charlotte ("Liberty"; 1909– ; born in England; daughter of Nathaniel Charles Rothschild; Walter's niece), 320(l) caricature by, 40

Rothschild, EMMA Louise von ("Lady Rothschild"; 1844–1935; born in Frankfurt; daughter of Mayer Carl von Rothschild; married her first cousin, Nathan Mayer (Natty) Rothschild; Walter's mother)
aged 18, 3
aged 91, 54
appearance, 11, 12
Charles, relations with, 13, 22–3, 59, 89, 90
death, 3, 139, 305, 306
a dog lover, 207
English and other languages, 6, 13
family and friends, 4–5, 12–13, 13–16, 30–1, 238, 239n., 250n.
Frankfurt, 65
gold pendant given to Natty, 10
good works, 224, 321(3), 339(5), 341(28)
health, 12
literature, reading, 30–1, 320(16)
marriage, 9–11
memory, 285

Natty, relations with, 10, 11–12, 22, 32, 38, 50, 50n., 104
palm, lines on, 12
a perfectionist, 13
personality, 1, 4, 9, 13, 16, 21, 22, 31, 97
private papers destroyed by, 297
in public life, 217–18, 236–7, 245, 250, 251
resembling Miss Lipschitz, 19
Rozsika, relations with, 97
scientific studies, 58–9
straightlaced, 11n., 22, 23
a tree lover, 324(36)
and Tring Park, 3–4, 4–7, 8, 13–14, 19–21, 287, 288–9
Walter, relations with, 18, 21–2, 23–4, 52–4, 55, 56, 58–9, 60, 65–6, 70, 74, 98, 104, 107, 110, 112, 139, 207n., 215, 216, 221, 222, 249, 286, 288–90, 292, 293, 304, 313
her Will, 306; 56
mentioned, 7n., 29, 49, 61, 68, 72, 81, 210, 240n., 266, 272n.

Rothschild, Charlotte Louise Adela EVELINA (1873–1947; born in England; daughter of Nathan Mayer (Natty) Rothschild; married Clive Behrens; Walter's sister), 5, 6, 16, 49, 54, 58, 61, 97, 108; 12, 13

Rothschild, EVELINA de (1839–1866; born in England; daughter of Baron Lionel de Rothschild; married Ferdinand James Anselm de Rothschild; Walter's aunt), 10

Rothschild, EVELYN Achille de (1886–1917; born in England; son of Leopold de Rothschild; Walter's first cousin; killed in action in Palestine during World War 1), 266

Rothschild, FERDINAND James Anselm ("Baron Ferdinand"; 1839–1898; born in Germany; son of Anselm Salomon von Rothschild; naturalised Englishman; M.P. for Aylesbury (Liberal); married Evelina, Walter's paternal aunt), 5, 100, 110, 198, 225, 226, 340(5)

Rothschild, GUSTAV Samuel James de (1829–1911; born in France; son of James Mayer de Rothschild; married Cécile Anspach; partner in Rothschild Frères, Paris), 39, 326(28), 332(4), 340(14)

Rothschild, GUTLE (nee Schnapper; 1753–1849; married Mayer Amschel

Rothschild; mother of ten [surviving] children of whom the five brothers founded banks in Germany, England, Italy, France and Austria), *1*

Rothschild, GUY Édouard Alphonse Paul de (1909– ; born in France; son of Édouard Alphonse James de Rothschild; married (1st) Alix Schey von Koromla, (2nd) Marie-Helene van Zuylen de Nyevelt; Walter's cousin; senior partner in Rothschild Frères, Paris), 208*n.*

Rothschild, HANNAH (1783–1850; married Nathan Mayer de Rothschild; Walter's great grandmother), 17

Rothschild, HANNAH de (1851–1890; born in England; daughter of Mayer Amschel de Rothschild; married 5th Earl of Rosebery; Walter's cousin), 9, 15–16, 37, 110

Rothschild, HANNAH Mary (1962– ; born in England; daughter of Nathaniel Charles Jacob Rothschild; Walter's cousin), 320(16)

Rothschild, HENRI James Nathaniel Charles de ("Baron Henri"; 1872–1946; born in France; son of Nathan James Édouard de Rothschild and Emma's sister Thérèse; married Mathilde Sophie Henriette de Weisweiler; Walter's maternal first cousin; paediatrician, playwright, inventor), 83, 110, 208, 208–10, 239*n.*, 295*n.*, 332(9), 347(3)

Rothschild, JAMES Armand Edmond de (1878–1957; born in France; naturalised British subject; son of Edmond James de Rothschild; married Dorothy Matilde Pinto; Walter's cousin; Liberal M.P., leading Zionist), 39*n.*, 239*n.*, 242, 246–7, 248, 250, 251, 252, 267, 279, 281–2, 285, 340(5)

Rothschild, Nathan JAMES Édouard de ("Baron James"; 1844–1881; born in London, but opted to be French; son of Nathaniel de Rothschild; married Laura Thérèse de Rothschild; Walter's uncle by marriage; lawyer, bibliophile), 9, 58, 110, 111, 333(13); memory, 111*n.*

Rothschild, JEANNE Charlotte Louise Marthe de (1874–1929; daughter of Nathan James Édouard de Rothschild; married Barone Abram David Leonino; Walter's maternal first cousin), 209

Rothschild, JULIANA de (née Cohen; 1831–1877; born in England; married Mayer Amschel de Rothschild; Walter's great aunt by marriage), 15

Rothschild, LEOPOLD de ("Leo"; 1845–1917; born in England; son of Baron Lionel de Rothschild; married Marie Perugia; Walter's uncle; partner in N. M. Rothschild & Sons), 11, 12, 14, 34–5, 57, 60, 319(5)

Rothschild, LIONEL Nathan de ("Baron Lionel"; 1808–1879; born in England; son of Nathan Mayer de Rothschild; married Charlotte von Rothschild; Walter's paternal grandfather; senior partner in N. M. Rothschild & Sons), 5, 28, 47*n.*, 225, 296, 340(5)

Rothschild, LIONEL Nathan de (1882–1942; born in England; son of Leopold de Rothschild; married Marie-louise Beer; Walter's first cousin; senior partner in N. M. Rothschild & Sons), 279, 340(5)

Rothschild, LOUIS Nathaniel de (1882–1955; born in Vienna; son of Salomon Albert Anselm von Rothschild; married Countess Hildegard Johanna Caroline Marie Auersperg; Walter's cousin; senior partner in S. M. Rothschild, Vienna), 18, 91*n.*

Rothschild, LOUISE de (1820–1894; born in England; daughter of Nathan Mayer Rothschild; married Mayer Carl von Rothschild (Frankfurt); Walter's maternal grandmother and paternal great aunt), 9–10, 13, 20, 25*n.*, 68

Rothschild, LOUISE de (née Montefiore; 1821–1910; born in England; daughter of Abraham Montefiore; married Sir Anthony de Rothschild; Walter's great aunt, by marriage; Constance's mother), 14, 240*n.*

Rothschild, MARGARETHA (1855–1905; married the Duc de Gramont), 321(19)

Rothschild, MARIE de (née Perugia; 1862–1937; born in Austria; daughter of Achill Perugia and Nina Landauer; married Leopold de Rothschild; Walter's aunt by marriage), 254–5, 266*n.*, 297

Rothschild, Hannah MATHILDE von (1832–1924; born in Frankfurt; daughter of Anselm Salomon von

Rothschild; married Wilhelm Carl von Rothschild; Walter's cousin), 17

Rothschild, MATHILDE Sophie Henriette de (née Weisweiler, 1874–1926; married Baron Henri), 332(l)

Rothschild, MAURICE Edmond Charles de (1881–195 7; born in France; younger son of Edmond James de Rothschild; married Noémie Claire Alice Palmyre Halphen; Walter's cousin), 11n., 338(9)

Rothschild, MAYER Amschel de ("Uncle Muffy"; 1818–1874; born in England; son of Nathan Mayer de Rothschild; married Juliana Cohen; Walter's paternal great uncle), 5, 64, 100, 110, 208, 340(5)

Rothschild, MINNA Caroline von (1857–1903; born in Frankfurt; daughter of Wilhelm Carl von Rothschild; married Maximilian Benedikt Heyum Goldschmidt; Walter's cousin), 50

Rothschild, MIRIAM Caroline Alexandrine de (1884–1965; born in Paris; daughter of Baron Edmond James de Rothschild; married Albert Max von Goldschmidt-Rothschild; Walter's cousin), 17, 94, 110

Rothschild, MIRIAM Louisa (1908– ; born in England; daughter of Nathaniel Charles Rothschild; married George Henry Lane; Walter's niece and author of his biography), 21n., 45–6, 49, 59n., 100, 103, 109n., 143, 205, 247, 290–1, 294, 299, 300, 301, 309, 310, 320(16), 333(19), 364

Rothschild, N. M., & Sons ("New Court") Charles at, see under Rothschild, Charles Hungary, loans to, 96

importance of, in financial world, 28, 297

Natty at, see under Rothschild, Nathan Mayer

Natty's brothers at, 12, 233n., 243

in recession of the 1930s, 18

Russian Government loan to, refused, 33, 226, 340(14)

Walter at, see under Rothschild, Walter

Rothschild, NATHAN MAYER (1777–1836; born in Frankfurt; son of Mayer Amschel Rothschild; married Hannah Barent Cohen; Walter's great grandfather; began trading at Manchester in 1820; founder of N. M. Rothschild & Sons, London), 4, 17, 123, 125, 225, 295

Rothschild, Nathan Mayer (known as "Nathaniel Mayer", "NATTY"; 1st Baron Rothschild of Tring; 1840–1915; born in England; son of Baron Lionel de Rothschild; married Emma Louise von Rothschild; Walter's father; the most widely respected and most influential member ofthe family; senior partner of N. M. Rothschild & Sons; M.P. for Aylesbury; first Jewish peer), 4, 51

abduction, threats of, 97

appearance, 12, 347(3)

Balfour, relations with, 31–2, 37–8, 38, 242

and the bank, 12, 28, 33, 115, 226, 230, 297, 340(14)

at Cambridge, 26–7, 41, 74–5, 328(16)

capabilities, 16, 25–6, 66, 111, 154, 234

caricatures of, 39

Charles, relations with, 22, 86–7, 233

and children, 50

with Randolph Churchill at Tring, 35

consulted on financial matters, 28, 29, 31–2, 39, 45, 83–4, 91, 111, 296–7

death, 38, 45, 234

dock strike, 39, 301

a dog lover, 207

Emma, relations with, 10, 11–12, 22, 32, 38, 50, 50n., 104

funeral, 48, 49

gold pendant given by Emma to, 10

grandchildren, relations with, 47–9, 50

health, 12, 25

and Jewish affairs, 26–30, 32, 33–6, 38–9, 154, 217–18, 226, 236–7, 238–40, 249, 259, 313, 321(3), 340(14)

languages, 25, 295n.

letters, letter-writing, 47n., 47–8, 50

letters to Rosebery, 68, 297

marriage, 9–12

memory, 71, 111, 158, 285

military service, 81

an M.P., 5, 225, 340(5)

a nature lover, 6

parents, relations with, 5, 25, 66

personality, 10, 13, 22, 25, 26, 38, 124n., 280n., 295

private papers destroyed by, 296–7

in public life, 27–30, 31–2, 33–6,

37–8, 40, 41–7, 51, 81–4, 85, 91,
225, 226, 229, 234, 236–7, 238–40,
296–7
real ruler of England", 229
with "Snip", *22*
a social reformer, 41, 47, 154
Sultan of Turkey's gift to, 39
taste, lapses of, 7–8
Tring Park bought for, by father, 5–8
Tring Park, management of, 7, 19, 20,
44, 58, 106, 210–12, 321(5), 324(36)
at Tring shoot, *34*
with baby Walter, *1*
Walter, relations with, 12, 22, 32,
50–1, 58, 60–1, 64–5, 67–8, 72, 80,
84, 87, 92, 97, 100, 104, 107, 108,
220, 221, 222, 227, 230–2, 242, 313
his Will, 87
women, relations with, attitudes to,
7*n*., 15, 17, 225
mentioned, 21, 31, 56*n*., 71, 97, 206–7,
224, 266, 287, 319(1), 339(5),
347(5), 348(13)
Rothschild, NATHANIEL Mayer (1836–
1905; born in Vienna, son of Anselm
Salomon), 332(4)
Rothschild, NATHANIEL Nathan (1812–
1870; born in England; son of Nathan
Mayer Rothschild; married Charlotte
de Rothschild (Paris); Walter's paternal
great uncle), 58, 111
Rothschild, ROZSIKA (née Wertheimstein;
1870–1940; born in Hungary; daughter
of Alfred von Wertheimstein; married
Nathaniel Charles Rothschild; Walter's
sister-in-law), 18
aged 20, *22*
in 1935, *59*
appearance, 95
and Charles's ghost, 175
and Charles's illness, 93, 174–5, 243,
247*n*., 288
collecting trips with Charles, 174,
337(10)
and death of Charles, 147, 288
and Sister Claire, 307–8
Emma, relations with, 4, 11*n*., 13, 50,
97
and death of Emma, 305, 306
and Jewish affairs, Zionism, 238,
239*n*., 241, 242, 243–7, 250, 251,
253, 254–5, 261, 262, 266*n*., 267,
272, 279
marriage, 94–8

Natty, relations with, 47, 49
personality, 22, 47, 94, 207*n*., 245
private papers destroyed by, 170, 297
her reading, 96
self-educated, 18, 96
with two sisters, *38*
and lawn tennis, 96
and Tring Museum, 299, 304, 309–10,
310, 337(10)
Tring Park managed by, 106–7, 287–8,
289, 305
Walter, relations with, 87, 92, 93, 183,
221, 227*n*., 232, 249*n*., 268–9, 287,
291–2, 298, 300
Weizmann, relations with, 245–7, 251
and World War I, 243–4
mentioned, 23, 207, 233, 334(13)
Rothschild, LAURA THÉRÈSE de
("Baroness James"; 1847–1877; born
in Frankfurt; daughter of Mayer
Carl von Rothschild; married Nathan
James Édouard de Rothschild (Paris);
Walter's maternal aunt), 9, 110,
208–10, 321(19, 21), 333(12)
Rothschild, Nathaniel Mayer VICTOR
(3rd Baron Rothschild; 1910– ; born
in England; son of Nathaniel Charles
Rothschild; married (1st) Barbara
Judith Hutchinson, (2nd) Teresa
Georgina Mayer; Walter's nephew),
103, 110, 291, 298, 309, 320(13, 16)
Rothschild, Lionel WALTER (2nd
Baron Rothschild; 1868–1937; born
in England; son of Nathan Mayer
(Natty) Rothschild and Emma Louise
von Rothschild; zoologist, addressee
of the Balfour Declaration), 52,
313–14
as a baby, *4*
aged 3–4, *5*
aged 3½, *7*
aged 5, *6*
aged 64, *53*
in 69th year, with Monné, *60*
abstemious, 71
animals, live, collected by, 61, 62, 63,
102–4, 197–8
animals, love of, feeling for, 21–2, 24,
53, 54–5, 57, 62, 123–4, 137, 141–2,
197–8, 206, 212–14; albinos, 63,
138; large animals, 197, 212, 250*n*.
animals, reports of, lists made of, 67,
68, 71, 81

animals and plants named after,
212–13, 317, 339(21), 364
animosity of other scientists to, 132,
188 appearance, 12, 23, 52, 53–4,
65, 70, 98, 184, 210, 348(3)
art, his views on, 292
attention to detail, 113, 167
avant-garde ideas, 224
and Balfour Declaration, Zionism
(*qq. v.*) 234–86 *passim*
and the bank, 52, 75, 100–101, 116,
220, 227, 229–31, 301, 316–17
Barmitzvah, 88
birth, 1
blackmailed, 59, 92, 98, 116, 139,
205, 220, 221–2, 230, 242, 243, 293,
297–8, 302
at Bonn, 65, 67–9; *16*
and botany, 173, 212
in Bucks Yeomanry, 61, 80–1, 84–5; *20*
butterflies (*q.v.*) his main
entomological interest, 148–9
butterfly hunt at Ashton, *46*
at Cambridge, 57, 67, 68, 70, 74–5,
206
caricature of, by niece Liberty, *40*
Charles, relations with, 22–3, 86–7,
91, 92, 93, 146, 167*n.*, 221–2, 230,
262, 286, 290, 309
with Charles and tortoise, *23*
childhood, 19, 21–2, 25, 52–6, 58
city life, hatred of, 101
clumsiness, 80
collections, 2, 67, 71–3, 102, 121,
135–8, 139, 149, 152, 201
collecting expeditions organised by,
73, 76–8, 138–9, 154–61, 170–80,
185–96, 197–203, 293; cost of, 166
collecting expeditions undertaken
by, 93, 116–17, 124, 127, 159–60,
161–7, 222, 233
collecting, hazards of, 179, 198;
yellow fever, 193–4
collectors, numbers of, 155–7, 187
committee work, 293
the common touch, 101–2, 123
conservative, 224–5
contribution to science, 213; Jordan's
view of, 151
correspondence lost, 301
correspondence stored in laundry
baskets, 91–3, 220, 221, 242,
297–8
death, 307, 312

debts, 92
delegation of work, 62
at a public dinner, *51*
distinction, 24
and dogs, 206, 307, 308
Emma, relations with, 18, 21–2, 23–4,
52–4, 55, 56, 58–9, 60, 65–6, 70,
74, 98, 104, 107, 110, 112, 139,
207*n.*, 215, 216, 221, 222, 249, 286,
288–90, 292, 293, 304, 313
eye for colour, 73, 137
eyesight, 73–4
with King Ferdinand, *52*
finance, 87, 92, 100, 118
an F.R.S., 132
German, command of, 55–6, 74, 131
Germany, links with, 65–6
ghost, 119
handwriting, 52, 92*n.*, 215–16, 250*n.*
with Hartert and Jordan at Tring, *45*
health, 25, 52–3, 53–4, 63, 124, 243
at Home Farm, 305, 306–7, 308–9
honorary degree, after receiving, *38*
horseman and driver, 80
illness, final, 311–12
information about him, dearth of, 296,
297–8, 300–1, 314
islands leased, for conservation, 114,
197, 205
Jewish affairs, concern for, 33, 110,
226, 234 (*see also under* Balfour
Declaration; Zionism)
Jewish state, 237, 241
his laugh, 99
leg injury, 305–6
library, books bought by, 59, 68, 108,
114, 125–6, 298
a good listener, 21, 144
manhandled by unemployed workmen,
224
meals, silent at, 142, 292–3
memory, 53, 71, 73, 78, 109, 113–14,
120, 123, 155, 158, 159, 166, 248
mistresses, 11*n.*, 92, 98, 216–19,
220–1, 222, 223
an M.P., 110, 116, 220, 224, 225–7,
249
mother, *see* Rothschild, Emma
museum, decides to start, 1 (*see also*
Tring Museum)
museum, gift *inter vivos* of Tring
museum, 310, 315; memorandum to
Trustees, 315–16
Natty, relations with, 12, 22, 32, 50–1,

58, 60–1, 64–5, 67–8, 72, 80, 84, 87,
   92, 97, 100, 104, 107, 108, 220, 221,
   222, 227, 230–2, 242, 313
at Natural History Museum with King
   George V, *47*
nature, love of, 212
new species described by, *see* family
   tree on end-paper.
nieces and nephews, awkward with,
   290–2, 294
personality, eccentricities,
   indiscretions, carelessness, secrecy,
   21, 22, 23–4, 52, 60, 63–5, 71, 75,
   101, 110, 119, 120–2, 131, 148–9,
   191*n.*, 215–16, 220, 227, 247–8,
   249, 280, 312, 314
and the press, 229
publications, 2, 73, 131, 147, 153, 205,
   220, 228, 233, 286, 293–4 (*see also*
   *Avifauna of Laysan*; *Extinct Birds*;
   *Novitates Zoologicae*; *Revision of*
   *the Sphingidae*)
records destroyed, 297
riding lessons, 60, 207*n.*, *7*
rows, 93, 100, 107, 204, 228–9
and salerooms, 114
scar, 126–7
schooling, tutors, 54, 55–6, 61–2; 27,
   *16*; no formal education, 55, 64, 177
scientists, relations with other, 76–9,
   122–3, 132, 138, 167–9, 203–4,
   228–9 (*see also under* Günther;
   Hartert; Jordan)
shooting, interest in, skill at, 60–1, 63,
   63–4, 80, 213
speech impediment, 1, 60, 81, 86, 205,
   250
staff, his magic touch with, 119, 143,
   155, 166–7, 191
theatre, interest in, 23, 74–5, 104
timekeeping by, 114–15, 217, 293
his white top hat, 224; *41*
and Tring Council, 244
Tring Park, move to, 6, 8
at Tring Park Agricultural Show, *44*
at Tring after Charles's death, 287–93
at Tring shoot, *34*
a Trustee of British Museum, 228–9
unmarried, 52
Weizmann and Sokolow, relations with,
   236, 241, 243, 248, 260–6, 271–6,
   286
his Will, bequests, 232, 299, 309, 310
women, relations with, 61–2, 62, 68–9,

98, 104, 216 (*see also* blackmailed
   *and* mistresses, *above*)
writing, style and mannerisms, 64,
   67, 70
Zionism and Jewishness, 234–5, 236–8,
   239–40, 240–3, 249–53, 256–8,
   259–60, 279, 283
zoological knowledge, work, 24, 53,
   54, 57, 67, 71–2, 120–1, 124, 130–4,
   205, 291 (*see also* publications,
   *above, and* Tring Museum,
   importance of)
Rothschild, Walter, his nieces and
   nephews (general and anonymous
   references; see also individual names),
   9, 12, 13, 23*n.*, 47–9, 80, 95*n.*, 115,
   142, 181, 207, 210, 223, 232, 234, 269,
   286, 287, 289–90, 290–3, 306, 310
Rothschild, WILHELM Carl von ("Uncle
   Willy"; 1828–1901; born in Frankfurt;
   married Hannah Mathilde von
   Rothschild; Walter's cousin and great
   uncle (by virtue of Wilhelm's sister
   marrying Baron Lionel); partner in M.
   A. Rothschild & Sohne, Frankfurt-
   am-Main), 50
Rothschild family
   animals, affinity with, 207–14
   animals named after, 212–13, 317,
   339(21), 365
   bonfire, 295*n.*, 301
   breeding winners, 208
   collections formed by, 110, 331 (note)
   established in society by Rosebery
   marriage, 37
   fabulous stories concerning, 295,
   297–8
   Frankfurt home in Jew St, *2*
   French and English branches, 39*n.*,
   340(14)
   incestuous, 9
   information about, dearth of, 296–301
   as M.P.s, politics, 225
   plants named after, 212, 317, 339(18),
   365
   praise of, fulsome, 36
   schism, 25 3–4
   strangers' letters to, 216
   women-folk, 16–18, 321(3)
Rotumah (giant tortoise), 202–3, 315
Roumania, Jews in, 33, 252, 255
Rowlands, Maldwyn, 300
Royal Entomological Society, 293
Royal Horticultural Society, 212

Royal Society, 132, 163*n*., 205, 294
Ruckbiel (collector), 155
Ruskin, John, 14
Russell, Hon. William Odo Theophilus Villiers (1870–1951), 238, 243, 245, 246
Russia, Jews in, 32–6, 226, 252, 255, 259, 275–6 (Odessa)
Ruwenzori Expedition (1906), 166

Sacher, Harry, 252, 253, 261
Sage and Co., 112
St. Bernard dog, Walter's, *see* Bonny
Salisbury, Lady, 29
Salisbury, Lord, 297*n*.
Salisbury, Sir Edward (botanist), 128
Samuel, Herbert Louis (1st Viscount Samuel; 1870–1963; High Commissioner for Palestine 1920–5), 236*n*., 237*n*., 239–40, 250, 252, 253, 267, 273, 279, 280, 285
Samuel, Stuart, 258*n*., 273, 279, 344(47)
San Blas, 186
Sandwich Islands (now Hawaiian Islands), 73, 76–8
Sanford, Dr. L. C., 303
Save the Children Fund, 245
Schama, Simon, 234*n*., 241
science, controversies in, 128
Sclater, Dr. P. L. (ornithologist), 128, 132, 158–9, 163, 228
Scott, C. P., 246*n*.
sea elephant, 11–12; 52
sea otter, 187
Sebag-Montefiore, Clarice, *see* Rothschild, Clarice
Seebohm, Henry (ornithologist), 128
Settima, 179
Sharpe, Mr. (beneficiary), 288
Sharpe, Bowdler (ornithologist), 128, 132, 228; *46*
Shaw, George Bernard, with Walter at a public dinner, *51*
Sheals, Dr. J. G. (zoologist), 349(6)
Sheard, Mr., 187
Shelduck, Ruddy, 54
Sherfield, Lord (Roger Makins), 323(8)
Silver, Rabbi, 281
Smit, F. G. A. M. (entomologist), 119
Smith, Sir Drummond, 319(2)
Smith, Frank, 299, 348(20)
Smuts, Jan Christiaan (statesman), 45
Snip (Natty's dog), 206–7
"snobs", the, 224

Snow, Dr. David William (ornithologist), 125, 138, 140
Snowden, Philip (Chancellor of the Exchequer, 1924, 1929–31), 45
Snowden, Mrs. Philip, 285
Society for the Promotion of Nature Reserves, 293
Socorro (Benito) Group Isles, 186
Sokolow, Nahum (leading Zionist), 236
    and drafting of Zionist declaration, 251, 260*n*., 261, 265*n*., 267, 283
    Walter, relations with, 241, 243*n*., 260*n*., 261, 263, 265*n*., 266, 267, 271, 279*n*., 301
    mentioned, 252, 267*n*., 281, 284, 285, 344(47)
South Sea Islands, *see* Pacific Islands
species and subspecies
    first clear description and definition of biological species, 45, 153
    series needed for study, 137–8
    speciation, 141, 144, 186, 308*n*.
    subspecies, 272
    trinomial system, 121, 128–9, 132–4
    Walter's manner of describing, 124
Spy, caricature by, *41*
*Standard*, 36
Stanley, Lord, 41
Staudinger (dealer), 157
Stein, Georg (collector), 155
Stein, Leonard, 235–6, 241*n*., 242, 256, 263*n*., 281, 344(38)
Steinbach (skinner), 166
Stoddart, D. R. (zoologist), 205
Stolypin, P. A., 36
Stresemann, Erwin (ornithologist), 129, 131, 139–40, 336(11)
Sudan, collecting in, 170–3
swallowtails, *see* Papilios *under* butterflies
Swaythling, Lord, 29
swifts, 141
    Alpine, 85
Switzerland, Alps, collecting in, 116, 148, 160
Sykes, Sir Mark (6th Baronet 1879–1919), 241*n*., 251–2, 253, 261*n*., 267, 270, 276*n*., 279*n*., 283
    Sykes-Picot agreement (1916), 275, 278
systematics, systematists, 72, 139–40, 148, 150, 291, 308, 316

Tariff Reform, 32, 225, 227

Teleki, Countess, 94*n*
Teleski, Count, 330(6)
Tenderson, Lizzie, *see* Ritchie, Lizzie
Théâtre Pigalle, 154, 210, 239*n*.
Thomas, Oldfield (zoologist), 212
Thomas, Phyllis, 59, 101, 109, 115*n*.,
    119, 131, 141, 184, 299, 302, 303
Thomas, William Morris, 109
*Times, The*, 31, 256–7, 320(2)
Times Lending Library, 320(16)
Tite, Gerald, 115
tits, 140
tortoises, giant, 103, 113, 114, 185,
    197–205, 232, 314–15; *23*
Trafford, W. H., *35*
Tree Kangaroos, 74
trees, 30
Trevelyan, George (historian), 90
Tring (town and neighbourhood), 19–20,
    44, 224, 244, 347(4)
Tring Bowling Club, 293
Tring Museum
  bird collection, *see* birds
  British Museum, given to, 103, 117,
      309–11, 315–16
  built for Walter's 21st birthday, 104
  childhood beginnings, 1, 57, 62–3, 71,
      72, 124; *14*
  clean, 113
  collectors, collecting for, 102, 137,
      187 (*see also under* Hartert Jordan;
      Rothschild, Charles; Rothschild,
      Walter; *and under* birds; butterflies)
  curators, 121, 302 (*see also* Hartert
      Jordan; Minall)
  early display cases, *28*
  electric light, 109
  entomological collections, *see* beetles;
      butterflies; fleas; insects
  exchange, specimens acquired by,
      113–14
  exhibition cases, handled personally by
      Walter, 108, 111–12
  extent of, 2, 52, 71, 112, 140, 145
  importance of, 122–3, 139–41, 149–50,
      228, 315
  ledger from, 299–300; *55*
  library, 59, 68, 108, 114, 125–6, 298
  live animals associated with, 102–4,
      108–9
  lively nature of, 101–2, 137, 152
  new buildings, after Great Row, 112,
      115–16, 222

  opened to public, public interest in,
      101–2, 108, 123, 316; *26*
  publications, *see* revisions *and under*
      Hartert Jordan; Rothschild, Walter
  purchases, finances, 76, 105–7, 114,
      118, 222, 243
  rebuilding (1960s), 299
  records destroyed, 63, 185, 298–300
  reorganised after Great Row, 92, 112,
      115–16, 222
  theft from, 181–4
  two sections, public and study, 108
  visitors, numbers of, 101, 154, 316
  Walter at work in, *43*
  Walter's commitment to,
      responsibility for, 81, 100, 101,
      109, 112, 118–19, 120–1, 123–4,
      137, *and passim*
  wartime difficulties, 243
  working conditions, 141, 145–6
  zebra cases, *29*
  mentioned, 76, 212, 250*n*.
Tring Park, 3–8; *8*
  in 1760, *5*
  alterations to, 1, 5, 7, 109
  bequeathed by Natty to Charles, 87
  disintegrates after death of Charles,
      287–9
  dogs' cemetery, 207, 298
  Emma's love for, 3–4, 5–7, 8
  estate management: farm stock, 58,
      210–12, 338(note); welfare of staff
      and tenants, 20–1, 44
  interior, *9*
  offered to British Museum, refused,
      103
  records destroyed, 63, 298, 338(note)
  schoolroom group, *13*
  shoots at 63–4
  trees at, 324(36)
  Walter at Agricultural Show, *44*
  Walter's museum and, 62, 106–7
  a Wren house, 4
Trinity College, Cambridge, 27*n*.
trinomial system, *see under* species
Tropical Disease Prevention Society,
    279–80
Tschlenow, Dr., 267*n*.
Turkey, Sultan of, 33
  gift to Natty, 39
Types, definition of, 182–4
Tyrrell, Sir William (later Lord Tyrrell),
    238, 246

Ullswater, Lord, 293

Vansittart, Lord, 285
Vereté, Mayir, 238, 255*n*., 283
Verrall, G. H., 109*n*.
Victoria, Lake, 312
Victoria, Queen, 17, 23, 27, 28*n*., 29; *36*
Victoria Club, Aylesbury, 293
Vincent, Professor John, 27

Waddesdon Manor, 5, 110, 332(4)
Wales, Prince of, *see* Edward VII
walrus, 187
Walsingham, Lord, 213
Walters, Marie, 98, 101, 116    (*see* Fredensen)
water lilies, blue, 173, 312
Waterloo, Battle of, 295
Waterstrade, John, 187
Watkins and Doncaster, 157
Webb (collector), 154*n*.
Webster, Charles, 270
Webster, Frank Blake, 185–96 *passim*, 198, 201, 299, 365–9
Wedgwood, Josiah, 280, 282, 285
Weiske, Emil, 155
Weizmann, Dr. Chaim (1874–1952; Zionist
  Balfour, relations with, 26, 38, 241–2, 261, 272
  Balfour Declaration not addressed to, 236
  decisive role of, 246
  and drafting of Zionist document, 251, 267
  England, passion for, 38
  Feisal, negotiations with, 272–3, 274
  and Meinertzhagen, 276
  memory imperfect, 236, 276
  ostrich collecting for Walter, 271–2
  and Palestine boundaries, 277
  President of English Federation of Zionists, 236
  Rothschild family, relations with, 239; Charles, 239–40, 243; Dorothy, 240; Edmond, 239*n*.; James, 246, 248; Rozsika, 242, 243–7, 251; Walter, 235, 241, 245, 246–8, 251–2, 253, 258, 260–1, 262–4, 265, 266–7, 271–4, 277, 279, 281, 285–6, 300–301
  on Stuart Samuel, 258*n*.
  simplicity of mind, 64
  at Thanksgiving Meeting, 267

mentioned, 39, 281, 283, 284, 344(47)
Wemyss, Lady, 38
Wertheimer, J., 30
Wertheimstein, Aranka, 323(8); *21*
Wertheimstein, Charlotte, *21*
Wertheimstein, Heinrich, 98
Wertheimstein, Joseph von, 95
Wertheimstein, Sarolta, 23, 98
Wertheimstein, Victor, and duelling, 98
Wertheimstein family, 95–6
Whitney family, 302
Wilde, Oscar, 223
Wilkins (herdsman), 211
Williams, Roland Edmund Lomas Vaughan, 94
Wilson, Scott, 77, 78
Wilson, President Woodrow, 264
Wingate, Sir Reginald, 27–8
Wolf, Lucien, 41, 124*n*., 235, 255, 256, 258, 262*n*., 347(5)
Wollaston, A. F. R., 157, 169, 170–3, 174–6
Wollaston, Nicholas, 175, 336(3), 348(21)
Wombwell, Mrs., 11
Woods (valet), 289, 305
World War I, 45, 233, 243–4

Young, Fred, 111, 115, 119, 181–3, 293, 298–9; *55*
Yunnan, collecting in, 157

Zangwill, Israel, 25*n*., 39*n*., 239*n*., 267*n*.
zebras, 61, 102, 108–9, 122; *29, 37*
Zeitlyn, Elsley, 259
Zimmerman, E. C., 144
Zionism, 38, 234–86
  American Zionist organisation, 264, 272
  Arabs and, 272–3, 274, 275
  Balfour's conversion to, 38
  Bible, statesmen and others influenced by, 237
  Britain, supposed advantages accruing to, 237–8, 240, 249, 250–2
  British Government declaration in favour of, *see* Balfour Declaration
  document summarising Zionist views prepared, 251, 253
  English Federation of Zionists, 236
  impractical?, 239, 249–50
  Jewish opposition to, 253–60, 262–3, 265, 279

Natty's conversion to, 239, 240, 249, 259
political complexities, 241
Rozsika and, 238, 241, 242, 243–7, 250, 251, 253, 279
Walter's attitude to, 234–5, 236–8, 239–40, 240–3, 249–53, 256–8, 259–60, 279, 283
World Zionist Organisation, Conference, 236, 279, 281
Zoo, London, 205
Zoril, 212